Negation

Perspektiven der Analytischen Philosophie
Perspectives in Analytical Philosophy

Herausgegeben von
Georg Meggle und Julian Nida-Rümelin

Band 7

Walter de Gruyter · Berlin · New York

1996

Negation

A Notion in Focus

Edited by
Heinrich Wansing

W
DE
G

Walter de Gruyter · Berlin · New York

1996

♾ Printed on acid-free paper which falls within the guidelines
of the ANSI to ensure permanence and durability.

Library of Congress Cataloging-in-Publication Data

Negation : a notion in focus / edited by Heinrich Wansing.
p. cm. – (Perspectives in analytical philosophy = Perspektiven der analytischen Philosophie ; Bd. 7)
"Proceedings volume of an interdisciplinary workshop held during the conference Analyomen 2 of the Gesellschaft für Analytische Philosophie (GAP), in Leipzig, September 7–10, 1994" – Pref.
Includes indexes.
ISBN 3-11-014769-6 (alk. paper)
1. Negation (Logic)–Congresses. I. Wansing, H. (Heinrich). II. Gesellschaft für Analytische Philosophie. III. Series: Perspectives in analytical philosophy ; Bd. 7.
BC199.N4N44 1996
160–dc20 96-3567
CIP

Die Deutsche Bibliothek – CIP-Einheitsaufnahme

Negation : a notion in focus / ed. by Heinrich Wansing. – Berlin ; New York : de Gruyter, 1996
(Perspektiven der analytischen Philosophie ; Bd. 7)
ISBN 3-11-014769-6
NE: Wansing, Heinrich [Hrsg.]; GT

Printed in Germany
Printing: Arthur Collignon GmbH, Berlin
Binding: Lüderitz & Bauer, Berlin
Cover design: Rudolf Hübler, Berlin

Contents

Preface

The notion of negation is one of the fundamental concepts of Philosophy, Logic, Linguistics, and Knowledge Representation. Often logical systems are characterized in terms of properties of their negation operator(s). Moreover, investigations into the nature of negation are of central importance for classical themes in the Philosophy of Language and Philosophical Logic, such as the semantics and pragmatics of presupposition, the syntax and semantics of pronominal anaphora, and the semantic paradoxes.

The present collection of research papers is the proceedings volume of an interdisciplinary workshop held during the conference *Analyomen 2* of the Gesellschaft für Analytische Philosophie (GAP), in Leipzig, September 7–10, 1994. The aim of the workshop was to focus on various aspects of negation within the interface of Philosophical Logic, Linguistics, and Artificial Intelligence. The main topics addressed during the workshop, as well as in the present volume, are:

- the problem of a general, abstract characterization of the notion(s) of negation,
- the analysis of certain negation phenomena in natural language and the use of negative information in logical syntax,
- the role of negation and the interplay of various negation operators in information processing and representation.

Obviously, these views on negation are not independent of one another. The linguistic analyses of negation mechanisms in natural languages, for example, are relevant to any investigation into the role of negation in theoretical and applied Knowledge Representation, while the logical characterization of negation operators is essential for systematic linguistic analysis. The approaches thus interact in clarifying the impact of negation in semantic representations and philosophical problems. This interaction of research interests, insights and problems is vividly reflected in the papers of the present volume, which may be viewed as a striking indication of the recent interest in negation across several related disciplines.

I wish to thank the authors for contributing to *Negation. A Notion in Focus*, as well as all participants of the workshop for engaging in fruitful discussions and creating a friendly and stimulating atmosphere during the conference. Special thanks go to Georg Meggle and Siegfried Gottwald for supporting the idea of a workshop on negation. Moreover, I gratefully acknowledge financial support from the Center for Advanced Studies of the University of Leipzig (to be precise, the Geistes- und Sozialwissenschaftliches Zentrum), GAP, and the Institute of Logic and Philosophy of Science at the University of Leipzig. Thanks are also due to Ingolf Max for repeated LaTeXnical assistance.

Leipzig, November 1995 *Heinrich Wansing*

Part I
Negation in Philosophical Logic

Generalized Ortho Negation*

J. Michael Dunn

Abstract. There are various more or less familiar ways of defining negation in non-classical logics, e.g., the Kripke definition of negation for intuitonistic logic, and Goldblatt's definition of ortho negation (in a weak version of quantum logic). This last uses an orthogonality ("perp") relation. In earlier papers I have already examined some relationships of these to each other and to some other definitions. In this paper I partly survey this earlier work and extend and systematize it, concentrating on the perp-style definition of negation. Things are made complicated by the fact that there can be various underlying order-structures (poset, semi-lattice, lattice, distributive lattice). For simplicity we shall stress the first and last of these.

1 Introduction

My purpose here is to discuss relationships among some of the main properties of negation as they emerge in classical logic and in various non-classical logics.[1]

There are various more or less familiar ways of giving the semantics of negation in non-classical logics. Perhaps most familiar is the Kripke definition of negation for intuitionistic logic, wherein $\neg\varphi$ is true at a given evidential state α, just in case φ is false, not just at that state (as would be the requirement for classical negation), but as well at all other accessible states β.

The Routleys' (1972) definition of De Morgan negation in relevance logic (the mathematical aspects of which were anticipated by Białynicki-Birula and Rasiowa (1957)) uses an operator $*$ on states, and $\neg\varphi$ is true at a given state α just in case φ is false, not at that state, but at its "twin" α^*.

The Birkhoff-von Neumann-Goldblatt definition of ortho negation in quantum logic uses a relation of incompatibility (variously called "orthogonality" or just "perp") between states, and $\neg\varphi$ is true at a given state α just in case every state at which φ is true is incompatible with α. This "perp" definition of negation is the semantical treatment that emerges from "gaggle theory" (short for "ggl" or "generalized galois logics"), which aims to give a uniform semantical treatment of various non-classical logics. Cf. (Dunn 1990).

---

*I wish to thank my research assistant Steve Crowley for helpful comments concerning this paper.

[1]There are undoubtedly other important semantical treatments of negation besides those I choose to discuss. I think particularly of Nelson's "constructive falsity," which appears not to fit within the framework I set here.

In an earlier paper (Dunn 1993b) I studied some relationships of the "perp semantics" for negation to that of the Routleys' (1972) "star semantics" for the De Morgan negation of relevance logic. In the present paper I focus on the perp semantics and extend and systematize the work begun on it in (Dunn 1993b). In yet another paper (Dunn 1996) I will survey this work, and also investigate relationships among these familiar definitions of negation and the four-valued semantics of Dunn and Belnap (cf. (Anderson/Belnap/Dunn 1992) for references).

2 Varieties of Negation

Let $\mathfrak{P} = (P, \leq)$ be a poset with *bounds* $\mathbf{0}$, $\mathbf{1}$, i.e., $\forall x(\mathbf{0} \leq x)$ and $\forall x(x \leq \mathbf{1})$. The elements of P are to be thought of as propositions, and $a \leq b$ is to be read "the proposition a implies the proposition b." So we shall call $\mathfrak{P}$ a *consequence poset* when we have this interpretation in mind. If the reader would like to think in terms of sentences instead of propositions, and read "$\varphi \vdash \psi$" instead of "$a \leq b$", no harm is done. But to save the trouble, we recommend that the reader who prefers sentences just assume that the logic has been "divided out" so that deductively equivalent sentences have been identified into equivalence classes. We are already postulating of consequence ($\leq$) that it is a pre-order, i.e., a reflexive and transitive relation, and anti-symmetry then becomes part of the definitional framework. $\mathbf{0}$ is a proposition that implies every proposition and $\mathbf{1}$ is a proposition that is implied by every proposition.[2] We will also from time to time assume that the poset is outfitted with the lattice operations $\wedge$ and $\vee$ (thought of as conjunction and disjunction) satisfying the following laws, which say of $a \wedge b$ that it is a greatest lower bound of a and b, and of $a \vee b$ that it is the least upper bound of a and b:

(1) $a \wedge b \leq a, \qquad a \wedge b \leq b$

(2) $a \leq b$ and $a \leq c$ implies $a \leq b \wedge c$

(3) $a \leq a \vee b, \qquad b \leq a \vee b$

(4) $a \leq c$ and $b \leq c$ implies $a \vee b \leq c$.

We will also sometimes assume the distributive law:

(5) $a \wedge (b \vee c) = (a \wedge b) \vee (a \wedge c)$.

[2]Such propositions are sometimes naively thought to be out of the spirit of relevance logic, but even in relevance logic such propositions can be explicitly defined, at least once propositional quantifiers are available: $\mathbf{0} = \forall p(p), \mathbf{1} = \exists p(p)$.

Hazen (199+), following a suggestion of I. L. Humberstone, has investigated what he dubs *subminimal negation*, which viewed as an operation on a consequence poset simply satisfies contraposition. In our framework this can be characterized by requiring order-inversion:

(6) $a \leq b$ only if $\neg b \leq \neg a$ (*contraposition*).

Hazen also has conjunction, disjunction, and nested implications, characterized by Hilbert's axioms for "positive logic." Došen's (1986) system N differs from Hazen's system by the addition of:

(7) $\neg a \wedge \neg b \leq \neg(a \vee b)$.

This last requires conjunction and disjunction for its statement. It would appear that the negation/consequence fragments of Hazen's and Došen's systems coincide with each other and with the system we are examining in terms of a poset and a negation operator satisfying contraposition. In section 6 we shall examine extensions with $\wedge$ and $\vee$, and study subminimal negation enriched with (7), calling it *preminimal negation.*

Given two partially ordered sets $(P, \leq)$ and $(Q, \leq')$, a pair of functions $(\neg, -)$ is called a *Galois connection between P and Q* (cf. Birkhoff) iff $\neg : P \longmapsto Q$, $- : Q \longmapsto P$, and

(8) $x_1 \leq x_2$ implies $\neg x_2 \leq' \neg x_1$

(9) $y_1 \leq' y_2$ implies $-y_2 \leq -y_1$,

(10) for $x \in P$, $x \leq - \neg x$,

(11) for $y \in Q$, $y \leq' \neg - y$.

It is easy to see that an equivalent definition[3] would be to require

(12) for $x \in P$, $y \in Q$, $x \leq -y$ iff $y \leq' \neg x$ (*Galois property*).

Here we are most interested in the special case of a Galois connection when the two posets are identical (there is one underlying set of propositions), and we speak then of *Galois (connected) negations,* or *split negation.*[4] It is confusing to have two negations, and in most of the paper the two operations are in fact identified with each other. But it is sometimes illuminating to view them as separate.[5]

---

[3]Birkhoff cites J. Schmidt with no reference or proof; cf. (Dunn 1993b) for a proof.

[4]I owe this latter term to my former student Chrysafis Hartonas.

[5]It is interesting to note that if $\neg$ is part of a Galois connection, then one can prove (7). Thus $\neg a \wedge \neg b \leq \neg(a \vee b)$ follows by the "Galois property" from $a \vee b \leq -(\neg a \wedge \neg b)$.

Clearly Galois negations are each subminimal negations, with the additional *Galois double negation* properties: $x \leq -\neg x$, $x \leq \neg - x$.

The familiar *minimal negation* of Johansson (1936) arises from subminimal negation by adding to *contraposition*,

(13) $a \leq \neg\neg a$ (*constructive double negation*),

or alternatively requiring simply

(14) $a \leq \neg b$ only if $b \leq \neg a$ (*constructive contraposition*).

Clearly constructive contraposition could be stated equivalently as a biconditional, and thus stated is similarly a special case of the Galois property when $\neg = -$. Minimal negation is simply a pair of Galois negations where $\neg = -$.

Intuitionistic negation (Brouwer-Heyting) is obtained by adding to minimal negation a property that in effect says that contradictions imply everything, *i.e.*, $a \wedge \neg a = \mathbf{0}$. Since we do not necessarily have conjunction (lattice meet) this is defined officially as follows:

(15) $a \leq b$ and $a \leq \neg b$ only if $a = \mathbf{0}$ (*absurdity*).

De Morgan negation is a departure from minimal negation in another direction from intuitionistic negation. It comes about by adding:

(16) $\neg\neg a \leq a$ (*classical double negation*).

Since minimal negation already satisfies the converse, this amounts to requiring that $\neg\neg a = a$ (*period two*), and is equivalent to requiring that $\neg$ is a 1-1 order-inverting operation (*dual automorphism*) of period two. Clearly $\neg$ is also onto in virtue of $a = \neg\neg a$.

Alternatively, we can define De Morgan negation by requiring that it satisfy *constructive contraposition* and its dual:

(17) $\neg a \leq b$ only if $\neg b \leq a$ (*classical contraposition*).

Finally to get *ortho negation* we add to minimal negation both the properties of absurdity and classical double negation. If the underlying consequence poset is a distributive lattice, then $\neg$ is just Boolean negation (complement), but ortho negation is studied more generally in the (non-distributive) framework of quantum logic. In a framework where we have only negation and binary consequence there is really no way to distinguish the more general ortho negations from the mere Boolean ones.

We can show this last by showing that $a \leq -(\neg a \wedge \neg b)$ (and similarly for b). But this follows by the Galois property from $\neg a \wedge \neg b \leq \neg a$. This is relevant to the discussion of (41) in section 6 below.

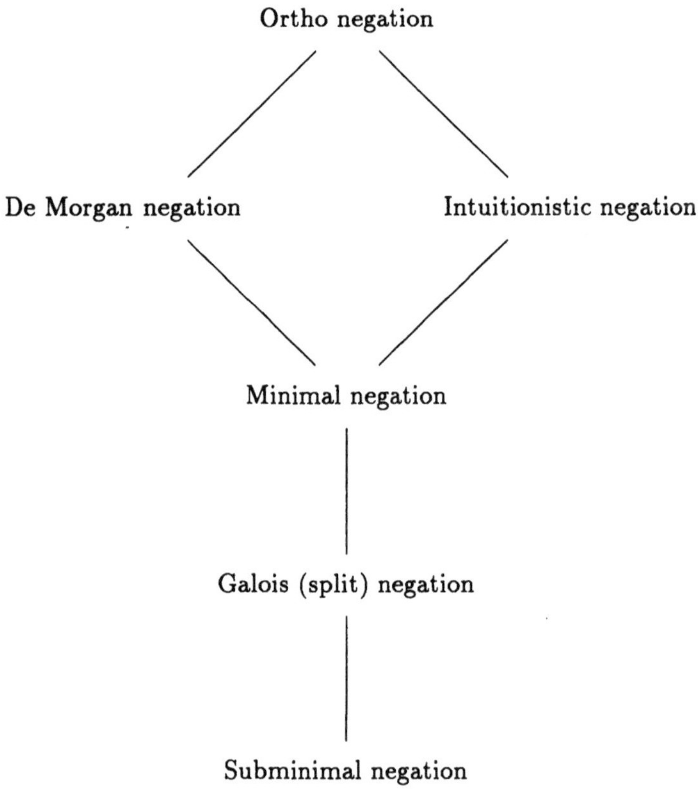

Fig. 1: The Kite of Negations

We summarize the negations discussed above in Figure 1. We do not at this point place preminimal negation in the diagram, since its role will not be fully sorted out until section 6.

3 States and Propositions

We shall adopt the by now common point of view that we can work with a set U of "states". Depending on how "fine grained" they are, the elements of U might be thought of as possible worlds, and we get something like the usual framework of "possible world semantics". But it is often more appropriate to think of states as partial worlds. Also though it seems odd to talk of impossible states, especially if we are thinking of these as *physical* states, from a formal point of view there is no reason that the elements of U need be thought of as consistent. We shall thus think of the elements of U as *states of information,* since there is no requirement that information be either consistent or complete.

It has become common to identify a proposition with a set of states, or equivalently as a mapping from states into the two truth values. We shall call such a concept of a proposition a *UCLA proposition.*[6] We can think of a valuation v as assigning to each sentence such a "proposition": $\|\varphi\|_v = A \subseteq U$.

The standard treatment of the logical operations in this framework is to understand conjunction as intersection, disjunction as union, and negation as set-theoretical complement (relative to U). This of course gives a Boolean algebra.

It is sometimes good for various reasons to consider an "information order" $\sqsubseteq$ on the set of states, where $\alpha \sqsubseteq \beta$ means that the information in the state β contains the information in the state α. As is standard, we assume that $\sqsubseteq$ is a partial order. When working with possible worlds the relation $\sqsubseteq$ becomes mere identity, since possible worlds are both consistent and complete.[7] But when working with partial and/or inconsistent informational states, it is quite reasonable that the information associated with one state can strictly contain the information associated with another. For this reason we do not take a "proposition" A to be a mere set of states, but require that A be *hereditary* in the sense that if $\alpha \in A$ and $\alpha \sqsubseteq \beta$ then $\beta \in A$. We accordingly build hereditariness into our concept of *(UCLA) proposition.*

When we deal with propositions as hereditary subsets of U, we can still understand conjunction and disjunction as in the classical case, because hereditary sets are closed under both intersection and union. But negation cannot be understood as complement, because it is easy to construct examples of hereditary subsets whose complements are not hereditary[8] (though they are always "dual hereditary," i.e., if $\alpha \in A$ and $\beta \sqsubseteq \alpha$ then $\beta \in A$). This gives a

[6]The name honors the university that has had both R. Carnap and R. Montague on its faculty, since in modern times they (together with others, e.g., S. Kripke and R. Stalnaker) have been proponents of this construction. But the idea actually originates with Boole, who suggested thinking of propositions as "sets of cases," thereby forging a link between the two interpretations of his calculus, as an algebra of sets or as an algebra of propositions. Cf. (Dipert 1978).

[7]Cf. (Dunn 1996). This explains why in the usual Kripke semantics for modal logic, propositions can be taken to be mere sets of possible worlds without worrying about heredity, whereas in the usual Kripke semantics for intutionistic logic there is need to impose heredity (see below).

[8]Let U have just two distinct elements α, β with $\alpha \sqsubset \beta$. Then $\{\beta\}$ is a hereditary subset, whereas its complement $\{\alpha\}$ is not.

kind of "explanation" of why in the Kripke semantics for intuitionistic logic negation is defined by:

(18) $\neg A = \{\chi : \forall\alpha(\chi \sqsubseteq \alpha \text{ implies } \alpha \notin A\}$.

It is easy to see that this operation preserves hereditary subsets.

We shall next examine definitions of negation that depend on "incompatibility," and ultimately come back to the Kripke definition of intuitionistic negation to see how it can be accommodated in such a framework.

4 Incompatibility ("Algebraic Version")

Early in this century there was a famous debate between Demos and Russell over the existence of negative facts. Demos held that if the apple is not red, this is because there is some other positive fact, e.g., the apple is green, which is incompatible with the apple being red (cf. (Demos 1917)). Russell reasonably replied that incompatibility is itself negative in character, and so there is still the negative (higher-order) fact of the one fact being incompatible with the other, but we will overlook this here, for at the very least, even granting this point, it is interesting that one can define negation in terms of one primitive relation of incompatibility.[9]

There are two ways of thinking of negation in such a metaphysical framework. We shall call these the "algebraic" and the "semantic," and explore them in turn.

For the algebraic, it is convenient to think of "facts" as propositions, forming a complete join semi-lattice, together with a relation of "incompatibility" I, i.e., as a structure $(A, \leq, \bigvee, I)$ where $\leq$ and I are binary relations on A and $\bigvee$ is an operation on subsets of A. The relation $\leq$ is a partial order (reflexive, transitive, and anti-symmetric), and every $B \subseteq A$ has a least upper bound $b \in A$, i.e., $\forall x(\forall y \in B(y \leq x)$ implies $b \leq x)$. The l.u.b of a set is unique and is denoted by $\bigvee B$. We think of $a \leq b$ as the fact a implies the fact b, and $\bigvee B$ can be thought of as the (infinite) disjunction of B. It is also required (*antitonicity of incompatibility*) that if $a \leq b$ and $b\,I\,x$, then $a\,Ix$, and similarly if $a \leq b$ and $x\,I\,b$, then $x\,Ia$. There are of course various requirements that one might reasonably put on I (irreflexivity and symmetry come quickly to mind), but we shall postpone these and see what can be done on the basis of the framework so far.

[9]There is also at least a hint of this definition in Plato's *Sophist* (257b,c): "When we are told the negative signifies the opposite, we shall not admit it; we shall admit only that the particle "not" indicates something different from the words to which it is prefixed, or rather from the things denoted by the words that follow the negative." Cf. (Horn 1989).

Recall that traditional logic distinguishes between *contradictories* (literal negations) and *contraries.* An example of a contrary of "the apple is red" is "the apple is green". Contraries cannot both be true, i.e., they are incompatible. However, "the apple is green" is not the weakest proposition contrary to (incompatible with) "the apple is red". For it does not merely deny that the apple is red; rather, it goes on to say specifically what other (incompatible) color the apple has. The proposition "the apple is red" has many contraries, including "the apple is green", but also "the apple is puce", "the apple is violet", etc. To deny that the apple is red is to say that the apple has some other color, which might be understood as the (infinite) disjunction "the apple is green, or the apple is puce, or the apple is violet, or ...". Thus the negation of a proposition is the infinite disjunction of all of its contraries.

Assuming that disjunction behaves like the join operation of lattice theory, another way of expressing this is to say that the *negation* of a proposition a is the weakest proposition contrary to (incompatible) with a. This allows us to characterize $\neg$, even in the absence of $\bigvee$, by:

(19) $x \leq \neg a$ iff $x \, I \, a$.

It is immediate from (19), substituting $\neg a$ for x, that $\neg a$ is at least a contrary of a:

(20) $\neg a \, I \, a$.

And the right-to-left of (19) says that $\neg a$ is weaker than any contrary x.

From (20) we can prove contraposition. Thus suppose $a \leq b$, then (by (19)) $\neg b \leq \neg a$ if $\neg b \, I \, a$. But this last follows from $\neg b \, I \, b$ (20) and our assumption $a \leq b$ by the antitonicity of incompatibility.

Ignoring for the moment the natural assumption that the relation I is symmetric, we can define another negation operation as follows:

(21) $x \leq -a$ iff $a \, I \, x$.

From (19), (21) it is easy to establish that we have Galois negations:

(22) $a \leq \neg b$ iff $b \leq -a$.

Thus using (19) and (21), we obtain: $a \leq \neg b$ iff $a \, I \, b$ iff $b \leq -a$.

The two negations are identified when I is symmetric. Thus since $\neg a \, I \, a$, by symmetry we obtain $a \, I \, \neg a$, and so from (21) we obtain $\neg a \leq -a$. A perfectly "symmetric" arguments yields $-a \leq \neg a$. We leave to the reader the proof that conversely I is symmetric when the two negations are identified.

What happens when I is made irreflexive? One might naturally think that this leads to "absurdity", but this turns out to be true in more senses that one. Let us suppose the hypotheses of the absurdity postulate: $a \leq b$, $a \leq \neg b$. Then $a\,I\,b$ follows from (19) and the second hypothesis. But then from the first hypothesis and the antitonicity of incompatibility, it follows that $a\,I\,a$, which is impossible given the assumption that I is irreflexive. So using the principle that from the impossible, anything follows, we obtain that $a = \mathbf{0}$ as required for the principle of absurdity. Before we say "Q.E.D." and move on, let us observe that we have really proven too much. Not only can we prove that $a = \mathbf{0}$, we can also prove that $a = \mathbf{1}$ or $a =$ **anything**. We have in fact shown that under the hypothesis of irreflexivity, there are no contradictions in the sense of propositions that imply both of two contradictory propositions. This suggests that there is a deficiency in the algebraic version of negation as incompatibility. Almost every logic, even relevance logic, admits contradictions. Relevance logic would be distinguished from classical logic and intuitionistic logic by denying that such a proposition $= \mathbf{0}$, there would still be such propositions. Indeed there would be propositions that imply their own negation, e.g. $a \wedge \neg a$, whereas (19) has the immediate effect of denying this (plug in a for x and apply irreflexivity).

We could go on to explore how one might get De Morgan negation by looking at "closed" propositions (those of the form $a = \neg\neg a$), but there would still be the problem in getting ortho negation that there could be no absurd propositions. Perhaps it is time to switch gears and examine a "semantic" (model-theoretic) rendering of the intuition that negation can be founded on incompatibility.

5 Incompatibility ("Semantic Version")

The second, "semantic," approach leads to the idea of a structure $(U, \perp)$, where U is a set of "states" and $\perp$ (pronounced "perp" and thought of as incompatibility) is a binary relation on U. We shall also assume the presence of an "information order" $\sqsubseteq$, where $\alpha \sqsubseteq \beta$ means that the information in α is included in the information in β. The information order is not always necessary and in fact sometimes gets in the way of exposition, and so we sometimes ignore it.

As in (Dunn 1993b), by a *perp frame* we mean a structure $(U, \perp, \sqsubseteq)$, where $\perp$ isotonic in each of its positions with respect to $\sqsubseteq$:[10]

(23) $\alpha \perp \beta$ and $\alpha \sqsubseteq \alpha'$ implies $\alpha' \perp \beta$,

[10] It is worth noting that the isotonicity requirement on $\perp$ plays almost no role in the sequel. This is in sharp contrast to the antitonicity requirement on I in the previous section. The only role of isotonicity is to allow us to show that if A is hereditary, then both $A^{\perp}$ and ${}^{\perp}A$ are hereditary. The reader who wants to disregard the isotonicity requirement can do so with little or no harm.

(24) $\alpha \perp \beta$ and $\beta \sqsubseteq \beta'$ implies $\alpha \perp \beta'$.

By a model we shall mean a structure $(U, \perp, \sqsubseteq, \vDash)$, where $(U, \perp, \sqsubseteq)$ is a perp frame and $\vDash$ is a relation between the states in U and sentences, which relation satisfies certain conditions which we shall specify. The first of these is the *hereditary condition* which requires that for every atomic sentence p, if $\alpha \vDash p$ and $\alpha \sqsubseteq \beta$ then $\beta \vDash p$. This simply expresses our understanding of $\sqsubseteq$ as an order of increasing information.

We have been a bit coy about what sentences we have, but surely we will have for each sentence φ its negation $\neg\varphi$. A sentence $\neg\varphi$ will be made true by a state $\chi \in U$ just in case χ is incompatible with all states α that make φ true. In symbols,

(25) $\chi \models \neg\varphi$ iff $\forall\alpha(\alpha \models \varphi$ implies $\chi \perp \alpha)$.

We shall label this 'perp negation'.

The seminal source for perp negation is in (Birkhoff/von Neumann 1936) with their "quantum logic."[11] They suggested that propositions can be identified with certain sets of states of a Hilbert space (the closed subspaces). There is a binary relation $\perp$ of orthogonality on the states of a Hilbert space, and Birkhoff and von Neumann suggested that given a proposition A, its negation might be defined as the set of states orthogonal to every state in A. Spelling this out, let us say that $\chi \perp A$ iff $\chi \perp \alpha$ for every $\alpha \in A$. We can then define ${}^{\perp}A = \{\chi : \chi \perp A\}$. Clearly there is the dual definition where the negation would be the set of states such that every state in A is orthogonal to them. Writing $A \perp \chi$ iff $\alpha \perp \chi$ for every $\alpha \in A$, we can denote this dual negation by $A^{\perp} = \{\chi : A \perp \chi\}$. Since orthogonality is symmetric, $A^{\perp} = {}^{\perp}A$, and so in the setting of quantum logic we have a distinction without a difference. But let us bear in mind that in other contexts we might have a non-symmetric relation analogous to orthogonality.

Goldblatt has used a somewhat abstract version of this style of definition in producing a semantics for orthologic. The idea is to consider a frame $(U, \perp)$, where $\perp \subseteq U^2$ is an irreflexive, symmetric relation. Goldblatt in effect defined a valuation v to be an assignment of truth or falsity to sentences that behaves in the following recursive manner:

(26) $\chi \models_v \varphi \wedge \psi$ iff $\chi \models_v \varphi$ and $\chi \models_v \psi$
(27) $\chi \models_v \neg\varphi$ iff for all $\alpha \in U, \alpha \models_v \varphi$ implies $\chi \perp \alpha$.

If Goldblatt had stopped here, the law of double negation would be violated, so he had to put on another condition. First he defined $X \subseteq U$ to be $\perp -$ *closed* iff for all $\alpha \in U, \alpha \notin X$ only if for some χ, $\chi \perp X$ and not $\alpha \perp \chi$

[11]The actual history of this is complicated. Cf. (M. Jammer 1974).

(in Goldblatt's framework the converse always holds). It is hard to put this definition into words, but perhaps it helps to say that if $\alpha \notin X$ this is because α is "indistinguishable" from some state χ (α is compatible with χ) that cannot be in X because it is "different from" (incompatible with) every state in X. Goldblatt then requires that the set of states that satisfy an atomic sentence be $\perp$–closed.

It is easiest to explain the requirement of $\perp$–closure by writing $\|\varphi\|_v$ instead of $\{\chi \in U : \chi \models_v \varphi\}$,and rewriting the semantical clauses for conjunction and negation as:

(28) $\|\varphi \wedge \psi\| = \|\varphi\| \cap \|\psi\|$,

(29) $\|\neg\varphi\| = {}^{\perp}\|\varphi\|$.

Now the requirement of $\perp$–closure can be explained as requiring for each atomic proposition p, that ${}^{\perp\perp}\|p\| = \|p\|$. It turns out that an easy induction shows that for complex sentences as well, $\|\varphi\|$ is always a $\perp$–closed set.

This gives the idea of Goldblatt's representation of ortholattices, where conjunction and negation are understood as intersection and perp respectively. Disjunction can then be defined *via* the De Morgan law $\varphi \vee \psi = \neg(\neg\varphi \cap \neg\psi)$. It turns out then that disjunction is evaluated not as union, but as something smaller than the union, and this means that in general conjunction does not distribute over disjunction.

Goldblatt makes the plausible assumptions that the perp relation $\perp$ is irreflexive as well as symmetric. If one reads $\perp$ as "incompatibility" it is difficult not to make these assumptions. It would seem that a state cannot be incompatible with itself, and if one state is incompatible with another, surely the reverse seems true. But let us think of an information state α being incompatible with an information state β as meaning that α contains a piece of information that β denies.[12] As soon as one allows the presence of inconsistent information states, as must be done to accommodate relevance and other "paraconsistent logics", we must give up irreflexivity.

Giving up symmetry is perhaps more difficult to motivate. But consider the following example due to C. Hartonas[13] — see (Hartonas/Dunn 1993). My son's practising his saxophone prevents my reading (a technical paper), but my reading does not one wit prevent his practising the saxophone. So the state of his playing the saxophone causes sound, which I hear, and which

[12]The reading actually more closely fits the idea of β being incompatible with α, but this gets confusing since it reverses the direction. I think nothing hangs on this "ordinary language" point, at least for our purposes.

[13]It should be stressed that Hartonas made up this example, and while it is true that my son does play the saxophone, it is not true that this actually interferes with any of my work.

interferes with my concentration. On the other hand my concentration in no way has causal effects that interfere with his playing the saxophone, excluding of course the unusual circumstance that I get really irritated and yell at him.

There is a rejoinder to this. It goes "Look, you are saying that if S then $\neg C$, but by contraposition, if C then $\neg S$. That is to say if you can't concentrate when he plays the saxophone, then he can't be playing the saxophone if you are concentrating."

This response is the type familiar to anyone of us who has taught elementary logic, and tried to motivate various principles of formal logic in relation to purported natural language counter-examples. But I think things get very subtle here, in a way that is reminiscent of, if not the very same phenomenon as, the distinction between necessary and sufficient conditions. It is a necessary condition of my reading that my son *not* play the saxophone, but my reading is not a sufficient condition for my son's not playing the saxophone (would sometimes that it were).

In a private communication, Greg Restall has suggested that the "dual" of *prevents* is *precludes*, and points out that while my reading does not *prevent* my son from playing the saxophone, it does *preclude* him from doing so. I am not sure about all of the nuances of the words, but there seems something right about there being some distinction along these lines. Restall also suggests that all this has something to do with the direction of the causal relation, and again there seems something right about that.

Notice that if $\perp$ is not required to be symmetric, we can in effect get two negations: $A^{\perp}$ and ${}^{\perp}A$. It is easy to check that they satisfy the following law:

(30) $A \subseteq B^{\perp}$ iff $B \subseteq {}^{\perp}A$.

This is the Galois property (12).

So we have two ways of playing with the conditions on the perp relation: we can make it irreflexive, symmetric, or both. Making it symmetric glues the split negations into one. It is easy to see that whether the two negations have been glued together or not, that we get the following ("constructive") forms of double negation:

(31) $A \subseteq {}^{\perp}(A^{\perp})$; $A \subseteq ({}^{\perp}A)^{\perp}$.

When $\perp$ is symmetric these of course reduce to the single form of constructive double negation:

(32) $A \subseteq A^{\perp\perp}$.

One can wonder whether there is a natural (first-order) condition on $\perp$ that would assure full (classical) double negation ($A^{\perp\perp} = A$). It appears that there is not.[14] However the usual way to assure double negation is simply to restrict attention to those sets of states that are closed in just the sense that $A^{\perp\perp} \subseteq A$, i.e., $A = A^{\perp\perp}$.

Let us suppose now that we have a perp frame $(U, \perp, \sqsubseteq)$,and let us consider a collection of subsets of U that are *hereditary* in the sense that if $\alpha \in A$ and $\alpha \sqsubseteq \beta$ then $\beta \in A$. The reader can easily verify that by putting the following conditions on the perp relation and/or requiring closure, we obtain the following conditions on the poset, defining $\neg A = A^{\perp}$ and $-A = {}^{\perp}A = A^{\perp^{-1}}$.

Negation	**Poset**	**Perp Semantics**
Subminimal	$x \leq y \Rightarrow \neg y \leq \neg x$	$\neg A = A^{\perp}$
Galois	$x \leq \neg y \Leftrightarrow y \leq -x$	$-A = {}^{\perp}A$
Minimal	$x \leq \neg y \Rightarrow y \leq \neg x$	$\perp$ symmetric
Intuitionistic	Minimal + $x \leq y, \neg y \Rightarrow x = \mathbf{0}$	$\perp$ irreflexive, symm.
De Morgan	Minimal + $\neg\neg x \leq x$	$A^{\perp\perp} = A$, symm.
Ortho	Intuitionistic + $\neg\neg x \leq x$	$A^{\perp\perp} = A$, irrefl., symm.

A question occurs about the completeness of the postulates on the poset in relation to the corresponding semantical conditions. This question is answered by a representation result from (Dunn 1993b), which we shall now describe.

Let $\mathfrak{P} = (P, \leq, \neg, -)$ be a bounded poset with two order-inverting mappings $\neg$ and $-$. We shall call this a *subminimal two negation poset.* This is a somewhat artificial construct in terms of most of the applications, but it is convenient in stating the general representation theorem. In order to get a single subminimal negation it is permitted to assume that the two operations $\neg$ and $-$ are the same (so then the more natural construct is to assume just a single operation). For Galois negations the two operations might well be different, but for minimal negation they are actually required to be the same.

By a *two perp frame* we mean a structure $(U, \perp_1, \perp_2, \sqsubseteq)$,where each of $(U, \perp_1, \sqsubseteq)$ and $(U, \perp_2, \sqsubseteq)$ is a perp frame. Every two perp frame gives rise to a subminimal two negation poset quite naturally, where we let P be the set of hereditary subsets of U, we let $\leq$ be $\subseteq$ among those hereditary subsets,

[14]Indeed Greg Restall, at The Australian National University, has informed me that he has a proof, but I have not seen it.

and for $A \in P$, we define $\neg A = A^{\perp_1}$, and we define $-A = A^{\perp_2}$. Of course $\mathbf{0}$ is the empty set and $\mathbf{1}$ is U. Let us call this the *full two negation poset determined by the frame.* It is easy to verify that this is a subminimal two negation poset.

When $\perp_1$ and $\perp_2$ are converses (as is needed for Galois negations) and $\perp_1$ is symmetric (as is needed for minimal negation) then $\perp_1 = \perp_2$. We can then collapse a two perp frame to just a perp frame, and similarly we can collapse a two negation poset determined by the frame to simply a (single) negation poset.

Theorem 1 (Representation) *Let $\mathfrak{P} = (P, \leq, \neg, -)$ be a subminimal two negation poset. Then there is a two perp frame so that $\mathfrak{P}$ can be isomorphically embedded in the full subminimal two negation poset determined by that frame.*

Proof. By a *cone* of $\mathfrak{P}$ we mean a set $C \subseteq P$ that is closed upwards under $\leq$, i.e., if $a \in C$ and $a \leq b$ then $b \in C$. Think of the cones as "theories' closed under implication $\leq$. A cone C is *proper* when C is not all of P. Proper cones correspond to absolutely consistent theories. We define the *canonical two perp frame determined by* $\mathfrak{P}$ as follows. U_c = the set of proper cones of $\mathfrak{P}$. For cones C_1 and C_2 we define $C_1 \perp^{\neg} C_2$ iff $\exists x \in C_1\ (\neg x \in C_2)$, and we define $C_1 \perp^{-} C_2$ iff $\exists x \in C_1\ (-x \in C_2)$. And finally we define $C_1 \sqsubseteq_c C_2$ iff $C_1 \subseteq C_2$. We can look at the structure $(U_c, \perp^{\neg}, \perp^{-}, \sqsubseteq_c)$ as a double perp frame. It is in effect shown in (Dunn 1993b) that the desired embedding is the *canonical isomorphism*:

(33) $h(a) = \{C : C \text{ is a proper cone of } \mathfrak{P} \text{ and } a \in C\}$.

We sketch the proof here, observing that we must show

$$h(\neg a) = {}^{\perp} h(a), \text{ i.e.,}$$

(34) $\forall C(\neg a \in C \text{ iff } \forall C'(a \in C' \text{ iff } C' \perp^{\neg} C)$,

and similarly for $-$.

Left-to-right is immediate. Right-to-left is proven by contraposition, so we assume that $\neg a \notin C$ and show that $\exists C'(a \in C'$ and not $C' \perp^{\neg} C)$. Setting C' = the principal cone determined by $a = [a) = \{x : a \leq x\}$, we cannot have $C' \perp^{\neg} C$. For otherwise there is some $x \geq a$ such that $\neg x \in C$. But then by contraposition $\neg a \geq \neg x$, and so $\neg a \in C$, contrary to our hypothesis.□

Corollary 2 *Let $\mathfrak{P} = (P, \leq, \neg, -)$ be a bounded poset with Galois negations $(\neg, -)$. Then there is a two perp frame $(U, \perp_1, \perp_2, \sqsubseteq)$, where $\perp_1$ and $\perp_2$ are converses, so that $\mathfrak{P}$ can be isomorphically embedded in the full two negation poset determined by that frame.*

Proof. It is easy to see that if $(\neg, -)$ are Galois negations, then $\perp^{-}$ is just the converse of $\perp^{\neg}$. Thus suppose $C_1 \perp^{\neg} C_2$ because $x \in C_1$ and $\neg x \in C_2$. Since $x \leq -\neg x$ then $-\neg x \in C_1$ and so $C_2 \perp^{-} C_1$. The argument for the other direction uses $x \leq \neg - x$.□

As noted above, when we make both of $\perp_1$ and $\perp_2$ symmetric, then if they are converses they are in fact the very same relation. This makes it unnecessary to distinguish $\neg$ from $-$, so in the remaining corollaries we work with structures $(P, \leq, \neg)$.

Corollary 3 *Let* $\mathfrak{P} = (P, \leq, \neg)$ *be a bounded poset with minimal negation* $\neg$. *Then there is a perp frame* $(U, \perp, \sqsubseteq)$, *with* $\perp$ *symmetric, so that* $\mathfrak{P}$ *can be isomorphically embedded in the full two negation poset determined by that frame.*

Proof. Reconfigure $(P, \leq, \neg)$ temporarily as a split negation poset $(P, \leq, \neg, -)$, where $\neg = -$. Then $\perp^{\neg} = \perp^{-}$, and we know from the previous corollary that they are converses of each other.□

Corollary 4 *Let* $\mathfrak{P} = (P, \leq, \neg)$ *be a bounded poset with intuitionistic negation* $\neg$. *Then there is a perp frame* $(U, \perp, \sqsubseteq)$, *with* $\perp$ *symmetric and irreflexive, so that* $\mathfrak{P}$ *can be isomorphically embedded in the full negation poset determined by that frame.*

Proof. Here we have to change slightly the set of canonical states. Instead of letting U be the set of all (proper) cones, we use U_{con} = the set of *consistent* cones, where a cone C is *consistent* just when for no $x \in C$ do we also have $\neg x \in C$. It is clear that this does the job of making $\perp^{\neg}$ irreflexive, for if to the contrary some $C \perp^{\neg} C$, then $\exists x \in C$ such that $\neg x \in C$. And the argument that $\perp^{\neg}$ is symmetric is still the same as for Corollary 3. The only remaining issue then is whether can we still get the Representation Theorem (Theorem 1). The reader can check that there are only two places in its proof where we must show that cones exist. In each case one can show that the required cone is consistent.

The first place is in showing that h is 1-1. But there we assume that $a \neq b$, assume without loss of generality that this is because $a \not\leq b$, and then consider $[a)$, the principal cone determined by a, which clearly is in $h(a)$ and not in $h(b)$. Clearly $[a)$ must be consistent, for otherwise for some $x, a \leq x$ and $a \leq \neg x$, and so (by absurdity) $a = \mathbf{0}$, which contradicts $a \not\leq b$.

The only other place where one must show the existence of a cone is in the argument for the right-to-left of (34), which we did above by contraposition, assuming $\neg a \notin C$, and then showing that the desired cone $C' = [a)$. Again $[a)$ must be consistent or else by absurdity, $a = \mathbf{0}$. In the latter case, for arbitrary $x, a \leq \neg x$, i.e., by constructive contraposition $x \leq \neg a$, and so $\neg a = \mathbf{1} \in C$, contrary to our assumption.

For the next corollary we have to change the notion of "proposition" so that it is more that just a hereditary subset. We further require that it be "closed" in the sense that $A^{\perp\perp} = A$. We shall call the collection of hereditary closed subsets the *De Morgan negation poset* determined by the frame, outfitting it of course with the definition of $\neg A = A^{\perp}$ and of $\leq$ as $\subseteq$. Note that it is easy to show (cf. e.g. Birkhoff) that $\neg$ so defined preserves (hereditary) closed subsets.

Corollary 5 *Let $\mathfrak{P} = (P, \leq, \neg)$ be a bounded poset with De Morgan negation $\neg$. Then there is a perp frame $(U, \perp, \sqsubseteq)$, with $\perp$ symmetric, so that $\mathfrak{P}$ can be isomorphically embedded in the full De Morgan negation poset determined by that frame.*

Proof. What needs to be checked is that $h(a)$ is always closed. But this is obvious since $h(a) = h(\neg\neg a) = (ha)^{\perp\perp}$.□

Corollary 6 *Let $\mathfrak{P} = (P, \leq, \neg)$ be a bounded poset with ortho negation $\neg$. Then there is a perp frame $(U, \perp, \sqsubseteq)$, with $\perp$ symmetric and irreflexive, so that $\mathfrak{P}$ can be isomorphically embedded in the full De Morgan negation poset determined by that frame.*

Proof. Combines the proofs of all of the corollaries above.□

6 Distributive Lattices

Let us next consider what happens when the underlying poset is a bounded distributive lattice $\mathfrak{D} = (D, \wedge, \vee)$ with least element $\mathbf{0}$ and greatest element $\mathbf{1}$. We interpret $\wedge$ as conjunction, $\vee$ as disjunction, and as usual we can define a partial order (interpreted as implication) so $x \leq y$ iff $x \wedge y = x$. In order to represent a distributive lattice in such a way that $\vee$ is interpreted as union, and of course $\wedge$ is interpreted as intersection, we change the nature of the canonical representation. A *prime filter* F may be defined as a non-empty subset of D satisfying the following conditions:

(35) $x \wedge y \in F$ iff $x \in F$ and $y \in F$,

(36) $x \vee y \in F$ iff $x \in F$ or $y \in F$.

A *proper prime filter* is a prime filter F that is not D.

Let $\mathcal{PF}(\mathfrak{D})$ = the set of proper prime filters of $\mathfrak{D}$. We then define the "canonical representation" $h(a) = \{F \in \mathcal{PF}(\mathfrak{D}) : a \in F\}$. It is well-know from (Stone 1937) that this is an isomorphism into the power-set of $\mathcal{PF}(\mathfrak{D})$, interpreting $\wedge$ as intersection and $\vee$ as union.

The question naturally arises as to what happens to our results concerning negation in this context. The simple answer is that everything carries over intact except for the results concerning subminimal negation. The problem is that certain parts of the De Morgan laws hold automatically in the perp semantics, but that not all of these follow from the mere properties of subminimal negation.

For this discussion it is convenient to consider an abstraction due to Urquhart.

Definition 7 *An* Ockham lattice *is a structure* $(D, \wedge, \vee, \neg)$, *where* $(D, \wedge, \vee)$ *is a distributive lattice and* $\neg : D \mapsto D$ *is a* dual endomorphism:

(37) $\neg(x \vee y) = \neg x \wedge \neg y$,

(38) $\neg(x \wedge y) = \neg x \vee \neg y$.

There is a more familiar structure, and one on which Urquhart based his notion of an Ockham lattice. A De Morgan lattice is just like an Ockham lattice, except for a De Morgan lattice it is required that "negation" be of period two. In footnote 15 of (Dunn 1993b) I said it would be nice to somehow fit the Ockham lattices into the framework of that paper, but that I could not figure out how to represent them. This note was less clear than it should have been, since of course Urquhart had already given a topological representation using a $*$ operator not assumed to be of period two. What I meant was that I did not know how to do a representation using $\perp$, and I still do not, as we shall see.

Let me go into this a little. The problem is that it is easy to see that the following bits and pieces of the De Morgan laws hold in the perp semantics:

(39) $A^{\perp} \cup B^{\perp} \subseteq (A \cap B)^{\perp}$

(40) $(A \cup B)^{\perp} \subseteq A^{\perp} \cap B^{\perp}$

(41) $A^{\perp} \cap B^{\perp} \subseteq (A \cup B)^{\perp}$.

Thus for (39), if either everything in A bears perp to χ, or everything in B does, then everything that is in both A and B (a smaller set) bears perp to χ. And for (40), if everything in the union of A and B bears perp to χ, then everything in A does, and everything in B does. Conversely, for (41), if everything in A bears perp to χ, and everything in B does, then so does everything that is in either A or B.

But the converse of (39),

(42) $(A \cap B)^{\perp} \subseteq A^{\perp} \cup B^{\perp}$,

does not hold. It may well be that everything in both A and B (perhaps a very small set, even empty) bears perp to χ, but this does not mean that either everything in A bears perp to χ, or that everything in B bears perp to χ.

The fit of the perp definition of negation with the De Morgan laws is awkward. The perp definition gives at one and the same time too much and too little. I say *too much* because (41) holds only for minimal negation (actually Galois negations) and not for subminimal negation (although (39) and (40) do). I say *too little* because not only does (42) not hold, it seems impossible to think of a natural condition on the perp relation that would validate it. As an example of an unnatural condition, we could simply require globally that we are looking at a collection of subsets of U that satisfy (42).

This might not seem much worse than what is needed to get double negation in the perp framework. There we just restrict ourselves to subsets A that are perp-closed, i.e., where $A^{\perp\perp} = A$. But in defining a model this can be made just a requirement on the UCLA propositions that are assigned to atomic sentences, and an induction on formulas shows this holds for the UCLA propositions assigned to complex sentences as well. But it seems this cannot be done with the condition (42).

Now for subminimal negation we can fix things by simply adding the requirement

(43) $\neg a \wedge \neg b \leq \neg(a \vee b)$.

In fact, as we noted in section 2, (Došen 1986) contains precisely this requirement, and we introduced the label *preminimal negation* for *subminimal negation* plus (43).

We will work through the proof that preminimal negation can be represented, but first we must present some necessary definitions. Let $\mathfrak{D} = (D, \wedge, \vee, \neg)$ be a distributive lattice with preminimal negation. Then its canonical information frame $\mathfrak{F}_c\ (\mathfrak{D})$ is the structure $(\mathcal{PF}(\mathfrak{D}), \perp_c)$, where for $F, G \in \mathcal{PF}(\mathfrak{D})$, we define $F \perp_c G$ iff for some $x \in F$, $\neg x \in G$. By the *canonical isomorphism* we mean the function $h(a) = \{F \in \mathcal{PF}(\mathfrak{D}) : a \in F\}$.

Given any information frame $\mathfrak{F} = (U, \perp, \sqsubseteq)$ (not necessarily canonical), by the *full set of propositions on* $\mathfrak{F}$ we mean the set of all hereditary subsets of U. It is easy to see that this forms a bounded distributive lattice.

Theorem 8 (Representation) *Every distributive lattice with preminimal negation* $\mathfrak{D} = (D, \wedge, \vee, \neg)$ *can be embedded in the full set of propositions of the canonical information frame.*

Proof. That the canonical isomorphism h is 1–1 and preserves $\wedge, \vee$ (interpreting them as intersection and union) follows from Stone's representation of distributive lattices. We must then show

(44) $h(\neg a) = (h(a))^{\perp}$, i.e.,

(45) $\neg a \in F$ iff $\forall G(a \in G$ only if $\exists x[x \in G$ and $\neg x \in F])$.

Left-to-right is immediate from the canonical definition of $\perp$. For right-to-left we prove the contrapositive, assuming that $\neg a \notin F$. We need to prove $\exists G$ so that a $\in G$ and $\forall x[x \in G$ only if $\neg x \notin F])$, i.e., G is "compatible" with F. We start by setting $G_0 = [a)$. It it is clear that $a \in G_0$ and we can show $\forall x[x \in G_0$ only if $\neg x \notin F]$. For this last, let us suppose that $x \in G_0$, i.e., $a \leq x$. By contraposition, $\neg x \leq \neg a$. So $\neg x \notin G_o$ since otherwise $\neg a \in G_0$, contrary to our main assumption.

But there is no reason to assume that G_0 is prime. The principal filter $[a)$ is prime only in the unusual circumstance that a is a "prime element," i.e., $a \leq x \vee y$ implies $a \leq x$ or $a \leq y$. But the following results deliver a prime filter H such that $a \in H$ and H is compatible with F.

Lemma 9 *Let $\mathfrak{D} = (D, \wedge, \vee)$ be a distributive lattice, let $\neg$ be a unary operator on D,and let $F \in \mathcal{PF}(\mathfrak{D})$. Then the set $\mathcal{F}$ of $G \in \mathcal{PF}(\mathfrak{D})$ such that G is compatible with F has a maximal member H, on the assumption that $\mathcal{F}$ is non-empty.*

Proof. Consider the set $\mathcal{F}$ of proper filters G that are compatible with F, ordering $\mathcal{F}$ by inclusion. We have assumed that $\mathcal{F}$ is non-empty. We call a set of filters $\mathcal{C}$ a *chain* just when for any two filters $G_1, G_2 \in C$, either $G_1 \subseteq G_2$ or $G_2 \subseteq G_1$. Now consider any chain $\mathcal{C} \subseteq \mathcal{F}$. We show $\bigcup \mathcal{C} \in \mathcal{F}$. It is routine to show that the union of a chain of filters is a filter, so the interesting part is to show that $\bigcup \mathcal{C}$ is compatible with $\mathcal{F}$. Suppose the contrary, i.e., for some $x \in \bigcup \mathcal{C}$, $\neg x \in F$. Then there must be some filter $G_1 \in \mathcal{C}$ with $x \in G_1$. This contradicts our picking $\mathcal{C}$ as a set of filters compatible with $\mathcal{F}$. We can now apply Zorn's Lemma (cf. e.g. Birkhoff) to show that $\mathcal{F}$ has a maximal member H. □

Lemma 10 *Let $\mathfrak{D} = (D, \wedge, \vee)$ be a distributive lattice and let $\neg$ be a preminimal negation operator on D. Let H be a filter that is maximal with respect to being compatible with some given filter F. Then H is prime.*

Proof. Suppose that $a \vee b \in H$ and yet $a, b, \notin H$. Let $[H, a)$ be the filter generated by $H \cup \{a\}$, and similarly for $[H, b)$. Since each of $[H, a)$ and $[H, b)$ properly extend H, each must be incompatible with F. Thus we can find $x \in [H, a)$ with $\neg x \in F$, and similarly for some y and $[H, b)$. By a routine argument it can be shown that for some $c \in H$, $c \wedge a \leq x$ and $c \wedge b \leq y$. Then (distribution)

(46) $\quad c \wedge (a \vee b) = (c \wedge a) \vee (c \wedge b) \leq x \vee y.$

Since both c and $a \vee b \in H$, $c \wedge (a \vee b) \in H$, and so $x \vee y \in H$.

But since $\neg x, \neg y \in F$, $\neg x \wedge \neg y \in F$. Since we have $\neg x \wedge \neg y \leq \neg(x \vee y)$, this means $\neg(x \vee y) \in F$, but since $x \vee y \in H$ this contradicts the fact that H is compatible with F.□

We leave to the reader the exercise of "fiddling" with the canonical frame so as to see that the various requirements below on the perp semantics correspond to the various requirements for negation on a distributive lattice. The arguments are similar to those needed for the similar correspondences regarding posets from the previous section, and again to accommodate Galois negations we must introduce two canonical perp relations $\perp^{\neg}$ and $\perp^{-}$ and show they are in fact converses.

One corollary gets nicer. It is easy to check that for intuitionistic negation a filter is proper iff it is consistent (the key point being that if $a, \neg a \in F$ then $a \wedge \neg a = \mathbf{0} \in F$). So there is no need to change the nature of the canonical frame, restricting the proper filters to be consistent filters, as correspondingly was done for cones in Corollaries 4 and 6.

In the Kite of Negations (Fig. 1), preminimal negation is the appropriate replacement for subminimal negation. Note also that ortho negation on a distributive lattice is just Boolean negation. The semantical equivalent of this is that in the canonical frame, $\perp$ becomes the relation of difference, i.e., $F \perp_c G$ iff F is not identical to G. Thus if $F \perp_c G$ then for some $x \in F$, $\neg x \in G$. We must then have $\neg x \notin F$, since if otherwise then $x \wedge \neg x = \mathbf{0} \in F$, and hence F would not be proper. Conversely, if F is not identical to G there is some element in one and not the other. Assume without loss of generality that $x \in F$ and $x \notin G$. Since $x \vee \neg x = \mathbf{1} \in G$, from the fact that G is prime we conclude $\neg x \in G$, and so $F \perp_c G$.

We summarize the results above in the following table.

Negation	**Distributive Lattice**	**Perp Semantics**
Subminimal	$x \leq y \Rightarrow \neg y \leq \neg x$	None
Preminimal	Submin. + $\neg x \wedge \neg y \leq \neg(x \vee y)$	$\neg A = A^{\perp}$
Galois	$x \leq \neg y \Leftrightarrow y \leq -x$	$-A = {}^{\perp}A$
Minimal	$x \leq \neg y \Rightarrow y \leq \neg x$	$\perp$ symmetric
Intuitionistic	Minimal + $x \leq y, \neg y \Rightarrow x = \mathbf{0}$	$\perp$ irreflexive, sym.
Ockham	De Morgan laws	None
De Morgan	Minimal + $\neg\neg x \leq x$	$A^{\perp\perp} = A$, symm.
Ortho = Boolean	Intuitionistic + $\neg\neg x \leq x$	$\perp$ is non-identity, so $A^{\perp} = U - A$

7 Intuitionistic Negation

The reader might be wondering how the perp semantics for intuitionistic negation relates to the Kripke semantics. This is discussed in more detail in (Dunn 1993b), but for the convenience of the reader, we again describe the connection here. The Kripke-style definition of negation in intuitionistic logic says that $\neg\varphi$ is true at an evidential state χ just when φ is refuted at χ, i.e. at all stronger evidential states φ is never true.

(47) $\neg A = \{\chi : \forall\alpha(\chi \sqsubseteq \alpha \text{ only if } \alpha \notin A)\}$.

If we contrapose this we get:

(48) $\neg A = \{\chi : \forall\alpha(\alpha \in A \text{ only if } (\chi \not\sqsubseteq \alpha))\}$,

which has the same form as the perp definition except we have $(\chi \not\sqsubseteq \alpha)$ instead of $\chi\perp\alpha$:

(49) ${}^{\perp}A = \{\chi : \forall\alpha(\alpha \in A \text{ only if } \chi\perp\alpha)\}$.

One should not think that $\chi\perp\alpha$ and $(\chi \not\sqsubseteq \alpha)$ mean the same, but there is a one-way implication:

(50) $\chi\perp\alpha$ implies $(\chi \not\sqsubseteq \alpha)$.

Thus if $\chi\perp\alpha$ then canonically there exists $x \in \chi$ such that $\neg x \in \alpha$. But then we cannot have $\chi \sqsubseteq \alpha$, for canonically this means $\chi \subseteq \alpha$ and this means then that both x , $\neg x \in \alpha$, thus α is inconsistent (see Corollary 4, Theorem 1).

But the converse does not hold, $(\chi \not\sqsubseteq \alpha)$ does not imply $\chi\perp\alpha$. One can easily imagine an evidential state χ that is not a subset of α because it affirms something that α does not. But why should this mean that χ is incompatible with α. Of course with classical logic, where the evidential states are complete, χ would be incompatible with α, for whatever proposition x that χ asserts that is not asserted by α, then $\neg x$ must be asserted by α. But in the semantics of intuitionistic logic, states can be partial. Indeed, we could have α a subset of χ.

But it turns out that although $\chi\perp\alpha$ is not equivalent to $(\chi \not\sqsubseteq \alpha)$, it is equivalent to something involving χ, α, and the failure of $\sqsubseteq$. What is needed is that χ and α have no common (proper) extension, i.e., not $\exists\beta(\chi, \alpha \sqsubseteq \beta)$. We have the following equivalence:

(51) $\chi\perp\alpha$ iff not $\exists\beta(\chi, \alpha \sqsubseteq \beta)$.

We will not go through the details here (see Dunn 1993b), but canonically it amounts to there being a (proper) prime filter that extends any pair of "compatible" filters (where filters F and G are *compatible* if there is no $x \in F$ such that $\neg x \in G$).

From this it is easy to establish the equivalence of the perp definition of negation with that of Kripke. In the following we use $\neg A$ for the Kripke definition and ${}^{\perp}A$ for the perp definition. Thus assume $\chi \in \neg A$. This means that $\forall\beta(\chi \sqsubseteq \beta$ only if $\beta \notin A)$. We show $\chi \in {}^{\perp}A$ by showing $\forall\alpha((\chi \not\perp \alpha)$ only if $\alpha \notin A)$. For this we assume $(\chi \not\perp \alpha)$. So by the equivalence above $\exists\beta(\chi, \alpha \sqsubseteq \beta)$. But then we cannot have $\alpha \in A$ for if we did, then $\beta \in A$ by heredity. But since $\chi \in \neg A$, this means that $\forall\beta(\chi \sqsubseteq \beta$ only if $\beta \notin A)$.

For the converse suppose $\chi \in {}^{\perp}A$, i.e., $\forall\alpha(\alpha \in A$ only if $\chi\perp\alpha)$. We show $\forall\alpha(\alpha \in A$ only if $(\chi \not\sqsubseteq \alpha))$ by assuming $\alpha \in A$. But this follows immediately from (50).

The reader is referred to a similar but more complicated result connecting the Routleys' definition of De Morgan negation in terms of a $*$ operator with the perp style treatment here. The key again is that $\chi\perp\alpha$ is equivalent in the canonical frame to the existence of a "consistent" filter β such that $\chi, \alpha \subseteq \beta$, the difference being that unlike the intuitionistic case "consistent" is not the same as "proper" since we do not necessarily have absurdity. So χ and α can be inconsistent even though β is required not to be.

8 Lattices

In this section we investigate briefly what happens when we are presented with a structure $\mathcal{L} = (L, \wedge, \vee, \neg)$ where $(L, \wedge, \vee)$ is a lattice but not necessarily a distributive one. If $\neg$ is of period two and satisfies contraposition, it can be easily checked that it satisfies all of the De Morgan laws. We then call $\mathcal{L}$ *involuted.*[15] Goldblatt investigated structures like these but where it was also required that $x \wedge \neg x \leq y$ (and hence $y \leq x \vee \neg x$). Such a negation is called *orthocomplement* and such structures are called *orthocomplemented lattices* or just *ortholattices.* Goldblatt in fact showed how one could represent ortholattices by a perp relation that is irreflexive and symmetric, defining $A^{\perp}$ as usual as $\{\chi : A\perp\chi\}$, restricting propositions to be the Galois closed sets $A = A^{\perp\perp}$, and defining $A \wedge B = A \cap B$, $A \vee B = (A^{\perp} \cap B^{\perp})^{\perp}$. It is a small extension of this result to accommodate involuted lattices by dropping the requirement that $\perp$ be irreflexive. In the canonical frame the states are arbitrary filters (not necessarily prime). Cf. (Dunn 1993a, b).

Our investigation of the behavior of negation on an arbitrary lattice is not complete. It would be good to investigate the behavior of weaker negations, corresponding to preminimal negation, Galois negations, etc. The idea would be to consider some "nice" representation of lattices, and show how these various negations could be represented using something analogous to the perp relation and the role it plays for representing negation on posets and distributive lattices. There are at least three candidates for the underlying

[15]One is strongly tempted to call this a "De Morgan lattice", but the literature reserves this term for the distributive case.

lattice representation: that of Urquhart (1978), or of Hartonas and Dunn (1995), or of Hartonas (1993). Hartonas (1993) has already begun such an exploration.

References

Anderson, A.R., Belnap, N.D., & Dunn, J.M., *et al.*: *Entailment: The Logic of Relevance and Necessity*, Vol. 2. Princeton 1992.

Białynicki-Birula, A., & Rasiowa, H.: On the Representation of Quasi-Boolean Algebras. *Bulletin de l' Academie Polonaise des Sciences* 5 (1957), 259–261.

Birkhoff, G.: *Lattice Theory.* Providence (American Mathematical Society), (1940, 1948, 1967).

Birkhoff, G., & von Neumann, J.: The Logic of Quantum Mechanics. *Annals of Mathematics* 37 (1936), 823–843.

Demos, R.: A Discussion of a Certain Type of Negative Proposition. *Mind* 26 (1917), 188–196.

Dipert, R.: *Development and Crisis in Late Boolean Logic: The Deductive Logics of Peirce, Jevons and Schröder.* PhD thesis, Indiana University, 1978.

Došen, K.: Negation as a Modal Operator. *Reports on Mathematical Logic* 20 (1986), 15–27.

Dunn, J.M.: Gaggle Theory: An Abstraction of Galois Connections and Residuation with Applications to Negation and Various Logical Operations. In: *Logics in AI, Proceedings European Workshop JELIA 1990*, LNCS 478, Berlin 1990, 31–51.

Dunn, J.M.: Partial-Gaggles Applied to Logics with Restricted Structural Rules. In: *Substructural Logics*, ed. by P. Schroeder-Heister & K. Došen, Oxford 1993a, 63–108.

Dunn, J.M.: Perp and Star: Two Treatments of Negation. In: *Philosophical Perspectives* (*Philosophy of Language and Logic*) 7 (1993b), ed. by J. Tomberlin, 331–357.

Dunn, J.M.: Gaggle Theory Applied to Modal, Intuitionistic, and Relevance Logics. In: *Logik und Mathematik: Frege-Kolloquium 1993*, ed. by I. Max and W. Stelzner, Berlin/New York 1995a.

Dunn, J.M.: A Comparison of Various Model-theoretic Treatments of Negation: A History of Formal Negation. Forthcoming in: *What is Negation*, ed. by D. Gabbay & H. Wansing, Oxford 1996.

Everett, C.J.: Closure Operators and Galois Theory in Lattices. *Transactions of the American Mathematical Society* 55 (1944), 514–525.

Goldblatt, R.I.: Semantic Analysis of Orthologic. *Journal of Philosophical Logic* 3 (1974), 19–35.

Girard, J.-Y.: Linear Logic. *Theoretical Computer Science* 50 (1987), 1–102.

Hartonas, C.: Lattices with Additional Operators: A Unified Approach to the Semantics for Substructural Logics. Indiana University Logic Group Preprint Series, IULG-93-27, 1993.

Hartonas, C.: *Semantic Aspects of Substructural Logic.* Doctoral Dissertation, Indiana University, 1994.

Hartonas, C., & Dunn, J.M.: Duality Theorems for Partial Orders, Semilattices, Galois Connections and Lattices. Indiana University Logic Group Preprint Series, IULG-93-26, 1993.

Hazen, A.: Subminimal Negation. Unpublished ms., 1994.

Horn, L.R.: *A Natural History of Negation.* Chicago 1989.

Jammer, M.: *The Philosophy of Quantum Mechanics.* New York 1974.

Johansson, I,: Der Minimalkakül, ein reduzierter intuitionistischer Formalismus. *Compositio Mathematica* 5 (1936), 119–136.

Jónsson B., & Tarski, A.: Boolean Algebras with Operators. *American Journal of Mathematics* 73-74 (1951-52), 891–939, 127–162.

Kripke, S.: Semantic Analysis of Intuitionistic Logic I. In: *Formal Systems and Recursive Functions*, ed. by J. Crossley & M. Dummett, Amsterdam, 1965, 92–129.

Meyer R.K., & Routley, R.: Algebraic Analysis of Entailment. *Logique et Analyse*, n.s. 15 (1972), 407–428.

Ore, O.: Galois Connexions. *Transactions of the American Mathematical Society* 55 (1944), 493–513.

Priestley, H.A.: Ordered Topological Spaces and the Representation of Distributive Lattices. *Proceedings of the London Mathematical Society* 24 (1972), 507–530.

Routley R., & Meyer, R.K.: The Semantics of Entailment I. In: *Truth, Syntax and Modality,* ed. by H. Leblanc, Amsterdam, 1973, 199–243; The Semantics of Entailment II-III. *Journal of Philosophical Logic* 1 (1972), 53–73, 192–208.

Routley, R., & Routley, V.: Semantics of First-Degree Entailment. *Noûs* 6 (1972), 335–359.

Stone, M.: Topological Representations of Distributive Lattices and Brouwerian Logics. *Casopsis pro Pestovani Matematiky a Fysiky* 67 (1937), 1–25.

Urquhart, A.: A Topological Representation Theorem for Lattices. *Algebra Universalis* 8 (1978), 45–58.

Urquhart, A.: Distributive Lattices with a Dual Homomorphic Operation. *Studia Logica* 38 (1979), 201–209.

ORDER-DUALITY, NEGATION AND LATTICE REPRESENTATION

Chrysafis Hartonas*

1 Introduction

The question we raise is whether in the absence of a suitable negation operator the fact of order-duality of logical operators such as conjunction and disjunction is sufficient to sustain a representation by sets of the Lindenbaum algebra of the calculus, thus leading to the construction of a canonical frame for the logic. The concept of order-duality becomes critical if another convenience, the distribution law, is dropped.

The significance of resolving the lattice representation problem for non-distributive lattices becomes apparent when considering logical systems with restricted structural rules where distribution is often not assumed. Order-duality of meets and joins becomes manifest when the lattice is endowed with a classical-type negation operator and it is expressed in the de Morgan identity $a \vee b = \neg(\neg a \wedge \neg b)$. We explore a more general setup in which a generalization of the de Morgan identity holds. In this setup negation is absent but there is what may be thought of as a generalized negation operator[1], a *duality* on a partial order (or between two distinct partial orders). With a variant of the de Morgan identity in place we may always ignore joins, since they can be recovered from meets and from the duality map which generalizes negation.

The idea for a representation of lattices is then to represent a lattice as a diagram of intersection semi-lattices with a duality between them. The duality operator serves to recover joins on the semi-lattices which of course do not coincide with unions as the original lattice is not assumed to satisfy the distribution law. We conclude with a statement of a representation of a general bounded lattice by subsets of an appropriate set X. The result is subsequently strengthened by imposing topological structure on the set X and by thereby obtaining a characterization of the family of subsets representing the original lattice as the family of *stable compact-open* subsets of X.

---

*J. Michael Dunn is acknowledged for many useful discussions.

[1]In (Hartonas & Dunn 1993) we propose a general concept of *split negation* and investigate some varieties of it.

From a logical point of view there are two cases where the semantics of conjunction and disjunction is well understood. That where distribution is assumed and that where an orthonegation operator is present.

In both cases the semantic analysis of logical operators can be based on a representation by sets of the Lindenbaum algebras: Boolean and distributive lattices or ortholattices (Goldblatt 1975, Priestley 1970, Stone 1937, Stone 1938). Propositions, the semantic counterparts of sentences, are modelled as certain kinds of sets (just any subsets of a fixed set for the distributive case, or only the *regular* subsets in the case of orthologic (Goldblatt 1974, 1975).

What we intend to do in this paper is to extend this framework to the case of a possibly non-distributive calculus, possibly also lacking an orthonegation operator. The solution we present can be extended to the case of logical systems with a variety of additional operators (Hartonas 1994b).

1.1 Negation and Duality

A lattice $\mathcal{L}$ is a partially-ordered structure $(L, \leq, \wedge, \vee)$ where the greatest lower bound (glb) $a \wedge b$ of the elements a, b and their least upper bound (lub) $a \vee b$ exist. $\wedge$ and $\vee$ are *order-dual* operators in the sense that inverting the order in the lattice results in switching the roles of glb's and lub's. Order-duality becomes clear when the lattice comes equiped with a dualizing map[2], that is to say an antitone map δ ($a \leq b$ implies $\delta b \leq \delta a$) of period two ($\delta\delta a = a$, for any element a). In such a case joins can be recovered from meets and from the dualizing map δ by the de Morgan identity $a \vee b = \delta(\delta a \wedge \delta b)$. The informal content of the identity is that to compute the join of two elements a and b we need to move over to the opposite lattice of $\mathcal{L}$ (where the order is inverted) and compute there the meet of the dual objects $\delta a, \delta b$ of a and b.

In fact it is not necessary that the lattice be equipped with such an orthonegation operator. More generally, if $\mathcal{L}$ and $\mathcal{K}$ are lattices and $\delta : \mathcal{L} \to \mathcal{K}$ is an antitone map such that for any element a in the lattice $\mathcal{L}$ we have $\delta\delta a = a$, then joins in $\mathcal{L}$ can be recovered by the de Morgan identity $a \vee_L b = \delta(\delta a \wedge_K \delta b)$. We may then as well consider $\mathcal{L}$ and $\mathcal{K}$ as *meet-semilattices* since the full lattice structure can be recovered via δ.

Algebraically speaking a negation operator is an instance of a *galois connection*. Given partial orders $(P, \leq_P)$ and $(Q, \leq_Q)$, a pair of antitone maps λ from P to Q and ρ from Q to P is a galois connection just in case for any p in P and q in Q we have $p \leq_P \rho q$ iff $q \leq_Q \lambda p$. This is equivalent to the

[2]Logically speaking such a map is of course an orthonegation operator.

conditions that $p \leq_P \lambda\rho p$ and $q \leq_Q \rho\lambda q$. In the case of negation $P = Q$ and the two maps coincide $\lambda = \rho = \neg$.

Galois connections will not in general satisfy the period-two condition. When they do, that is to say when for any p in P and q in Q we have $p = \rho\lambda p$ and $q = \lambda\rho q$, then we call the connection a *duality*.

A further generalization of the de Morgan identity is then possible. For suppose that $\mathcal{S}$ and $\mathcal{K}$ are meet semilattices with a duality $\lambda : \mathcal{S} \to \mathcal{K}$ and $\rho : \mathcal{K} \to \mathcal{L}$. Then again both $\mathcal{S}$ and $\mathcal{K}$ are in fact full lattices and joins can be recovered from meets using the duality maps as follows:

(De Morgan) $a \vee_S a' = \rho(\lambda a \wedge_K \lambda a')$ $b \vee_K b' = \lambda(\rho b \wedge_S \rho b')$

This is the general form of the de Morgan identity we will rely on in the sequel. The case of orthonegation is an obvious instance: $a \vee b = \neg(\neg a \wedge \neg b)$.

1.2 Boolean and Distributive Lattices

We think of a Boolean algebra (BA) as the Lindenbaum algebra of classical logic. Similarly for a distributive lattice (DL) and negationless classical logic. Stone's original idea for the representation of BA's has been to model an element a of the Boolean algebra (which we may think of as an equivalence class of provably equivalent sentences) by the set of "interpretations" that make it true. If $\mathbf{2}=\{1,0\}$ is the set of truth values, then $\mathbf{2}$ has an obvious BA structure. By an "interpretation" we mean a map h from the given Boolean lattice $\mathbf{B}$ into the truth value lattice such that all operators are respected, namely such that $h(a \wedge b) = ha \wedge hb$, $h(a \vee b) = ha \vee hb$ and $h(\neg a) = -(ha)$ (where of course $-0 = 1$ and $-1 = 0$). Such a map is a Boolean algebra *homomorphism*.

It is not hard to see that for each such BA homomorphism h the set of elements that it makes true $\{a | ha = 1\}$ is a *maximal filter* (ultrafilter)[3] of the BA. Conversely, every maximal filter x determines such a homomorphism via its characteristic function: $ha = 1$ if a is in x and $ha = 0$ otherwise. Hence rather than modeling an element a of the BA $\mathbf{B}$ by the set of BA-homomorphisms h such that $ha = 1$ we may alternatively consider the set of maximal filters x such that a is in x.

Semantically speaking then we set up a set of *worlds* (states, situations, setups, indices etc) where in the canonical frame a world is simply a maximal filter of the Lindenbaum algebra of classical logic and we interpret a sentence

[3]A *filter* of a lattice $\mathcal{L}$ is a subset x of L that is upwards closed (a in x and $a \leq b$ implies b is in x) and such that if a and b are both in x then so is their glb $a \wedge b$. A filter x is *prime* if whenever $a \vee b$ is in x then at least one of a or b is also in x. It is *maximal* (or an *ultrafilter*) if there is no filter properly between it and the whole lattice (the improper filter).

as the set of worlds that contain the equivalence class of the given sentence. This establishes a completeness theorem for the logic.

For distributive lattices (DL) essentially the same intuition leads to a similar construction of a canonical frame. The only difference is that in the absence of negation and given a DL-homomorphism h from a given distributive lattice $\mathbf{D}$ into the truth-value lattice $\mathbf{2}$ the inverse image $h^{-1}(1) = \{a|ha = 1\}$ is not a maximal but a *prime* filter, hence the worlds in the canonical model are taken to be the prime filters of the Lindenbaum algebra of the calculus. This intuition completely breaks down, however, in the case of general lattices as we explain below.

2 Lattice Representation

Our interest in lattice representation is logical, or rather semantical. There is today a host of logical systems that drop the distribution law, some coming with a classical-type negation operator, such as Classical Linear Logic (Girard 1987), and some not, such as Intuitionistic Linear Logic (Abrusci 1990, Troelstra 1992). The model-theoretic question that arises is then to find a systematic approach to the semantics of such systems, an approach that does not depend directly on what particular operators are assumed but rather on what abstract properties are assumed for each of these operators. The two basic conveniences for a solution of such a problem are, first, an assumption of a classical-type negation operator and, second, an assumption of distribution of conjunction over disjunction. In both cases, as we already mentioned, a Kripke-style semantics for the relevant logical systems can be developed on the basis of the existing representations of the underlying lattices (distributive or ortholattices). We examine here the situation for the case where both of these convenient assumptions are dropped, that is to say the case where we simply have a bounded lattice L.

The Stone idea of modeling an element of a lattice by the set of lattice homomorphisms that map it to 1 (true) fails in the present case. The reason is that, as the reader can easily verify, for a given lattice homomorphism $h : \mathbf{L} \to \mathbf{2}$ the pre-image of 1, namely the set $\{a \in L|ha = 1\}$ is still a *prime* filter.

We base our representation on the idea that in an appropriate setup we may only consider the meet-semilattice structure and ignore joins. Specifically, we will represent a lattice L as a diagram of two intersection semilattices with a duality between them. From our previous discussion it will follow that each of the two intersection semi-lattices is in fact a full lattice and we will verify that the original lattice is isomorphic to one of them (and dually isomorphic to the other).

We generalize Stone's approach and consider the meet-*hemimorphisms* $h : \mathbf{L} \to \mathbf{2}$, that is to say the maps from $\mathbf{L}$ to $\mathbf{2}$ that only preserve meets. If h is such a hemimorphism then the pre-image of 1 under h is a *filter* but not necessarily a prime one. Indeed if $ha = 1$ and $a \leq b$ then $1 = ha \leq hb$ and so $hb = 1$ as well. Also, if $ha = 1$ and $hb = 1$ then $h(a \wedge b) = ha \wedge hb = 1 \wedge 1 = 1$ and so $\{e|he = 1\}$ is a filter. Conversely, if x is a filter of L then the characteristic function of x is a meet hemimorphism. Indeed, if at least one of a, b is not in x, then $a \wedge b$ is not in x either hence $h(a \wedge b) = 0 = ha \wedge hb$ (since either $ha = 0$ or $hb = 0$). Otherwise both a and b are in x in which case also $a \wedge b$ is in x and then again $h(a \wedge b) = ha \wedge hb$.

Hence we may consider as the dual space of a lattice L the space X of its filters. The representation map is the natural map $H : L \to \mathcal{P}(X)$ sending an element a of the lattice to the set of filters that contain it (the set of hemimorphisms that map it to true). We write $X_a = Ha$ for this set: $X_a = \{x \in X |\ a \in x\}$. Letting also $X_* = \{X_a | a \in L\}$ it is not hard to see that X_* is an intersection semilattice, where $X_a \bigcap X_b = X_{a \wedge b}$.

Incidentally, Goldblatt's representation of ortholattices (Goldblatt 1975) is presented along the same lines. Having an orthonegation $\neg$ in the lattice Goldblatt then proceeds to define an *orthogonality* relation $\perp$ on filters by $x \perp z$ iff there is an element $a \in x$ such that $\neg a \in z$. Since $\neg$ is an orthonegation, it easily follows that $\perp$ is an irreflexive and symmetric relation. The relation $\perp$ induces then a galois connection on subsets of the set of filters. By symmetry of $\perp$ the two galois maps coincide and can be explicitly defined on a subset $U \subseteq X$ by $U^{\perp} = \{x | U \perp x\}$, where $U \perp x$ means that for all $z \in U$, $z \perp x$. A *regular* subset of X is then defined in (Goldblatt 1975) as a subset $A \subseteq X$ such that $A = A^{\perp\perp}$. The sets X_a, $a \in L$ are then shown to be regular and $(\ )^{\perp}$ serves to define joins of regular subsets by the de Morgan identity $A \bigvee B = (A^{\perp} \cap B^{\perp})^{\perp}$.

In the absence of an orthonegation on the lattice we will make use of the generalization of the de Morgan identity resulting from a duality between two appropriate intersection semilattices. To find an intersection semilattice which is dual to X_* we first notice that $a \in x$ iff $[a) = x_a \subseteq x$.[4] Hence $X_a = \{x | x_a \subseteq x\}$. Consider then also the sets $X^a = \{x | x \subseteq x_a\}$, where we have dualized the inclusion $x_a \subseteq x$ to obtain $x \subseteq x_a$. Let also $X^* = \{X^a | a \in L\}$. Again it is not hard to verify that X^* is an *intersection*-semilattice, where $X^a \bigcap X^b = X^{a \vee b}$. What we need to establish is that there exists a duality between X_* and X^*.

As pointed out in (Birkhoff 1949), every binary relation R on a set X induces a galois connection on its subsets. In the particular case where $R = \leq$ is

[4]We will always write x_a for the principal filter $[a)$ generated by a.

a partial order the resulting galois connection is known as the Dedekind-McNeile connection. We use inclusion of filters as a binary relation on the set X of filters and generate the Dedekind-McNeile maps, explicitly defined on subsets $U, V \subseteq X$ by

$$\lambda U = \{x | U \leq x\} \quad \rho V = \{x | x \leq V\}$$

where $U \leq x$ means that x is above every element of U (an upper bound of U) and similarly for $x \leq V$.

It can be now verified that λ and ρ form a duality between X_* and X^*. This follows from the identities $\rho X_a = X^a$ and $\lambda X^a = X_a$. We prove the first and leave verification of the second to the interested reader. If $x \in \rho X_a$ then x is a lower bound of X_a hence $x \subseteq x_a$, since $x_a \in X_a$. Hence $x \in X^a = \{x | x \subseteq x_a\}$. For the converse, suppose $x \in X^a$, hence $x \subseteq x_a$. If z is any filter in X_a, then $x_a \subseteq z$ and thereby $x \subseteq z$, hence x is a lower bound of X_a, that is to say $x \in \rho X_a$.

Hence we may use the generalized de Morgan identity to define joins in X_*, letting

$$X_a \bigvee X_b = \lambda(\rho X_a \bigcap \rho X_b)$$

Thus we can conclude with the following

Theorem 1 *For every bounded lattice L there is a set X and a collection X_* of subsets of X such that X_* is a lattice isomorphic to L.*

Proof: For the proof we take X to be the set of filters of L (including the improper filter L itself) and use the map H defined by $a \overset{H}{\mapsto} X_a$. H is a lattice homomorphism since $H(a \wedge b) = X_a \bigcap X_b$ and $H(a \vee b) = \lambda(\rho X_a \bigcap \rho X_b) = X_a \vee X_b$. It is also clear that H is a bijection, hence a lattice isomorphism. □

We now proceed to characterize the lattice X_* independently of the representation map as a collection of subsets of X having some properties definable from structure to be imposed on X. Referring again to (Goldblatt 1975), the collection of subsets of the space of filters of an ortholattice is characterized there as the collection of *regular, compact-open* subsets. The topology imposed on X is that generated by the subbasis[5]

$$\mathcal{S} = \{X_a\}_{a \in L} \bigcup \{-X_b\}_{b \in L}$$

[5]A topology on a set X is a collection $\mathcal{T}$ of subsets of X, called the *open sets*, such that $\emptyset$ and X are in $\mathcal{T}$ and $\mathcal{T}$ is closed under arbitrary unions and finite intersections. A family $\mathcal{S}$ is a *subbasis* for a topology $\mathcal{T}$ if $\mathcal{T}$ is the collection inductively constructed by taking arbitrary unions of finite intersections of members of $\mathcal{S}$. For later reference we recall also that a set $K \subseteq X$ is *compact* in the topology $\mathcal{T}$ just in case if $U_i, i \in I$ is a family of open sets such that $K \subseteq \bigcup_{i \in I} U_i$ then there is a finite subcollection $U_j, j \in J \subseteq^{fin} I$ such that $K \subseteq \bigcup_{j \in J} U_j$. The space X is compact if it is a compact set in the previous sense. A set U is *clopen* if both U and $-U$ (its complement in X) are open. The space

We recall from (Goldblatt 1975) that the filter space with the above topology is a *Stone space*. In the absence of orthonegation we have been able to substitute the operator $(\ )^{\perp}$ on subsets by the two maps λ and ρ. If we let $\Gamma = \lambda\rho$ be their composite then it is easy to verify that for any subsets $U, V \subseteq X$ we have $U \subseteq \Gamma U$, $\Gamma\Gamma U = U$ and if $U \subseteq V$ then $\Gamma U \subseteq \Gamma V$. In other words Γ is a *closure operator* on subsets of X. We call a subset $A \subseteq X$ such that $\Gamma A = A$ a *stable* set. Theorem 1 can be then strengthened to the following theorem, concluding our lattice representation:[6]

Theorem 2 *For every bounded lattice L there is a Stone space X and a closure operator Γ on subsets of X such that L is isomorphic to the lattice of stable compact-open subsets of X.*

Proof: The space X is of course the filter space with the Stone topology discussed above and where Γ is the closure operator we defined by composition $\Gamma = \lambda\rho$ of the Dedekind-McNeile maps.

Every set X_a is (stable and) clopen, by the way the topology was defined, hence compact-open since the space is compact and Hausdorff (the latter follows from total separation). For the converse, notice first that the stable sets are exactly the upper closures $x\uparrow$ of single points x in the space. Indeed, if U is a stable set of filters, let $x = \bigcap U = \bigcap\{z | z \in U\}$. The intersection is never empty since for any $z \in U$, $\top \in z$ (recall that we assume a lattice always comes with a top element). Then $y \leq U$ iff $y \leq x$. Hence $\rho U = x\downarrow$ and $U = \lambda\rho U = x\uparrow$.

Consequently, every stable set U can be written in the form $U = \bigcap_{a\in x} X_a$, for some $x \in X$. Indeed, if U is stable, then $U = x\uparrow$, for some $x \in X$. It is then clear that $U = \bigcap_{a\in x} X_a$, where x is the base point of U.
It follows then that $-U = \bigcup_{a\in x}(-X_a)$. Hence, if U is clopen, by compactness, let $a_1, \ldots, a_n \in x$ such that $-U$ can be written as the finite union of the sets $-X_{a_i}$, $i = 1, \ldots, n$. Then $U = \bigcap_{i=1}^{i=n} X_{a_i}$. Letting $a = a_1 \wedge \cdots \wedge a_n$, it follows that $U = X_a$, qed. □

3 Conclusions

A lattice L with an orthonegation operator is a particular instance of a diagram of meet semilattices with a duality between them. We generalized the de Morgan identity $a \vee b = \neg(\neg a \wedge \neg b)$ and pointed out that if λ and ρ is a duality connecting two meet semilattices then each of them is a full lattice

---

is *totally-separated* if for any distinct points $x, z \in X$ there is a clopen set C containing exactly one of x, z. A *Stone space* is a compact, totally separated space.

[6] A Stone space is compact and Hausdorff. It is well known that in a compact-Hausdorff space compact and closed sets are the same. Hence in a Stone space we may use "compact-open" and "clopen" interchangeably.

with joins recoverable from meets and from the duality in a de Morgan manner. We then showed that every lattice determines a diagram of two intersection semilattices with a duality between them and, using this fact, we obtained a representation by sets of a general bounded lattice.

Frames for logics on a non-distributive calculus of conjunction and disjunction are of the form $\mathcal{F} = \langle W, \prec, \ldots \rangle$, where $\prec$ is a binary accessibiltiy relation on W. The *propositions* of the frame are stable sets $A = \lambda\rho A$, where λ and ρ are the maps defined on subsets of W by

$$\lambda U = \{x | U \prec x\} \quad \rho V = \{x | x \prec V\}$$

where $U \prec x$ means that for every z in U we have $z \prec x$ (and similarly for $x \prec V$). The semantic clause for disjunction that arises from our representation is of course not the standard Kripke clause since this has a built-in assumption of distribution. The set of states x where $\phi \vee \psi$ is verified cannot be just the union of the sets of states verifying one of ϕ or ψ (as in the Kripke clause) but needs to be completed by including states that may not support directly the truth of either ϕ or ψ. This is captured in the representation by the closure operator Γ. If $\|\phi\|$ is the proposition modelling ϕ (the set of x such that $x \models \phi$) then $\|\phi \vee \psi\| = \Gamma(\|\phi\| \cup \|\psi\|) = \lambda(\rho\|\phi\| \cap \rho\|\psi\|)$. The resulting clause for disjuction is

$$x \models \phi \vee \psi \text{ iff } \forall y(\text{ if } y \prec \|\phi\| \cup \|\psi\|, \text{ then } y \prec x)$$

(where recall that for a set V, $y \prec V$ means that $y \prec z$ for all z in V). Reading $x \prec y$ as *x extends to y*, $\phi \vee \psi$ is verified at a state x just in case x extends every state y that can be extended to all states z where either ϕ or ψ is verified. It is clear that every state x such that $x \models \phi$ or $x \models \psi$ satisfies this condition.

In the case of logical systems with additional connectives the frames can be extended to structures $\langle W, \prec, R, S, T, \ldots \rangle$ where $R, S, T, \ldots$ are relations on W serving to interpret the logical connectives. The general case of algebraizable logics with normal operators is treated in (Hartonas 1994b) where we also prove a duality of algebraic and Kripke semantics for normal algebraizable logics.

References

Abrusci, V.M.: Non-Commutative Intuitionistic Linear Logic. *Zeitschrift für math. Logik und Grundlagen der Mathematik* 36 (1990), 297–318.

Allwein, G., & Dunn, J.M.: Kripke Models for Linear Logic. *Journal of Symbolic Logic* 58 (1993), 514–545.

Allwein, G., & Hartonas, C.: Duality for Bounded Lattices. *Indiana University Logic Group, Preprint Series,* IULG-93-25, 1993.

Allwein, G., Dunn, J.M., & Hartonas, C.: Priestley Duality for Distributive Lattices with Normal Operators, in progress.

Birkhoff, G.: *Lattice Theory.* Providence (American Mathematical Society), (1940, 1948, 1967).

Davey, B.A., & Priestley, H.A.: *Introduction to Lattices and Order.* Cambridge Mathematical Textbooks, Cambridge 1990.

Došen, K.: Negation as a Modal Operator. *Reports on Mathematical Logic* 20 (1986), 15–27.

Dunn, J.M.: Gaggle Theory: An Abstraction of Galois Connections and Residuation with Applications to Negation and Various Logical Operations. In: *Logics in AI, Proceedings European Workshop JELIA 1990*, LNCS 478, Berlin 1990, 31–51.

Dunn, J.M.: Perp and Star: Two Treatments of Negation. In: *Philosophical Perspectives (Philosophy of Language and Logic)* 7 (1993), ed. by J. Tomberlin, 331–357.

Dunn, J.M.: Partial-Gaggles applied to Logics with Restricted Structural Rules. *Indiana University Logic Group, Preprint Series*, IULG-93-22, 1993b.

Girard, J.-Y.: Linear Logic. *Theoretical Computer Science* 50 (1987), 1–102.

Goldblatt, R.I.: The Stone Space of an Ortholattice. *Bulletin London Mathematical Society* 7 (1975), 45–48.

Goldblatt, R.I.: Semantic Analysis of Orthologic. *Journal of Philosophical Logic* 3 (1974), 19–35.

Hartonas, C.: *Algebraic and Kripke Semantics for Substructural Logics.* Phd thesis, Departments of Mathematics and of Philosophy, Indiana University, 1994a.

Hartonas, C. & Dunn, J.M.: Duality Theorems for Partial Orders, Semilattices, Galois Connections and Lattices. *Indiana University Logic Group, Preprint Series*, IULG-93-26, 1993.

Hartonas, C.: Duality for Lattice-Ordered Algebras and for Normal Algebraizable Logics. Technical Report 1994/18, University of Leicester, Dept. of Mathematics and Computer Science, 1994b. To appear in: *Studia Logica.*

Hartung, G.: A Topological Representation of Lattices. *Algebra Universalis* 17 (1992), 273–299.

Johnstone, P.T.: *Stone Spaces.* Cambridge 1986.

Jónsson, B., & Tarski, A.: Boolean Algebras with Operators. *American Journal of Mathematics*, 73–74 (1951–2), 891–939, 127–162.

Ono, H., & Komori, Y.: Logics Without the Contraction Rule. *Journal of Symbolic Logic* 50 (1985), 169–201.

Ono, H.: Structural Rules and a Logical Hierarchy. In: *Mathematical Logic, Proceedings of the Summer Schools and Conference on Mathematical Logic, Heyting '88*, ed. by P.P. Petkov, New York 1990, 95–104.

Ono, H.: Algebraic Aspects of Logics without Structural Rules. *AMS, Contemporary Mathematics*, 131 part 3, 1992, 601–621.

Ono, H.: Semantics for Substructural Logics. In: Schroeder-Heister, P., & Došen, K., eds., *Substructural Logics*, Oxford 1993, 259–291.

Priestley, H.A.: Representation of Distributive Lattices by means of Ordered Stone Spaces. *Bulletin London Mathematical Society* 2 (1970), 186–90.

Routley, R., & Meyer, R.K.: The Semantics of Entailment I. In: *Truth, Syntax and Semantics*, ed. by H. Leblanc, Amsterdam 1973, 194–243.

Schroeder-Heister, P., & Došen, K., eds.: *Substructural Logics*, Oxford 1993.

Stone, M.H.: The Representation of Boolean Algebras. *Bulletin American Mathematical Society* 44 (1938), 807–16.

Stone, M.H.: Topological Representation of Distributive Lattices and Brouwerian Logics. *Casopsis pro Pestovani Matematiky a Fysiky* 67 (1937), 1–25.

Troelstra, A.S.: *Lectures on Linear Logic.* Center for the Study of Language and Information, CSLI Lecture Notes No 29, 1992.

Urquhart, A.: A Topological Representation Theorem for Lattices. *Algebra Universalis* 8 (1979a), 45–58.

Urquhart, A.: Distributive Lattices with a Dual Homomorphic Operation I, II. *Studia Logica* 38, 40 (1979b, 1982), 201–209, 391–404.

Wansing, H.: *The Logic of Information Structures*, LNAI 681. Berlin 1993.

NECESSARY CONDITIONS FOR NEGATION OPERATORS*

Wolfgang Lenzen

1 Introduction

In recent years, various logics have been developed which deviate from the classical propositional calculus (**PC**) typically with regard to negation. Thus, not only the well-known calculus of intuitionistic logic, but also less-known systems of paraconsistent logic basically adopt the "uncritical" operators of conjunction, $\wedge$, and disjunction, $\vee$, while rejecting the seemingly all too simple, two-valued operator of negation, $\neg$. In what follows '$\sim$' will be used as a symbol for arbitrary other negation operators. Furthermore, given a certain logic **L**, we use '$\vdash_L$' to denote the syntactic relations of *provability* and of *deducibility* (or derivability) in **L** as they are determined by the axioms and rules of deduction of **L**. Thus '$\vdash_L p$' means that p is provable in **L**, while '$p_1, \ldots, p_n \vdash_L q$' expresses that, in **L**, the conclusion q can be derived from the premises $p_1, \ldots, p_n$. The corresponding semantic relations '$\{p_1, \ldots, p_n\}$ *logically imply* q' and 'p is *logically true*' (in **L**) are formalized by '$p_1, \ldots, p_n \models_L q$', and '$\models_L p$', respectively. The subscript 'L' will often be dropped for convenience when it is clear which logic **L** we are talking about. In what follows we presuppose an old-fashioned or "classical" understanding of logical implication in the sense of the subsequent condition:

(IMP) Whatever the details of the semantics of **L** may be, '$p_1, \ldots, p_n \models_L q$' holds if and only if the consequent q necessarily becomes true or gets assigned some other "distinguished" true-like value once all of the antecedent formulae $p_1, \ldots, p_n$ are true or have been assigned some other "distinguished" truth value.

---

*This paper is the partial outcome of a research on non-classical negation that was begun already in the late 80's and that was mainly carried out during the author's sabbatical leave in summer 1990. For several reasons, the first draft of the study "On Negation" remained a torso for quite a long time. A first, relatively small part of the material (former chapter IV) was improved and transformed into a paper on intuitionistic logic, (Lenzen 1991). Although the sequel was optimistically announced there as forthcoming in 1992, the bulk of the material stayed basically untouched until Heinrich Wansing invited me to participate in the *Analyomen*-workshop in September 1994. During the conference I presented an amended version of former chapter I. The present paper has profited quite a lot from the lively discussion during the workshop. The remaining material (especially former chapter V on paraconsistent logic) still needs to be elaborated and improved before appearing somewhen elsewhere. My special thanks are to Heinrich Wansing both for his constructive criticism and for encouraging me to bring this paper into a form that is hopefully worth while publishing.

Thus we do not restrict our considerations only to logics which can be characterized by a classical two–valued semantics. But if one employs a three–, or four–, or ... n–valued or any other kind of semantics, we still insist on the core–idea of logical implication by requiring (i) that one retains at least a counterpart – "true" – of the classical truth value 'true', and (ii) that a proposition q logically follows from a set of propositions $\{p_1, \ldots, p_n\}$ just in case that q cannot fail to have this distinguished value "true" if all of the p_i are "true". This requirement entails that, for each logic **L** to be considered here, the relation of logical implication must in particular be *reflexive*, $p \models_L p$, and *monotonic*, i.e. whenever a conclusion q logically follows from a certain set of premises $\{p_1, \ldots, p_n\}$, the same conclusion follows also from any larger set $\{p_1, \ldots, p_n, p_{n+1}, \ldots, p_m\}$. Clearly, if q is "true" whenever all sentences from $\{p_1, \ldots, p_n\}$ are "true", then a fortiori q must be "true" if all sentences from $\{p_1, \ldots, p_n, p_{n+1}, \ldots, p_m\}$ are "true". Since in a complete and sound logic **L** a consequent formula q can be syntactically derived from a set of antecedent formulas $\{p_1, \ldots, p_n\}$ if and only if the conclusion q logically follows from $p_1, \ldots, p_n$, the deducibility relation $\vdash_L$ of every logic also has to satisfy the laws of reflexivity and monotonicity:

(REFL) $p \vdash p$

(MONO) If $p_1, \ldots, p_n \vdash q$, then $p_1, \ldots, p_n, p_{n+1}, \ldots, p_m \vdash q$.

What does this restriction, however, mean for so–called "non–monotonic logics" as they have recently been developed mainly in the field of Artificial Intelligence? Now, a proponent of **MONO** need not deny that there are meaningful forms of non–monotonic *inferences* or non–monotonic *reasoning*. For instance, the basic idea of "default reasoning" may be described as follows. From the premises $p_1, \ldots, p_n$ one may conclude by way of default reasoning that q, if, in the absence of further contrary evidence, q can "normally" be expected to be true if the premises $p_1, \ldots, p_n$ all are known to be true. But when certain "unnormal" counter–evidence $p_{n+1}, \ldots, p_m$ turns up, it may no longer be reasonable to expect that q; instead it may then be "normal" to expect that $\neg q$.[1] Hence the "logic" of default reasoning, $\mathbf{L}^*$, is non–monotonic in the strong sense that $p_1, \ldots, p_n \vdash_{L^*} q$ is compatible with $p_1, \ldots, p_n, p_{n+1}, \ldots, p_m \vdash_{L^*} \neg q$. An advocate of the old–fashioned concept of logical implication therefore need not protest against the very construction of self–consistent calculi $\mathbf{L}^*$, in which the inference relation $\vdash_{L^*}$ is non–monotonic; and he may perhaps also not protest if such non–monotonic calculi nevertheless are called "logics". After all there is no copyright for the word 'logic', and during the long history of philosophy many strange theories have come to be called by that name. A proponent of the semantic principle

[1] A somewhat worn–off example is the default inference from the only premise p_1 = 'Tweety is a bird' to the conclusion q = 'Tweety can fly', which no longer remains valid once the additional premise is added p_2 = 'Tweety is a penguin'.

IMP and of its syntactic corollary **MONO** must, however, insist on one decisive point, namely that *the inference relation of a non-monotonic "logic" never constitutes a relation of logical implication*! This may be trivial, and in the past logicians usually appear to have been well aware of this fact. For instance, when philosophers like Hans Reichenbach or Rudolf Carnap constructed their systems of "inductive logic", they would never have been mislead to think that inductive inferences might be *logical* inferences in the sense of *deductive* inferences, after all. Hopefully contemporary advocates of default "logic" and other systems of non-monotonic reasoning will keep in mind that their "deductions" are not properly *logical* consequences either, because these inferences don't have the character of necessarily making the conclusion "true" once all the premises are assumed to be "true". Anyway, in what follows, we consider only logical systems **L** the deducibility relation of which satisfies besides **REFL** and **MONO** also the so-called "cut-rule":

(**CUT**) If $\Delta, p \vdash_L q$, and $\Gamma \vdash_L p$, then $\Delta, \Gamma \vdash_L q$

plus the trivial structural rules of permutation and contraction:

(**PER**) If $\Delta, p_1, p_2 \vdash_L q$, then $\Delta, p_2, p_1 \vdash_L q$

(**CON**) If $\Delta, p, p \vdash_L q$, then $\Delta, p \vdash_L q$.[2]

Now, in order to determine whether a certain monadic operator $\sim$ of such a logic **L** really is a *negation*, one will primarily consider characteristic axioms or rules of deduction which can be formulated without the help of additional operators. Some well-known examples are the principles of double negation introduction and elimination:

(**DNI**) $p \vdash \sim\sim p$ (**DNE**) $\sim\sim p \vdash p$

and the so-called *ex falso quodlibet* which is, however, more appropriately referred to as '*ex contradictorio quodlibet*':

(**ECQ**) $p, \sim p \vdash q$.

If the logic **L** contains only a small number of such characteristic principles, it becomes very difficult if not impossible to decide whether $\sim$ really is a negation. For instance, if $\sim$ were characterized in **L** just by the two double negation principles, then nobody could tell whether $\sim$ represents a negation or instead an affirmation, because **DNI** and **DNE** hold true both under the interpretation $\sim p = \neg p$ and under the opposite interpretation $\sim p = p$. Therefore it may be necessary to investigate also logical implications that hold between "negated" formulas and certain other formulas containing further operators besides $\sim$. For convenience we presuppose that each logic **L** contains at least the following elementary part of the usual theory of conjunction and disjunction:

---

[2] Δ and Γ are arbitrary finite or infinite sets of formulas. As usual, if Δ is infinite, $\Delta \vdash q$ is taken to hold iff there exists a finite subset Δ^* of Δ such that $\Delta^* \vdash q$.

(AND) $(p \wedge q) \vdash p$
$(p \wedge q) \vdash q$
If $p \vdash q$ and $p \vdash r$, then $p \vdash (q \wedge r)$

(OR) $p \vdash (p \vee q)$
$q \vdash (p \vee q)$
If $p \vdash r$ and $q \vdash r$, then $(p \vee q) \vdash r$.[3]

The availability of these laws would not be absolutely essential for the main aim of our paper, i.e. for the development of certain conditions of adequacy that any proposed monadic operator has to satisfy in order to be rightly called a negation. It will turn out below that the decisive conditions can in general be formulated without the help of additional operators. However, given the theory of $\wedge$ and $\vee$ in the background of the respective logic **L**, it is much easier to illustrate the import of these conditions by means of simple examples.

2 Conditions of Adequacy

In this section three types of negation principles will be discussed: *Unacceptable* principles which a logic **L** must *never* satisfy, if its "negation" operator $\sim$ is to rate as a real negation; *dispensable* principles which, though they are valid principles of classical negation, need not necessarily be satisfied by arbitrary other negations; and *indispensable* principles which a logic **L** always has to satisfy if its monadic operator $\sim$ is to count as a genuine negation. Let us begin by considering some unacceptable principles. If **L** contains the elementary theory of conjunction, we would certainly be reluctant to accept a monadic operator $\sim$ as a real negation, if **L** were to contain an axiom or rule of deduction according to which the conjunction of two propositions entails the "falsity" of, say, its left conjunct:

(UN 1) $(p \wedge q) \vdash \sim p$.

Instead it seems reasonable to require that, in any plausible sense of the word 'false', the falsity of one conjunct entails the falsity of the entire conjunction. Similarly, if the logic **L** contains the classical theory of disjunction, then the monadic operator $\sim$ would not rate as a real negation, if **L** were to contain an axiom or rule of deduction according to which the disjunction of two propositions entails the "falsity" of one of the disjuncts, say:

(UN 2) $(p \vee q) \vdash \sim p$.

[3]In addition **L** may but need not contain a theory of implication which can be either material implication, $\supset$, or strict implication, $\rightarrow$. In this case we assume the implication operator to represent the metalinguistic relation of deducibility in the sense that $\vdash_L (p \supset q)$ or $\vdash_L (p \rightarrow q)$ iff $p \vdash_L q$.

If **L** contains only a part of **AND** and **OR**, we could otherwise derive either from **UN 1** or from **UN 2** the clearly unacceptable principle

(**UN 3**) $p \vdash \sim p$.

Although a self-consistent logic **L** may in general contain *some* propositions p which logically entail $\sim p$, it would be very strange indeed if *every* proposition could be proven to entail its own "falsity". This would evidently mean that the logic **L** itself is inconsistent, or that the negative formulae $\sim p$ does not really express the *negation* of p. Since **UN 3** is unacceptable, so is – a fortiori – the following principle according to which every proposition p would be provably "false" (in the sense of the non-classical negation operator $\sim$):

(**UN 4**) $\vdash \sim p$.

To be sure, if a certain logic $\mathbf{L}^*$ is *inconsistent*, i.e. if for every proposition q one has $\vdash_{L^*} q$, then the result **UN 4** would trivially hold no matter how the negation operator $\sim$ is interpreted. But it seems uncontroversial to postulate that a *self-consistent* logic must never satisfy any of the principles **UN 1**, **UN 2**, **UN 3**, or **UN 4**. If it is observed that **UN 1** – **UN 3** are deductively equivalent to each other (provided that **L** contains some most elementary laws of conjunction and disjunction)[4], it is sufficient to state the *first condition of adequacy* as follows:

(**ADQ 1**) If **L** is a self-consistent logic, and if $\sim$ is a real negation operator, then **L** must not satisfy **UN 3**, i.e. not every proposition p may entail its own negation.

In other words: If a self-consistent logic **L** contains a unary operator $\sim$ such that, for every proposition $p, \sim p$ logically follows from p, then $\sim$ does not represent a real negation of **L**.

Next let us consider some characteristic laws of classical negation which seem to be *dispensable* for arbitrary negations. In the literature several non-classical negations $\sim$ have been suggested for which *tertium non datur*

(**TND**) $\vdash (p \vee \sim p)$

does not necessarily hold. Let us – preliminarily – refer to such operators as *strong* negations. From the point of view of classical two-valued semantics, the non-validity of **TND** for strong negation operators $\sim$ might be described as follows. The fact that p is *not true* does not always guarantee that p is *false* in the strong sense of $\sim p$, because the truth of $\sim p$ may require something more than the mere (classical) falsity of p. Intuitionistic negation appears to represent an example of such a strong negation. The intuitionistically

[4]Suppose **UN 1** to hold in **L**, then we have in particular $(p \wedge p) \vdash \sim p$ from which one obtains **UN 3** via the conjunction law $p \vdash (p \wedge p)$; conversely, if **UN 3** holds in **L**, then one obtains **UN 1** by means of $(p \wedge q) \vdash p$. The reader may verify in a similar way that **UN 2** is equivalent to **UN 3**, too.

negated formula $\sim p$ roughly says that p is *provably false* or that p can be shown to entail a contradiction. A related kind of strong negation operator might be defined in the framework of ordinary modal logic with the help of classical negation by:

(Def. 1) $\sim_s p := \Box\neg p$.

If the necessity–operator $\Box$ satisfies the usual "truth axiom" ($\Box p \supset p$), then strict or necessary falsity in the sense of **Def. 1** entails but is not conversely entailed by ordinary (plain) falsity, $\neg$. In this sense a strongly negated formula is stronger than a classically negated formula. Therefore strong negations in general do not satisfy **TND**. Similarly, the principle of double negation elimination, **DNE**, fails to hold for $\sim_s p$, since $\sim_s\sim_s p$, i.e. $\Box\neg\Box\neg p$ or $\Box\Diamond p$, does not generally entail p.[5]
Other non–classical negations $\sim$ have been suggested for which in particular **ECQ** does not necessarily hold. Let us – again preliminarily – refer to them as *weak* negations. The negation operators that have been put forward in some systems of paraconsistent logic appear to represent such weak negations. Another particularly simple example of a weak negation might be defined in the framework of classical modal logic by

(Def. 2) $\sim_w p := \Diamond\neg p$.

Such a weakly negated proposition only maintains that p is *possibly false*; this, of course, is fairly compatible with the assumption that p is actually true. Since, then, the conjunction of p and $\sim_w p$ will in general be self–consistent, principle **ECQ** does not hold for weak negation operators. Similarly, it may be shown that the principle of double negation introduction, **DNI**, fails to hold for the particular operator $\sim_w p$. In view of **Def. 2**, a doubly negated proposition $\sim_w\sim_w p$ "says" that p is possibly necessary. However, there exist some normal modal calculi where the actual truth of p does not entail $\Diamond\Box p$.[6] In sum, then, the consideration of strong and weak negation operators suggests that none of the classical principles **TND**, **ECQ**, **DNI** or **DNE** should be postulated as a general condition of adequacy for arbitrary negations.[7]
Let us now turn to the most important category of principles which are *indispensable* for any real negation operator. As was mentioned earlier in

[5]If a modal system would contain ($\Box\Diamond p \supset p$) as a theorem, then by the usual "rule of necessitation" also $\Box(\Box\Diamond p \supset p)$ would be provable. By the law of $\Box$–distribution, $\Box(p \supset q) \supset (\Box p \supset \Box q)$, one would thus obtain ($\Box\Box\Diamond p \supset \Box p$). But in **S5** $\Box\Box\Diamond p$ follows already from $\Diamond p$ so that here in sum the clearly unvalid principle ($\Diamond p \supset \Box p$) would become provable. For basic facts of modal logic cf., e.g., (Lemmon 1977).

[6]Observe that, with $\neg p$ substituted for p, the principle ($p \supset \Diamond\Box p$) entails ($\neg p \supset \Diamond\Box\neg p$) or, by contraposition, ($\neg\Diamond\Box\neg p \supset \neg\neg p$), i.e. ($\Box\Diamond p \supset p$), which was shown above not to be modally valid.

[7]Some further dispensable principles will be discussed in section 4.

connection with **UN 1**, it seems reasonable to require that in any plausible sense of 'false', the falsity of one conjunct entails the falsity of the entire conjunction: $\sim p \vdash \sim(p \wedge q)$. Similarly it seems indispensable to require that the falsity of a disjunction entails the falsity of each of the disjuncts, say: $\sim(p \vee q) \vdash \sim p$. As a matter of fact, these two inferences are only special instances of a much more general principle which any monadic operator $\sim$ apparently has to satisfy in order to be justifiably called a negation. Just as a negative number -x is the smaller the greater the positive number x is itself, so also – one will want to say – a negative or negated proposition $\sim p$ is "the falser" "the truer" the positive proposition p is itself. Somewhat more exactly: If p is "at most as true" as q, i.e. *if p logically entails q*, then conversely q is "at most as false" as p, i.e. the *"falsity" of the latter proposition q logically entails the "falsity" of the former proposition p*:[8]

(CP 1) If $p \vdash q$, then $\sim q \vdash \sim p$.

Before stating some reasons for accepting **CP 1** as an absolutely indispensable principle for negation, let me briefly consider a few argument that have been raised in the literature against contraposition. Da Costa and Wolf (1980, 199) criticize the "axiom of contraposition: $(p \supset q) \supset (\sim q \supset \sim p)$", because when it is "coupled with double negation [it] leads to a collapse" of their "dialectical" logic into classical **PC**. First, however, within the framework of paraconsistent logic this collapse should better be taken as an argument to give up douple negation rather than contraposition. Second, contraposition plus double negation does not automatically lead to classical logic. In a logic containing the operator of material implication, however, **CP 1** must not be formally represented by the quoted axiom $(p \supset q) \supset (\sim q \supset \sim p)$ or by a corresponding rule of deduction such as $(p \supset q) \vdash (\sim q \supset \sim p)$, but rather by the following rule: If $\vdash (p \supset q)$, then $\vdash (\sim q \supset \sim p)$. Independently of the issue of double negation, the former turns out to be much stronger than the latter because in conjunction with the ordinary implication principle $q \vdash (p \supset q)$ it entails $q \vdash (\sim q \supset \sim p)$ and hence by way of modus ponens also the following variant of **ECQ**: $q, \sim q \vdash \sim p$.

D. Pearce (1992, 67) similarly argues contraposition not to be indispensable for arbitrary negations since a particular type of negation which he refers to as "hard negation [...] cannot be contrapositive". This claim is then substantiated by a counterexample of a logic which fails to satify the specific principle: If $\sim p \vdash_L q$, then $\sim q \vdash_L p$. As will be proven below in section 4, however, this contraposition principle is much stronger than our basic **CP 1**, because the former unlike the latter entails double negation elimination,

[8]This core of the traditional theory of "conversio per contrapositionem" has been formulated by one of the forerunners of modern logic, G.W. Leibniz, in (1984, 522) as follows: "Si ex propositione L [...] sequitur propositio M [...] tunc contra ex falsitate propositionis M sequitur falsitas propositionis L."

DNE. Unfortunately, Pearce was practically bound to overlook the difference between **CP 1** and the criticized principle of contraposition because he based his investigations on the full theory of double negation.

Let me now present some constructive arguments in favor of **CP 1**. In ordinary discourse there exist many different forms of affirmation and also many different forms of negation. To maintain, to claim, to believe, to assert, etc. all belong to the former category; and to deny, to doubt, to reject, to disbelieve, to disprove, etc. to the latter. What appears to be common to all these forms of negation is that they are somehow *opposite* to (a corresponding way of) *affirmation*. Sometimes this opposition can be made explicit by transforming the negation into an affirmation with a negative or negated content; sometimes the negation can be analyzed as representing the direct (classical) negation or denial of some affirmative expression. For instance, making use of the formal apparatus of epistemic logic developed in (Lenzen 1980), the negative attitude of some person a who *doubts* that p can be paraphrased by the expression $B(a, \neg p)$ (to abbreviate 'a believes that $\neg p$'), while a's disbelieving that p presumably has to be interpreted as $\neg B(a,p)$, i.e. 'a does not believe that p'.

This is not the place to investigate the linguistic relations between arbitrary negative and affirmative expressions in greater detail. Rather I want to inquire in an abstract and admittedly speculative way into the logical relationships between affirmations and negations as they are suggested by analogy to negative and positive expressions in arithmetic. Elementary mathematics teaches us that the product of two positive numbers, $(+x) \times (+y)$, as well as the product of two negative numbers, $(-x) \times (-y)$, yields a positive number, while $(-x) \times (+y)$ and $(+x) \times (-y)$ always gives a negative number. Similarly, the affirmation of an affirmative expression and the negation of a negative expression appear to yield an affirmation, while the affirmation of a negation as well as the negation of an affirmation normally represents a negation.[9] In particular, it seems rather safe to claim that the classical negation of an affirmative expression itself constitutes a negation, while the classical negation of an arbitrary negation yields an affirmation. E.g., if $\sim p$ is taken to mean that p is "false", or that p is "not true", or that p is "impossible", or that p is "unlikely to be true", or what not, then the classically

[9]Cf., e.g., Leibniz's sequence of definitions and corollaries in (1982, 86–88): "4) *Affirmatio* est enuntiatio de enuntiatione, dicens eam esse veram. 5) *Negatio* est enuntiatio de enuntiatione, dicens eam non esse veram. [...] 10) Si affirmatio est vera, negatio non est vera. 11) Si affirmatio non est vera, negatio est vera [...] 13) Si negatio est vera, affirmatio non est vera. 14) Si negatio non est vera, affirmatio est vera." The omitted article 12 contains the following version of the basic principle of contraposition (which Leibniz often refers to by the medieval term 'regressus'): "Si positis enuntiationibus sequatur nova et haec sit falsa, etiam aliqua ex illis erit falsa. Hoc est axioma." For more details of Leibniz' logic, cf. (Lenzen 1990).

negated expression $\neg\sim p$ either says that p is not false, or that p is true, or that p is not impossible, or that p is not unlikely to be true, etc. All these expressions evidently represent some sort of affirmation.

Now, for any affirmative operator A, the premise that p logically entails q generally seems to warrant that $A(p)$ logically entails $A(q)$, too. To be sure, I have no proof of this very strong principle. In order to lend it at least some credibility, however, let it be pointed out that this law holds not only for each so-called affirmative modality in alethic modal logic (and in related systems for other modal operators), but also, e.g., for probabilistic notions saying that p is likely, that p is probable, that p is certain, etc. If it could thus be taken for granted that affirmations are generally closed under logical implications, **CP 1** might be "proven" as follows. Let p, q be propositions such that p logically entails q and q is "false" in the sense of the operator $\sim$, i.e. let $\sim q$ be true. Suppose further that - contrary to **CP 1** - p would not be "false" in the sense of $\sim$; i.e. suppose that $\sim p$ does not hold. This can be expressed metalinguistically by means of classical negation as $\neg\sim p$. Hence we have in sum the assumption that $p \vdash q$; that $\sim q$; and that $\neg\sim p$. Now if $\sim$ really is a *negation* operator - no matter what particular kind of "negation" one has in mind - then $\neg\sim$ certainly will not be a negation itself but instead expresses some sort of *affirmation*. Thus by our assumption that affirmations are closed under logical implication, $\neg\sim p$ logically entails $\neg\sim q$; and hence we would arrive at the classical contradiction that, on the one hand, $\sim q$, and on the other hand, $\neg\sim q$.

The discussion during the Workshop has shown that the "proof" of **CP 1** just given, even if its main idea is basically sound, remains problematic insofar as the central assumption of the logical closure of arbitrary affirmations is in need of justification at least to the same degree as the principle that it purports to justify. Therefore let me sketch another argument in favor of **CP 1** which does without the latter assumption. As was argued above in connection with principle **IMP**, a logical implication $p \vdash_L q$ must always be understood as holding only if the conclusion q cannot fail to be "true" once the entailing proposition p itself is assumed to be "true". Or, to give a somewhat more formal paraphrase: If $p \vdash_L q$, then, necessarily, (if p is "true", then q is "true"). So, by classical contraposition: If $p \vdash_L q$, then, necessarily, (if q is not "true", then p isn't "true" either). But to say that some proposition p is not "true" (either in the sense of classical, two-valued semantics or in the sense of some other distinguished true-like value) appears to be tantamount to *negating* p (in some way or another). Thus one obtains the following version of **CP 1**: If $p \vdash q$, then, necessarily (if q is "false", then p must be "false", too). I hope these reflections have lent some plausibility to the claim that **CP 1** really is an indispensable principle for negation and that therefore a *second condition of adequacy* may be stated as follows:

(ADQ 2) A unary operator $\sim$ is a negation of the logic **L** only if it satisfies **CP 1**: If $p \vdash_L q$, then $\sim q \vdash_L \sim p$.

CP 1 represents a rather weak form of contraposition only. In section 4 below, several stronger principles of contraposition will be discussed which do not hold for arbitrary negations. For the moment suffice it to point out that because of **CP 1** each negation, although not necessarily being a *truth-functional* (or extensional) operator, is at least a *propositional* (or intensional) operator in the following sense: If p and q are logically equivalent sentences thus expressing the same proposition, then $\sim p$ and $\sim q$ have to be logically equivalent, too:

(EQUI) If $p \dashv\vdash q$, then $\sim p \dashv\vdash \sim q$.

Now, even in conjunction with **ADQ 1**, the *necessary* condition **ADQ 2** is not *sufficient* for negation operators because **CP 1** is satisfied, among others, both by the "tautology-operator", $\sim_t$, and by the "contradiction-operator", $\sim_c$, which might be defined in the framework of classical **PC** as follows:

(Def. 3) $\sim_t p := (p \vee \neg p)$ **(Def. 4)** $\sim_c p := (p \wedge \neg p)$.

These operators certainly cannot be rated as genuine negations. As a matter of fact, the "tautological negation" according to **Def. 3** is not a negation because, for any proposition p, the formula $\sim_t p$ is provable in **PC**[10]. This violates our condition **ADQ 1** according to which $\sim$ is a genuine negation of a logic **L** only if not every proposition is provably "false". In a similar way the "contradictory negation" according to **Def. 4** should not be regarded as a genuine negation because otherwise no proposition, not even the impossible ones, could be proven to be "false". One might therefore think of postulating another condition of adequacy as follows:

(ADQ 3) A unary operator $\sim$ is a negation of the logic **L** only if there is at least one proposition p such that $\vdash_L \sim p$.

However, this condition appears somewhat problematic because the language of **L**, even if it contains the theory of conjunction and disjunction, may nevertheless lack the means for expressing a particular proposition p which is "sufficiently" false so that $\sim p$ *must be provable* in **L**. To be sure, if **L** would contain the conceptual means for expressing a *classical* contradiction such as $(p \wedge \neg p)$, then we were certainly justified in requiring that *this* proposition, which according to the familiar law of consistency is provably false in classical logic, $\vdash_{PC} \neg(p \wedge \neg p)$, must also be provably "false" in any

[10] Cf. also (Wansing 1993, 142), fn. 11 where it is similarly argued that the "tautology operator" should not be accepted as a negation.

other logic, $\vdash_L \sim(p \wedge \neg p)$, no matter what particular interpretation is given to the negation operator $\sim$. However, in **L** one can in general formulate only the corresponding conjunction $(p \wedge \sim p)$ which perhaps expresses no real contradiction at all. Therefore it is not clear whether the "non–classical law of consistency": $\vdash_L \sim(p \wedge \sim p)$, represents an indispensable principle for arbitrary negations. A more detailed discussion of this principle will be postponed to the end of the next section.

As regards the aforementioned scruples concerning **ADQ 3**, however, note that when p is a *theorem* of a certain logic **L**, then, whatever the details of the semantics of **L** may be, p is necessarily "true" and hence p cannot be "false" in any reasonable sense of the word. So if p is a theorem of **L**, and if the operator $\sim$ really represents a *negation*, then $\sim p$ expresses sort of a *contradiction* in **L** and must therefore itself be classified as necessarily "false". This observation justifies to postulate the following principle of *double negation introduction for theorems*:

(DNT) If $\vdash_L p$, then $\vdash_L \sim\sim p$.

This principle is weaker than the usual law of double negation introduction, **DNI**, which was argued above not to be indispensable for arbitrary negations. Semantically speaking, the inference $p \vdash \sim\sim p$ amounts to the condition that, necessarily, if p is "true", so will be $\sim\sim p$. In contrast **DNT** only requires that, necessarily, when p is *necessarily* "true", $\sim p$ must be necessarily "false", and hence $\sim\sim p$ must be necessarily "true", too.

Now, each logic **L** "normally" does contain *some* theorems p. This is trivial whenever **L** contains an operator of material or strict implication. Then at least some formulas such as $(p \supset p)$ or $(p \rightarrow p)$ will be provable in **L**. And if **L** contains the theory of conjunction, the rules of inference **AND** can be transformed with the help of '$\supset$' or '$\rightarrow$' into corresponding theorems such as $\vdash ((p \wedge q) \supset p)$ or $\vdash ((p \wedge q) \rightarrow q)$, etc. So, in view of **DNT**, there "normally" exist in **L** some theorems of the form $\sim\sim p$, and a fortiori there also exists some q (namely $q = \sim p$) such that $\vdash_L \sim q$. In other words, **ADQ 3** will "normally" be satisfied after all. Yet we cannot absolutely exclude the possibility that in an "unnormal" $\{\wedge, \vee, \sim\}$–logic **L** neither the "positive" formula $(p \vee \sim p)$ nor the "negative" formula $\sim(p \wedge \sim p)$ nor any other formula is strictly provable[11]. Therefore instead of **ADQ 3** only the subsequent *fourth condition of adequacy* will be postulated:

[11]The "minimale Aussagenlogik" of Kutschera (1985, 30) represents such a counterexample: "In ihr ist insbesondere kein einziger Satz beweisbar". Interestingly, the failure of $(p \supset p)$ as a theorem is motivated by the observation: "wegen der Definition der Implikation durch die Disjunktion wäre $[p \supset p]$ mit dem Prinzip tertium non datur $[\sim p \vee p]$ äquivalent". Now if someone wants to give up **TND**, it would seem to be much more natural to reject the definition of implication in terms of disjunction rather than to dispense with the reflexivity of the implication operator.

(ADQ 4) A unary operator $\sim$ is a negation of the logic **L** only if for every proposition p: If $\vdash_L p$, then $\vdash_L \sim\sim p$.

3 Gabbay's "Necessary and Sufficient" Condition

In a paper of (1988), Dov Gabbay investigated the question "What is Negation in a System?", and he suggested the following answer[12]:

(GAB) "$\sim$ is a form of negation [...] iff for any Δ and any p the following holds: $\Delta \vdash \sim p$ iff for some q such that $\vdash \sim q$ we have $\Delta, p \vdash q$."

The set Δ is taken by Gabbay to contain the "unwanted" formulas of the respective logic **L**, but this heuristic interpretation seems to be irrelevant for the import of the quoted criterion where Δ just functions as a variable for arbitrary sets of sentences. In order to be able to compare the general criterion **GAB** with our own conditions of adequacy, consider first the following necessary conditions which are obtained from **GAB** by setting $\Delta = \emptyset, \Delta = \{\sim p\}$, and $\Delta = \Delta^* \cup \{r\}$, respectively:

(GAB 1) $\vdash \sim p$ iff there is a q such that $\vdash \sim q$ and $p \vdash q$;

(GAB 2) $\sim p \vdash \sim p$ iff there is a q such that $\vdash \sim q$ and $\sim p, p \vdash q$;

(GAB 3) $\Delta^*, r \vdash \sim p$ iff there is a q such that $\vdash \sim q$ and $\Delta^*, r, p \vdash q$.

The equivalence **GAB 1** may be split up into two implications. The first one: 'If $\vdash \sim p$ then there is a q such that $\vdash \sim q$ and $p \vdash q$' holds trivially – just set $q = p$! The converse implication:

(GAB 4) If there is a q such that $\vdash \sim q$ and $p \vdash q$, then $\vdash \sim p$

says that if p logically entails a proposition q which is provably false, then p must be provably false, too. This principle of contraposition is slightly weaker than the basic requirement **CP 1**. It will turn out soon, however, that **GAB** entails another principle of contraposition which is even stronger than our **CP 1**.

In view of the trivial consequence $\sim p \vdash \sim p$, it follows from **GAB 2** that there exists a proposition q such that $\vdash \sim q$ and $\sim p, p \vdash q$, or – by the laws of conjunction – $(p \wedge \sim p) \vdash q$. Because of **GAB 4**, then, $(p \wedge \sim p)$ itself is provably false. As was already pointed out in (Wansing 1993, 143), one thus obtains the crucial "law of consistency"

(CONSIS 1) $\vdash \sim(p \wedge \sim p)$.

With the help of **CONSIS 1**, however, the following stronger principle of contraposition can be proven:

[12]Cf. (Gabbay 1988, 100); in the quotation I have tacitly replaced some of Gabbay's symbols by the logical terminology used throughout this paper. In particular I write simply '$\vdash p$' instead of Gabbay's '$\emptyset \vdash p$'.

(GAB 5) If $\Delta, p \vdash q$, then $\Delta, \sim q \vdash \sim p$.

For if $\Delta, p \vdash q$, then by the laws of conjunction $\Delta, p, \sim q \vdash (q \wedge \sim q)$; hence there exists an r, namely $r = (q \wedge \sim q)$, such that (by **CONSIS 1**) $\vdash \sim r$ and such that $\Delta, p, \sim q \vdash r$. **GAB 3** thus entitles us to conclude that $\Delta, \sim q \vdash \sim p$. **GAB 5** is stronger than our basic principle of contraposition, **CP 1**, because it not only contains the latter as a special instance for $\Delta = \emptyset$, but also entails the following principle which, in view of the parallel to so–called "disjunctive syllogism"

(DISSYL) $(p \vee q), \sim p \vdash q$,

is perhaps not inappropriately referred to as "conjunctive syllogism":

(CONSYL) $p, \sim(p \wedge q) \vdash \sim q$.

Clearly, since $p, q \vdash (p \wedge q)$, **CONSYL** immediately follows from **GAB 5** by the substitution $(\Delta/p; p/q; q/p \wedge q)$. Conversely strong **GAB 5** can be derived from weak **CP 1** in conjunction with **CONSYL** as follows. Let Δ be a finite set of formulas such that $\Delta, p \vdash q$, and let r abbreviate the conjunction of all elements of Δ. Hence the premise $\Delta, p \vdash q$ can be rendered as $(p \wedge r) \vdash q$, so that **CP 1** gives us $\sim q \vdash \sim(r \wedge p)$. By the laws of conjunction one further obtains that $(r \wedge \sim q) \vdash r \wedge \sim(r \wedge p)$. But, by **CONSYL**, $r \wedge \sim(r \wedge p) \vdash \sim p$, so that by transitivity of $\vdash$ we also get $r \wedge \sim q \vdash \sim p$, i.e. the desired conclusion $\Delta, \sim q \vdash \sim p$.

Next observe that the general criterion **GAB** is already satisfied if the special instances **CONSIS 1** and **GAB 5** hold. On the one hand, if a set of formulae Δ entails $\sim p$, then by the laws of conjunction $\Delta, p \vdash (p \wedge \sim p)$. Hence there exists a q, namely $q = (p \wedge \sim p)$, such that $\Delta, p \vdash q$ and such that – in view of **CONSIS 1** – $\vdash \sim q$. On the other hand, if there exists a q such that $\vdash \sim q$ and $\Delta, p \vdash q$, then – because of **GAB 5** – $\Delta, \sim q \vdash \sim p$. Since $\sim q$ is assumed to be provable, it follows that $\Delta \vdash \sim p$.

Summing up the results of the foregoing derivations, Gabbay's "necessary and sufficient" condition **GAB** turns out to be "equivalent"[13] to weak contraposition **CP 1**, the law of consistency, **CONSIS 1**, and conjunctive syllogism, **CONSYL**. On the one hand, this collection of principles is not entirely *sufficient* for arbitrary negations because **GAB** fails to warrant **ADQ 1**.[14] On the other hand, it is somewhat doubtful whether the two princi-

---

[13]'Equivalent' here and in related places is to be understood as 'equivalent within a logic **L** such that (1) **L** contains **AND** and **OR**; (2) the deducibility relation of **L** satisfies the requirements formulated in section 1; and (3) the ~–operator of **L** satisfies the conditions of adequacy formulated in section 2'.

[14]Consider, e.g., the particular logic $\mathbf{L}^*$ in which every negated formula $\sim p$ is provable but which nevertheless remains self–consistent because $\sim$ does not satisfy **ECQ** and thus not every unnegated formula q is provable in $\mathbf{L}^*$. As the reader may easily verify, the criterion **GAB** is then trivially satisfied, but the operator $\sim$ of this logic certainly should

ples **CONSYL** and **CONSIS 1** by which **GAB** surmounts our condition **ADQ** really are *necessary* conditions for negation operators. As regards the former, conjunctive syllogism certainly is *not* indispensable for arbitrary negations because it will be satisfied neither by strong negations à la **Def. 1** nor by weak negations à la **Def. 2**. To wit, if $\sim p$ is interpreted as $\Box\neg p$, then the premises of **CONSYL**, p and $\sim(p \wedge q)$, say that p is "true" and that $(p \wedge q)$ is necessarily "false". In order to infer that $\Box\neg q$, however, one would have to have the stronger premise that besides $\Box\neg(p \wedge q)$ also p is *necessarily* "true". Similarly, if $\sim$ is interpreted in the weak sense of **Def. 2**, then $p, \sim(p \wedge q) \vdash \sim q$ means as much as $p, \Diamond\neg(p \wedge q) \vdash \Diamond\neg q$, or, by classical contraposition, $p, \neg\Diamond\neg q \vdash \neg\Diamond\neg(p \wedge q)$; again, however, the conclusion $\Box(p \wedge q)$ follows from $\Box q$ only in conjunction with $\Box p$ and not just in conjunction with p simpliciter.

The question whether the law of consistency, **CONSIS 1**, also is dispensable for arbitrary negations or not, is much harder to decide. Several arguments might be given in favor of adopting $\sim(p \wedge \sim p)$ as an essential theorem for arbitrary negations. First, **CONSIS 1** is easily seen to follow from the subsequent weak version of *tertium non datur*:

(TNDF) $\vdash (\sim p \vee \sim\sim p)$.

Since $(p \wedge \sim p) \vdash p$ and $(p \wedge \sim p) \vdash \sim p$, one obtains $\sim p \vdash \sim(p \wedge \sim p)$ and $\sim\sim p \vdash \sim(p \wedge \sim p)$ by **CP 1**, hence by **OR** $(\sim p \vee \sim\sim p) \vdash \sim(p \wedge \sim p)$, and thus **CONSIS 1** follows from **TNDF**. However, **TNDF** itself may be regarded as *dispensable* for arbitrary negations since it is not satisfied, e.g., by strong negations à la **Def. 1**. To wit, the assumption $(\sim_s p \vee \sim_s\sim_s p)$ amounts to $(\Box\neg p \vee \Box\neg\Box\neg p)$, i.e. $(\Diamond p \supset \Box\Diamond p)$, and this principle holds only in **S5** but not in other modal calculi.

Second, as was already pointed out earlier, unlike the classical formula $(p \wedge \neg p)$, the conjunction $(p \wedge \sim p)$ does not always represent a real contradiction. In particular $(p \wedge \sim p)$ becomes perfectly self-consistent when $\sim p$ is interpreted in the weak sense of **Def. 2**. Therefore one cannot expect it to be provably false in the classical sense, $\neg(p \wedge \sim p)$. Still it seems plausible to require that $(p \wedge \sim p)$ should be provably "false" at least in the sense of the operator $\sim$ as it occurs within the formula $(p \wedge \sim p)$ itself. Yet this argument is not entirely conclusive either. After all, **CONSIS 1** will not necessarily

---

not be regarded as a genuine negation. Note incidentally that the remaining conditions of adequacy hold on Gabbay's approach. **ADQ 2**, i.e. **CP 1**, is trivial. **ADQ 3** immediately follows from **CONSIS 1**. Furthermore also **ADQ 4**, i.e. the principle of double negation introduction for theorems, **DNT**, follows from **CONSIS 1**. Take any q such that $\vdash q$; by **MONO** one has both $p \vdash q$ and $\sim p \vdash q$, so that elementary contraposition yields $\sim q \vdash \sim p$ and $\sim q \vdash \sim\sim p$. By the rules of conjunctions and another application of **CP 1** one obtains $\sim(\sim p \wedge \sim\sim p) \vdash \sim\sim q$; but according to **CONSIS 1** $\sim(\sim p \wedge \sim\sim p)$ is provable; hence $\vdash \sim\sim q$.

be satisfied by weak negations à la **Def. 2**. To wit, $\sim_w(p \wedge \sim_w p)$ "means" as much as $\Diamond\neg(p \wedge \Diamond\neg p)$, i.e. $\Diamond(p \supset \Box p)$, and this formula, though holding in *many* modal logics, is not a theorem of *each* such logic.[15] It might, however, be replied that the $\Box$–operator of such an "exotic" modal system which does not satisfy $\Diamond(p \supset \Box p)$ would better not be called a *necessity* operator at all and that the corresponding weak negation à la **Def. 2** therefore doesn't represent a serious counter–example to the validity of **CONSIS 1** either.

Anyway, the issue of the dispensability or indispensability of **CONSIS 1** need not be decided here once and forever. It will turn out below that **CONSIS 1** is satisfied if not by each, then at least by the most prominent "types" of negation operators, namely both by arbitrary weak and also by arbitrary strong negations. Since we are mainly interested in *necessary* conditions for negation operators, however, only *entirely unproblematic* principles will be adopted as conditions of adequacy here. Therefore, foregoing **CONSIS 1**, we obtain the following condition **ADQ** which summarizes the discussion of the preceding two sections:

(ADQ) Let **L** be a self–consistent logic such that not for every proposition q: $\vdash_L q$
Then the unary operator $\sim$ is a negation of **L** only if:
(1) not for every proposition $p : p \vdash_L \sim p$;
(2) for every proposition p: If $p \vdash_L q$, then $\sim q \vdash_L \sim p$;
(3) for every proposition p: If $\vdash_L p$, then $\vdash_L \sim\sim p$.

As was pointed out in connection with **ADQ 3** above, it follows from (3) that whenever **L** is a "normal" logic in the sense that, for at least one proposition $p, \vdash_L p$, then also the subsequent two principles hold:

(4) for at least one proposition p: $\vdash_L \sim\sim p$
(5) for at least one proposition p: $\vdash_L \sim p$.

In the subsequent section we will offer a coarse classification of the broad field of non–classical negation operators by examining some further principles which appear to be dispensable for arbitrary negations.

4 Weak and Strong Negation

From now on we presuppose for convenience that each logic **L** contains the basic theory of conjunction and disjunction and furthermore that **L** is "normal" so that for at least one p: $\vdash_L p$. In section 2 above, negation operators

[15]In particular $\Diamond(p \supset \Box p)$ can be falsified by the following Kripke–structure with three possible worlds i, j, k: Let the (irreflexive and non–transitive) accessibility relation R consist only of the pairs $\langle i, j\rangle$, and $\langle j, k\rangle$, and let p be true in j but false in k. Then in j the proposition $\Box p$, and hence also $(p \supset \Box p)$ is false, and therefore $\Diamond(p \supset \Box p)$ becomes false in world i!

$\sim$ have preliminarily been characterized as being weak or strong according to whether they fail to satisfy *ex contradictorio quodlibet*, **ECQ**, or *tertium non datur*, **TND**, respectively. Before redefining these notions in a slightly diverging way, let me first explain the intuitive idea behind this classification in some more detail. The most natural way of determining the "strength" of a negation $\sim$ would seem to proceed by way of *comparison with classical negation* $\neg$: A negation $\sim$ should evidently be considered as *strong* iff $\sim$ is "at least as strong as $\neg$" in the sense that $\sim p$ logically entails $\neg p$. Thus, if $\sim$ is a strong negation, then the conjunction $(p \wedge \sim p)$ logically entails the contradiction $(p \wedge \neg p)$ and must therefore itself be regarded as contradictory. In other words, if $\sim$ is a strong negation, then p and $\sim p$ can never be true together. Analogously $\sim$ should be considered as *weak* iff $\sim$ is "at most as strong as $\neg$", i.e. if $\sim p$ is logically entailed $\neg p$. Accordingly, if $\sim$ is a weak negation, then the disjunction $(p \vee \sim p)$ logically follows from the tautology $(p \vee \neg p)$ and hence is itself logically true. In other words, if $\sim$ is a weak negation, then at least one of the proposition p and $\sim p$ must always be true. Unfortunately such comparisons between $\sim$ and $\neg$ will normally not be possible within the logic **L** itself which usually does not contain $\neg$ as an additional operator besides $\sim$. One might, however, consider the "$\neg$-extension" $\mathbf{L}^*$ of **L** obtained by adding the theory of classical negation to **L**, and then define $\sim$ to be strong iff $\sim p \vdash^*_L \neg p$, or $\vdash^*_L (\sim p \supset \neg p)$[16]. Similarly, $\sim$ might be defined to be weak iff $\vdash^*_L (\neg p \supset \sim p)$. If this approach were chosen, the following basic facts could be proved: (1) If the negation $\sim$ of a logic **L** satisfies **ECQ**, then $\sim$ is strong. (2) If the negation $\sim$ of a logic **L** satisfies **TND**, then $\sim$ is weak.[17] Furthermore, from an intuitive point of view, the negation $\sim$ should also conversely come out as strong *only if* it satisfies **ECQ** and as weak *only if* it satisfies **TND**. However, the latter relations could seldom or never be *proved* because we have no guarantee that the deductive relations which hold in the $\neg$-extension $\mathbf{L}^*$ already hold in **L** itself. To be sure, if $\sim$ is a weak negation, then in $\mathbf{L}^*$ $(\neg p \supset \sim p)$ is provable, and hence by the classical laws for $\neg$ and $\supset$ also $\vdash^*_L (p \vee \sim p)$. But evidently there is no way to infer that $(p \vee \sim p)$ is already a theorem of **L**! Similarly, if $\sim$ is strong, then, in $\mathbf{L}^*$, $\sim p \vdash^*_L \neg p$ and hence by the laws of conjunction $(p \wedge \sim p) \vdash^*_L (p \wedge \neg p)$. Since **ECQ** holds for $\neg$, one further obtains $(p \wedge \sim p) \vdash^*_L q$, but how could we ever derive from this that already in **L** $(p \wedge \sim p) \vdash_L q$? In view of these

[16] Note that, since **L** is supposed to contain the theory of $\wedge$ and $\vee$, and since $\mathbf{L}^*$ contains in addition the theory of $\neg$, $(p \supset q)$ can be introduced in $\mathbf{L}^*$ by either of the definitions $(\neg p \vee q)$ or $\neg(p \wedge \neg q)$.

[17] *Proof*: (1) If **ECQ** holds in **L**, i.e. if $(p \wedge \sim p) \vdash_L q$, then in particular in $\mathbf{L}^* (p \wedge \sim p) \vdash^*_L (p \wedge \neg p)$ from which one obtains by classical contraposition $\neg(p \wedge \neg p) \vdash^*_L \neg(p \wedge \sim p)$. But, classically, one also has $\vdash^*_L \neg(p \wedge \neg p)$, so that in sum $\vdash^*_L \neg(p \wedge \sim p)$, i.e. $\vdash^*_L \neg(\sim p \wedge \neg\neg p)$ or $\vdash^*_L (\sim p \supset \neg p)$. (2) In $\mathbf{L}^*$ the following instance of disjunctive syllogism, $\neg p, (p \vee \sim p) \vdash^*_L \sim p$, holds; hence, if **TND** is provable in **L**, then $\neg p \vdash^*_L \sim p$.

problems we better drop the $\mathbf{L}^*$–approach and adopt instead the satisfaction of **ECQ** and of **TND** as *defining* marks of strong and weak negations:

(Def. 5) The negation operator $\sim$ of a logic **L** is a strong negation: $\Leftrightarrow (p \wedge \sim p)$ represents a contradiction in **L**, i.e. $p, \sim p \vdash_L q$.

(Def. 6) The negation operator $\sim$ of a logic **L** is a *weak* negation: $\Leftrightarrow (p \vee \sim p)$ represents a tautology in **L**, i.e. $\vdash_L (p \vee \sim p)$.

The "weakness" and "strength" of a negation $\sim$ can alternatively be characterized as follows:

Theorem 1

(1.1) A negation $\sim$ is weak if and only if it satisfies:
CONSIS 3 If $\sim p \vdash p$, then $\vdash p$;

(1.2) A negation $\sim$ is strong if and only if it satisfies:
CONSIS 4 If $p \vdash \sim p$, then $p \vdash q$.

Proofs of this theorem and of all subsequent theorems are given in the Appendix.

Now if a negation operator of a logic **L** is *both weak and strong*, then it is *classical* either in the sense that, in the $\neg$–extension $\mathbf{L}^*$ of **L**, $\sim p$ is deductively equivalent to $\neg p$, or in the sense that, in **L** itself, $\sim$ satisfies all characteristic axioms of classical negation:

(Def. 7) The negation operator $\sim$ of a logic **L** is a classical negation: $\Leftrightarrow \vdash_L (p \vee \sim p)$ and $p, \sim p \vdash_L q$.

Accordingly a negation is *genuinely strong* iff it is strong but not classical; and $\sim$ is *genuinely weak* iff it is weak but not classical. Furthermore, let us call a negation $\sim$ *semi–weak* and *semi–strong*, iff it satisfies the following special instances of the characteristic "weak" and "strong" principles **TND** and **ECQ**, respectively:

(Def. 8) The negation operator $\sim$ of a logic **L** is *semi–strong*:
$\Leftrightarrow \sim$ satisfies **ECQF**: $p, \sim p \vdash_L \sim q$;
The negation operator $\sim$ of a logic **L** is *semi–weak*:
$\Leftrightarrow \sim$ satisfies **TNDF**: $\vdash_L (\sim p \vee \sim\sim p)$.

Trivially a weak negation always is also semi–weak, and a strong negation a fortiori is semi–strong. Let us now see how these properties are related to the validity of certain other axioms and rules of deduction besides **ECQ** and **TND**.

To begin with, remember that according to our condition of adequacy **ADQ** every negation operator $\sim$ of a logic **L** which contains the basic theory **AND** and **OR** satisfies in particular the following De Morgan principles:

DM 1 $(\sim p \vee \sim q) \vdash \sim(p \wedge q)$

DM 2 $\sim(p \vee q) \vdash (\sim p \wedge \sim q)$.[18]

Next we find that the "law of consistency" which according to Gabbay is indispensable for arbitrary negations is satisfied at least *both by arbitrary weak and by arbitrary strong* negations. As a matter of fact, one can prove even the somewhat stronger result:

Theorem 2

Every *semi–weak* negation as well as every *semi–strong* negation satisfies:

CONSIS 1 $\vdash \sim(p \wedge \sim p)$[19]

This principle may thus be considered as "almost universally valid", after all. Anyway we want to call the negation $\sim$ of a logic **L** *almost classical* iff $\sim$ becomes classical as soon as **CONSIS 1** holds in **L**:

(Def. 9) The negation operator $\sim$ of a logic **L** is almost classical: $\Leftrightarrow$
If $\vdash_L \sim(p \wedge \sim p)$, then both $\vdash_L (p \vee \sim p)$ and $p, \sim p \vdash_L q$.

Next let us consider the laws of double negation introduction and elimination, the validity of which is largely independent of the "strength" or "weakness" of $\sim$[20] but instead turns out to be equivalent to the validity of certain De Morgan laws and certain laws of contraposition:

---

[18]**DM 1** and **DM 2** appear to be more or less universally accepted in the literature on non-classical negation. In particular they hold in the weak "direkte Aussagenlogik" of von Kutschera (1985). Furthermore also Pearce/Wagner (1991, 442) who otherwise reject a stronger version of contraposition accept at least the inferences from $\sim p$ and from $\sim q$ to $\sim(p \wedge q)$.

[19]As the reader may easily verify, **CONSIS 1** is deductively equivalent to the following variants:

CONSIS 1a If $p \vdash \sim p$, then $\vdash \sim p$

CONSIS 1b If $p \vdash q$ and $p \vdash \sim q$, then $\vdash \sim p$.

Note further that **CONSIS 1** entails

CONSIS 2 If $\sim p \vdash p$, then $\vdash \sim\sim p$

which in turn is deductively equivalent to doubly negated *tertium non datur*:

DNTND $\vdash \sim\sim(p \vee \sim p)$.

Proof: (a) **CONSIS 2** follows from **CONSIS 1** because, if $\sim p \vdash p$, then by elementary contraposition $\sim p \vdash \sim\sim p$ so that by **CONSIS 1a** $\vdash \sim\sim p$. (b) Since by **DM 2** $\sim(p \vee \sim p) \vdash \sim p$ and by **OR** $\sim p \vdash (p \vee \sim p)$, one obtains $\sim(p \vee \sim p) \vdash (p \vee \sim p)$ so that **DNTND** follows from **CONSIS 2**. (c) Conversely, **CONSIS 2** is obtained from **DNTND** as follows: If $\sim p \vdash p$, then because of the trivial $p \vdash p$ and by **OR** also $(p \vee \sim p) \vdash p$, hence by double application of elementary contraposition $\sim\sim(p \vee \sim p) \vdash \sim\sim p$ from which the conclusion of **CONSIS 2** is obtained via **DNTND**.

[20]In intuitionistic logic, however, **DNE** is equivalent to **TND**. Thus **TND** follows from **DNE** in conjunction with the intuitionistically valid **DNTND**; and theorem 4.45 of Heyting (1930) shows that **DNE** conversely follows from **TND**.

Theorem 3

(3.1) A negation $\sim$ satisfies **DNI** iff it satisfies one of the subsequent principles:
DM 3 $(p \wedge q) \vdash \sim(\sim p \vee \sim q)$;
DM 4 $(p \vee q) \vdash \sim(\sim p \wedge \sim q)$;
CP 4 If $p \vdash \sim q$, then $q \vdash \sim p$.

(3.2) A negation $\sim$ satisfies **DNE** iff it satisfies one of the subsequent principles:
DM 5 $\sim(\sim p \vee \sim q) \vdash (p \wedge q)$;
DM 6 $\sim(\sim p \wedge \sim q) \vdash (p \vee q)$;
CP 5 If $\sim p \vdash_L q$, then $\sim q \vdash_L p$.[21]

Some major results may now be summarized as follows:

Theorem 4

(4.1) A negation $\sim$ is strong whenever it satisfies disjunctive syllogism:
DISSYL $(p \vee q), \sim p \vdash q$;

(4.2) A negation $\sim$ satisfies **DNI** whenever it satisfies conjunctive syllogism:
CONSYL $p, \sim(p \wedge q) \vdash \sim q$;

(4.3) A negation $\sim$ is almost classical whenever it satisfies strong contraposition:
CP 6 If $\sim p \vdash_L \sim q$, then $q \vdash_L p$.

We conclude this investigation by the following corollary:

Theorem 5

(5.1) A negation $\sim$ is almost classical if it satisfies both **DNE** and **DNI**;

(5.2) A weak negation becomes classical if it satisfies **DISSYL**;

(5.3) A strong negation becomes classical if it satisfies **DNE**.[22]

---

[21]It may further be shown that if a negation satisfies **DNI**, then it also satisfies
DM 7 $(\sim p \wedge \sim q) \vdash \sim(p \vee q)$;
and if a negation satisfies DNE, then it satisfies:
DM 8 $\sim(p \wedge q) \vdash (\sim p \vee \sim q)$.
Proof: (a) $(p \vee q) \vdash (\sim\sim p \vee \sim\sim q)$ and $(\sim\sim p \vee \sim\sim q) \vdash \sim(\sim p \wedge \sim q)$ by **DNI** and **DM 1**; hence by **CP 1** $\sim\sim(\sim p \wedge \sim q) \vdash \sim(p \vee q)$ from which one obtains **DM 7** by another application of **DNI**. (b) Similarly, $\sim(\sim p \vee \sim q) \vdash (\sim\sim p \wedge \sim\sim q)$ and $(\sim\sim p \wedge \sim\sim q) \vdash (p \wedge q)$ by **DM 2** and **DNE**; hence by **CP 1** $\sim(p \wedge q) \vdash \sim\sim(\sim p \vee \sim q)$ from which **DM 8** follows by another application of **DNE**.

[22]The field of further principles and corresponding deductive relations is rather rich. Here only two additional results will be mentioned without proof:
(6.4) A weak negation is *semi-strong* if it satisfies double negation introduction, **DNI**.
(6.5) A strong negation is *semi-weak* if it satisfies de Morgan principle **DM 8** formulated in the preceding fn.

5 Appendix

Proof of Theorem 1:

1.1 (a) If $\sim p \vdash p$, then because of $p \vdash p$ by **OR** $(p \vee \sim p) \vdash p$; hence **TND** entails the desired conclusion $\vdash p$. (b) According to **DM 2** $\sim(p \vee \sim p) \vdash \sim p$, hence by **OR** also $\sim(p \vee \sim p) \vdash (p \vee \sim p)$, from which one infers by means of **CONSIS 3** that $\vdash (p \vee \sim p)$.

1.2 (a) If $p \vdash \sim p$, then because of $p \vdash p$ by **AND** $p \vdash (p \wedge \sim p)$; hence by means of **ECQ** also $p \vdash q$. (b) By **DM 1** $\sim p \vdash \sim(p \wedge \sim p)$, a fortiori $(p \wedge \sim p) \vdash \sim(p \wedge \sim p)$, so that it follows by **CONSIS 4** that $(p \wedge \sim p) \vdash q$, i.e. $p, \sim p \vdash q$.

Proof of Theorem 2:

(a) Let $\sim$ be semi–weak, i.e. let $(\sim p \vee \sim\sim p)$ be provable. Since by **DM 1** both $\sim p \vdash \sim(p \wedge \sim p)$ and $\sim\sim p \vdash \sim(p \wedge \sim p)$, it follows by **TNDF** that $\vdash \sim(p \wedge \sim p)$.

(b) If $\sim$ is semi–strong, then $(p \wedge \sim p) \vdash \sim q$; hence by elementary contraposition $\sim\sim q \vdash \sim(p \wedge \sim p)$. But **L** is supposed to be a normal logic so that there exists at least one q such that $\vdash q$. In view of **ADQ 3** it follows that $\vdash \sim\sim q$; thus we can derive $\vdash \sim(p \wedge \sim p)$.

Proof of Theorem 3:

3.1 (a) **DNI** entails **DM 3**: By **DM 1** we have $(\sim p \vee \sim q) \vdash \sim(p \wedge q)$, hence by **CP 1** $\sim\sim(p \wedge q) \vdash \sim(\sim p \vee \sim q)$ and by **DNI** $(p \wedge q) \vdash \sim(\sim p \vee \sim q)$.

(b) To see that **DM 3** entails **DNI**, just set $p = q$.

(c) **DNI** entails **DM 4**: Again by **DM 1** we have $(\sim\sim p \vee \sim\sim q) \vdash \sim(\sim p \wedge \sim q)$, hence by **DNI** also $(p \vee q) \vdash \sim(\sim p \wedge \sim q)$.

(d) Conversely **DNI** is obtained from **DM 4** by setting $q = p$!

(e) **DNI** entails **CP 4**: If $p \vdash \sim q$, then by **CP 1** $\sim\sim q \vdash \sim p$; hence by **DNI** also $q \vdash \sim p$.

(f) Conversely **DNI** follows from **CP 4** by the substitution $p/\sim p; q/p$!

Note, incidentally, that Došen (1988, 382–3) already proved **CP 4** to be deductively equivalent (within the framework of what he calls Hilbert system **BC**) to elementary **CP 1** plus **DNI**.

3.2 (a) DNE entails **DM 5**: By **DM 2** $\sim(\sim p \vee \sim q) \vdash (\sim\sim p \wedge \sim\sim q)$, hence by **DNE** also $\sim(\sim p \vee \sim q) \vdash (p \wedge q)$.

(b) Conversely **DNE** is obtained by **DM 5** by just setting $p = q$!

(c) DNE entails **DM 6**: Again by **DM 2** we have $\sim(p \vee q) \vdash (\sim p \wedge \sim q)$, hence by **CP 1** $\sim(\sim p \wedge \sim q) \vdash \sim\sim(p \vee q)$ and thus by **DNE** $\sim(\sim p \wedge \sim q) \vdash (p \vee q)$.

(d) Conversely **DM 6** entails **DNE** by the substitution q/p.

(e) DNE entails **CP 5**: If $\sim p \vdash q$, then by **CP 1** $\sim q \vdash \sim\sim p$, hence by **DNE** $\sim q \vdash p$.

(f) Conversely **DNE** follows from **CP 5** by the substitution $p/{\sim}p; q/p$!

Proof of Theorem 4:

4.1 Since trivially $p \vdash (p \vee q)$, the premises of **ECQ**, p and ${\sim}p$, immediately entail the premises of **DISSYL**, $(p \vee q)$ and ${\sim}p$, so that one obtains the desired conclusion $p, {\sim}p \vdash q$.

4.2 CONSYL implies **ECQF** because in view of the trivial ${\sim}p \vdash {\sim}(p \wedge q)$, the premises of **ECQF**, p and ${\sim}p$, entail the premises of **CONSYL**, p and ${\sim}(p \wedge q)$, so that one obtains the desired conclusion **ECQF**: $p, {\sim}p \vdash {\sim}q$. Furthermore if in **CONSYL** ${\sim}p$ is substituted for q, one obtains that $p, {\sim}(p \wedge {\sim}p) \vdash {\sim\sim}p$; but, according to Theorem 2, **ECQF** entails **CONSIS 1**: $\vdash {\sim}(p \wedge {\sim}p)$; hence in sum $p \vdash {\sim\sim}p$.

4.3 (a) If $\vdash {\sim}(p \wedge {\sim}p)$, then **CP 6** entails **ECQ**: If $\vdash {\sim}(p \wedge {\sim}p)$, then a fortiori ${\sim}q \vdash {\sim}(p \wedge {\sim}p)$, hence by **CP 6** $(p \wedge {\sim}p) \vdash q$.

(b) If $\vdash {\sim}(p \wedge {\sim}p)$, then **CP 6** entails **TND**: By **DM 2** one has ${\sim}(p \vee {\sim}p) \vdash ({\sim}p \wedge {\sim\sim}p)$; by **ECQ** also $({\sim}p \wedge {\sim\sim}p) \vdash {\sim\sim}(p \wedge {\sim}p)$, hence by transitivity ${\sim}(p \vee {\sim}p) \vdash {\sim\sim}(p \wedge {\sim}p)$, from which one obtains by **CP 6** that ${\sim}(p \wedge {\sim}p) \vdash (p \vee {\sim}p)$; but it is presupposed that $\vdash {\sim}(p \wedge {\sim}p)$; hence $\vdash (p \vee {\sim}p)$.

Proof of Theorem 5:

5.1 Almost trivially **DNE** and **DNI** together entail the strong contraposition principle **CP 6** so that by Theorem 4.3 the negation is almost classical.

5.2 If a weak negation $\sim$ satisfies **DISSYL**, then according to Theorem 4.1 it is also strong and hence classical.

5.3 If $\sim$ is a strong negation, then it is also semi–strong and hence by Theorem 2 it satisfies **CONSIS 1**: $\vdash {\sim}(p \wedge {\sim}p)$. As was shown in fn. 18, the negation $\sim$ thus satisfies **DNTND**: $\vdash {\sim\sim}(p \vee {\sim}p)$; so if $\sim$ also satisfies **DNE**, one immediately obtains $\vdash (p \vee {\sim}p)$, i.e. $\sim$ is weak, too, and hence in sum $\sim$ is classical.

References

da Costa, N.C.A. & Wolf, R.G.: Studies in Paraconsistent Logic I: The Dialectical Principle of the Unity of Opposites. *Philosophia* (Israel) 9 (1980), 189–217.

Došen, K.: Sequent Systems and Groupoid Models I. *Studia Logica* 47 (1988), 353–385.

Gabbay, D.M.: What is Negation in a System?. In: *Logic Colloquium '86*, ed. by F.R. Drake & J.K. Truss, Amsterdam 1988, 95–112.

Heyting, A.: Die formalen Regeln der intuitionistischen Logik. *Sitzungsberichte der Preußischen Akademie der Wissenschaften. Physikalisch–Mathematische Klasse*, 1930, 42–56.

Kutschera, F. v.: *Der Satz vom ausgeschlossenen Dritten – Untersuchungen über die Grundlagen der Logik.* Berlin 1985.

Leibniz, G.W.: De Vero et Falso, Affirmatione et Negatione, et de Contradictoriis. In: *Vorausedition zur Akademie-Ausgabe von Leibnizens Philosophischen Schriften*, ed. by Leibniz-Forschungsstelle der Universität Münster, fasc. 1 (1982), 86–88.

Leibniz, G.W.: Analysis Particularum. In: *Vorausedition zur Akademie-Ausgabe von Leibnizens Philosophischen Schriften*, ed. by Leibniz-Forschungsstelle der Universität Münster, fasc. 3 (1984), 514–532.

Lemmon, E.J.: *An Introduction to Modal Logic.* Oxford 1977.

Lenzen, W.: *Glauben, Wissen und Wahrscheinlichkeit – Systeme der epistemischen Logik.* Wien 1980.

Lenzen, W.: *Das System der Leibnizschen Logik.* Berlin 1990.

Lenzen, W.: What is (or at least appears to be) wrong with intuitionistic logic. In: *Advances in Scientific Philosophy – Essays in Honour of Paul Weingartner*, ed. by G. Schurz & G. Dorn, Amsterdam 1991, 173–186.

Pearce, D. & Wagner, G.: Reasoning with negative Information I: Strong Negation in Logic Programs. In: *Language, Knowledge, and Intentionality: Perspectives on the Philosophy of Jaakko Hintikka.*, ed. by L. Haaparanta, M. Kusch & I. Niiniluoto, (Acta Philosophica Fennica 49), Helsinki 1990, 430–453.

Pearce, D.: Reasoning with negative Information, II: Hard Negation, Strong Negation and Logic Programs. In: *Nonclassical Logics and Information Processing*, LNAI 619, ed. by D. Pearce & H. Wansing, Berlin 1992, 63–79.

Wansing, H.: *The Logic of Information Structures*, LNAI 681. Berlin 1993.

External, Restricted External, and Internal Negations in a Two-Dimensional Logic

Ingolf Max

Logicians usually divide negations into two main groups: (a) *external* (*classical*) negation and (b) *internal* (*non-classical*) negations. But some logicians, e.g. Stelzner,[1] argue for a threefold subdivision adding *restricted external* negations. Let M and N be any well-formed formal expressions, and let $\neg$ be the external negation. Then a restricted external negation n_r is definable using

$n_r M \equiv (\neg M \wedge N)$, with
(1) $M \vdash N$ (logical connection)
(2) *not* $(N \vdash M)$ (consistency)
(3) *not* $\vdash N$ (proper restriction)

It follows immediately that the *law of contradiction* holds with regard to the restricted external negation, and that the *law of excluded middle* does not:

$\vdash \neg(M \wedge n_r M)$, but not necessarily $\vdash n_r(M \wedge n_r M)$
not $\vdash (M \vee n_r M)$

Regarding restricted negations the law of excluded middle depends on the *truth of the restriction* describing an empirical situation:[2]

$\vdash (N \supset (M \vee n_r M))$

There are several reasons for introducing such restricted negations. Linguists, e.g., are very interested in negations whose scopes are not entire sentences but only well-defined parts of them.

If we look for the dual case it is very simple to characterize a weakened external negation n_w by:

$n_w M \equiv \neg M \vee N$, with
(1) $N \vdash M$ (logical connection)
(2) *not* $(M \vdash N)$ (non-triviality)
(3) *not* $\dashv N$ (proper weakening)

It is not surprising that now the *law of contradiction* does not hold, but the *law of excluded middle* holds:

not $\vdash \neg(M \wedge n_w M)$, but possibly $\vdash n_w(M \wedge n_w M)$,
additionally $\vdash ((M \wedge n_w M) \supset N)$ $\vdash (M \vee n_w M)$

[1] Cp. (Stelzner 1984, 79)
[2] Cp. condition (3).

It is clear that both restricted external and weakened external negations are functionally dependent on the external negation:

$$n_r M \vdash \neg M \qquad \neg M \vdash n_w M, \text{ therefore } \qquad n_r M \vdash n_w M.$$

The first equation above tells us that restricted external negations can also be named *strong* negations.

There are many 1–place functions (n_i) which are logically independent of the external/classical negation in the following sense:

$$\textit{neither } n_i M \vdash \neg M, \quad \textit{nor } \neg M \vdash n_i M$$

This does not exclude that there are a lot of interesting functional connections between such negations and the external one. But regarding logical dependency, negations of the former type are the only ones which I call *internal*. Therefore we get the following diagram (arrows are indicating the entailment relations here):

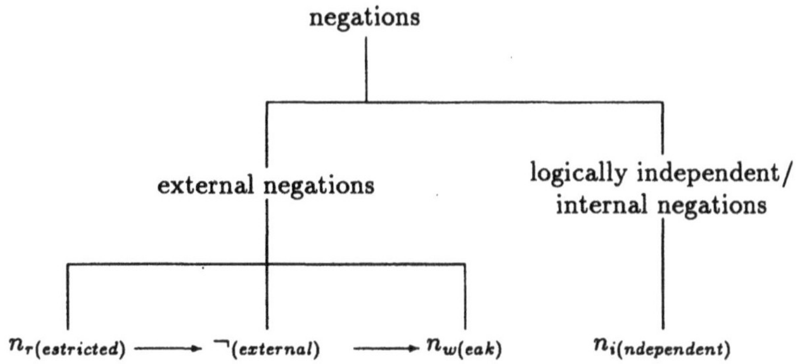

Neglecting 1–place tautological and inconsistent functions it holds that any 1–place function is either an external, strong, weak or logically independent negation in our schema. The reader could be astonished about the fact that he has no criteria at hand telling him which 1–place function is a negation at all. I think there are no a priori criteria deciding which 1–place function is a negation. My approach offers a schema with the external/classical negation at its core. My aim is to show that some negations are closer to the classical one than usually thought. I therefore advocate a *classical program*.

But there are a lot of negations discussed in the literature called *internal negations*. How do we handle them? First I want to show that several negations are not very properly characterized by calling them "internal". It would be a very interesting task to show that several "internal" negations are representable as, e.g., *restricted external* ones. In the case that there is no such reconstruction I intend to show that with respect to some restricted arguments M several of the internal negations can nevertheless be characterized as restricted external negations.

After introducing my system $\mathcal{N}$ — a classical propositional logic extended by n–place variable functors — and discussing the simplest case — types of product negations — I will demonstrate that Steinacker's *internal* negation can be constructed as a *restricted external* negation offering an S1–necessity–like variable functor as restriction N. Using several theorems stating equivalences between the system E_{fde} of tautological entailments and sublanguages of my system $\mathcal{N}$, I will show the possibility of representing a 4–valued *internal* negation in different ways. Depending on the *intuitive* interpretation of the 4 values — read two–dimensionally — one gets syntactically different negations. One of them is representable as a *restricted external* negation with respect to somehow restricted arguments M.

1 System $\mathcal{N}$

1.1 Primitive symbols

(1) $p, q, r, s, p_1, \ldots$ — propositional variables

(2) $\neg, \wedge, \vee, \supset, \equiv, \not\equiv$ — classical functors

(3) $\begin{bmatrix} \\ \end{bmatrix}$ — operator forming **pairs** of classical formulae

(4) $V_i^n \quad n \geq 1, 1 \leq i \leq 4^{4^n},$ — form of n–place variable functors

(5) $(\;)\ \Big(\;\Big)$ — parentheses

1.2 Formation rules

(1) A propositional variable standing alone is a formula of $\mathcal{N}$.

(2) If X, Y are formulae of $\mathcal{N}$, then $\neg X$, $(X \wedge Y)$, $(X \vee Y)$, $(X \supset Y)$, $(X \equiv Y)$, and $(X \not\equiv Y)$ are formulae of $\mathcal{N}$.

(3) If A, B are formulae of $\mathcal{N}$ formed without reference to the formation rules (3) and (4) (i.e. usual classical formulae), then $\begin{bmatrix} A \\ B \end{bmatrix}$ is a formula of $\mathcal{N}$.

(4) If X_1, X_n are formulae of $\mathcal{N}$ then $V^n X_1 \ldots X_n$ is a formula of $\mathcal{N}$.

(5) X is a formula of $\mathcal{N}$ iff its being so follows from (1)–(4)

1.3 Types of formulae

CL–formulae A, B, C, D (i.e. classical formulae) are those formulae which were exclusively formed by means of formation rules 1 and 2.

An **E–formula** $\mathcal{E}$ (i.e. elementary formula) is a formula of the form $\begin{bmatrix} A \\ B \end{bmatrix}$.

An **F–formula** $\mathcal{F}$ is a formula of the form $V_i^n \begin{bmatrix} A_1 \\ B_1 \end{bmatrix} \cdots \begin{bmatrix} A_n \\ B_n \end{bmatrix}$.

An **NC–formula** $\mathcal{Z}$ (i.e. a non–classical formula) is a formula of $\mathcal{N}$ which is neither (a) a CL–formula nor (b) an E–formula. Every F–formula is an NC–formula, but other examples are $\neg\neg \begin{bmatrix} p \\ q \end{bmatrix} \equiv \begin{bmatrix} p \\ q \end{bmatrix}$, $\neg V^3 \begin{bmatrix} p \\ q \end{bmatrix} \begin{bmatrix} r \\ s \end{bmatrix} \begin{bmatrix} p_2 \\ p_5 \end{bmatrix}$ etc.

1.4 Reduction rules

Roughly speaking, reduction rules should support a complete transformation of any non–classical formula $\mathcal{Z}$ to a formula of the form $\begin{bmatrix} A \\ B \end{bmatrix}$ (i.e. an *E–formula* of a special kind).
I use the following abbreviation of $X \Longrightarrow X[Y_1/Y_2]$: $Y_1 \Longrightarrow Y_2$.
Both "$X \Longrightarrow X[Y_1/Y_2]$" and "$Y_1 \Longrightarrow Y_2$" are read as "From X to infer $X[Y_1/Y_2]$", where by $X[Y_1/Y_2]$ we mean that formula which is the result of substituting any formula Y_2 for the formula Y_1 in all of its occurrences in X.

(1) *Reduction rules for classical functors:*

(1.1) $\neg \begin{bmatrix} A \\ B \end{bmatrix} \Longrightarrow \begin{bmatrix} \neg A \\ \neg B \end{bmatrix}$

(1.2) $\begin{bmatrix} A \\ B \end{bmatrix} \wedge \begin{bmatrix} C \\ D \end{bmatrix} \Longrightarrow \begin{bmatrix} A \wedge C \\ B \wedge D \end{bmatrix}$

$$\begin{bmatrix} A \\ B \end{bmatrix} \wedge C \Longrightarrow \begin{bmatrix} A \wedge C \\ B \wedge C \end{bmatrix} \qquad C \wedge \begin{bmatrix} A \\ B \end{bmatrix} \Longrightarrow \begin{bmatrix} C \wedge A \\ C \wedge B \end{bmatrix}$$

(1.3) Disjunction, implication, equivalence, and neg–equivalence as in 1.2.

(2) *Reduction rules for variable functors*

The *general form of substitution* is

SR $X \Longrightarrow X[\mathcal{F}/\mathcal{E}]$,

where by $X[\mathcal{F}/\mathcal{E}]$ I mean the result of substituting the *E–formula* $\mathcal{E}$ for the *F–formula* $\mathcal{F}$ in all occurrences of $\mathcal{F}$ in X.

The *special form* of **V–substitution** is

$$V^n \begin{bmatrix} A_1 \\ B_1 \end{bmatrix} \cdots \begin{bmatrix} A_n \\ B_n \end{bmatrix} \Longrightarrow \begin{bmatrix} \Phi^{2n} A_1 \ldots A_n, B_1 \ldots B_n \\ \Psi^{2n} A_1 \ldots A_n, B_1 \ldots B_n \end{bmatrix},$$

where Φ^{2n}, Ψ^{2n} are $2n$–place classical functors.[3]

1.5 Semantics

Validity of CL–formulae

Definition 1 *The classical formula A is* **classically valid** *(is a* **tautology**) *[symb.: $\vdash A$] iff the truth–value of A is 1 for all truth–values of the propositional variables.*

Validity of E–formulae

Definition 2 *The E–formula* $\begin{bmatrix} A \\ B \end{bmatrix}$ *is* **E–valid** *[symb.:* $\models \begin{bmatrix} A \\ B \end{bmatrix}$*] iff* $\vdash A$ *and* $\vdash B$.

Theorem 1 $\models \begin{bmatrix} A \\ B \end{bmatrix}$ *iff* $\vdash (A \wedge B)$.

Since all *NC–formulae* can be reduced to *E–formulae* this theorem means that validity in $\mathcal{N}$ is reducible to classical validity.

Validity of NC–formulae

Definition 3 *Let $\mathcal{Z}$ be any* NC–formula *and* $\begin{bmatrix} A_z \\ B_z \end{bmatrix}$ *that* E–formula *which is the result of the complete reduction of $\mathcal{Z}$, i.e. that all occurrences of variable functors and all occurrences of classical functors outside the scope of brackets are eliminated:*

$$\models \mathcal{Z} \text{ iff} \models \begin{bmatrix} A_z \\ B_z \end{bmatrix}.$$

1.6 Restricted external negations as variable functors

Now we are able to state more precisely the conditions for variable functors n_r to represent restricted external negations:[4]

---

[3]For the sake of simplicity and readability I take the well–known property of definability of the non–primitive classical functors for granted. Below, instead of $\Phi^{2n} A_1 \ldots A_n, B_1 \ldots B_n$ and $\Psi^{2n} A_1 \ldots A_n, B_1 \ldots B_n$ I will use any logically equivalent expressions.

[4]The conditions of variable functors representing weakened external negations can be obtained analogously.

$$\models n_r \begin{bmatrix} A \\ B \end{bmatrix} \equiv \neg \begin{bmatrix} A \\ B \end{bmatrix} \wedge \begin{bmatrix} \Phi^2 AB \\ \Psi^2 AB \end{bmatrix}, \text{ with}$$

(1) $\models \begin{bmatrix} A \\ B \end{bmatrix} \supset \begin{bmatrix} \Phi^2 AB \\ \Psi^2 AB \end{bmatrix}$ (logical connection)

(2) $\not\models \begin{bmatrix} \Phi^2 AB \\ \Psi^2 AB \end{bmatrix} \supset \begin{bmatrix} A \\ B \end{bmatrix}$ (consistency)

(3) $\not\models \begin{bmatrix} \Phi^2 AB \\ \Psi^2 AB \end{bmatrix}$ (proper restriction).

1.7 The simplest case: Types of product negations

Let us first introduce the notion of a *product operation* with respect to n–place variable functors:

Definition 4 *A variable functor of the form V^n is a* **product operation** *iff its reduction rule has the following form:*

$$V^n \begin{bmatrix} A_1 \\ B_1 \end{bmatrix} \cdots \begin{bmatrix} A_n \\ B_n \end{bmatrix} \Longrightarrow \begin{bmatrix} \Phi^n A_1 \ldots A_n \\ \Psi^n B_1 \ldots B_n \end{bmatrix},$$

i.e. such a variable functor acts separately like two different *n–place* functors in each dimension/line.

The two–dimensional approach offers the possibility to give each dimension (line) an interpretation of its own. One way of explicating this idea is to call any variable functor which has only an external/classical negation in its first reduction line an *n^1–negation*.

Definition 5 *A 1–place variable functor V^1 is an* **n^1–negation** *iff its reduction rule has the following form:*

$$V^1 \begin{bmatrix} A \\ B \end{bmatrix} \Longrightarrow \begin{bmatrix} \neg A \\ \Psi^2 AB \end{bmatrix}.$$

If we take the meet of these two definitions above we get the notion of an *n^1–product negation*:

Definition 6 *A 1–place variable functor V^1 is an* **n^1–product negation** *iff its reduction rule has the following form:*

$$V^1 \begin{bmatrix} A \\ B \end{bmatrix} \Longrightarrow \begin{bmatrix} \neg A \\ \Psi^1 B \end{bmatrix}.$$

Remembering the reduction rule for the external/classical negation we can now state that "$\neg$" is an n^1-product negation.[5] But there are others:

n^1-product negation as restricted external negation n_r^1

$$n_r^1 \begin{bmatrix} A \\ B \end{bmatrix} \Longrightarrow \begin{bmatrix} \neg A \\ B \wedge \neg B \end{bmatrix} \qquad \models n_r^1 \begin{bmatrix} A \\ B \end{bmatrix} \equiv \neg \begin{bmatrix} A \\ B \end{bmatrix} \wedge \begin{bmatrix} A \vee \neg A \\ B \end{bmatrix}$$

n^1-product negation as weakened external negation n_w^1

$$n_w^1 \begin{bmatrix} A \\ B \end{bmatrix} \Longrightarrow \begin{bmatrix} \neg A \\ B \vee \neg B \end{bmatrix} \qquad \models n_w^1 \begin{bmatrix} A \\ B \end{bmatrix} \equiv \neg \begin{bmatrix} A \\ B \end{bmatrix} \vee \begin{bmatrix} A \wedge \neg A \\ B \end{bmatrix}$$

n^1-product negation as internal negation n_i^1

$$n_i^1 \begin{bmatrix} A \\ B \end{bmatrix} \Longrightarrow \begin{bmatrix} \neg A \\ B \end{bmatrix} \quad \not\models n_i^1 \begin{bmatrix} A \\ B \end{bmatrix} \supset \neg \begin{bmatrix} A \\ B \end{bmatrix} \quad \not\models \neg \begin{bmatrix} A \\ B \end{bmatrix} \supset n_i^1 \begin{bmatrix} A \\ B \end{bmatrix}$$

This leads us to a more specific diagram (arrows are to indicate $\models$-valid material implications):

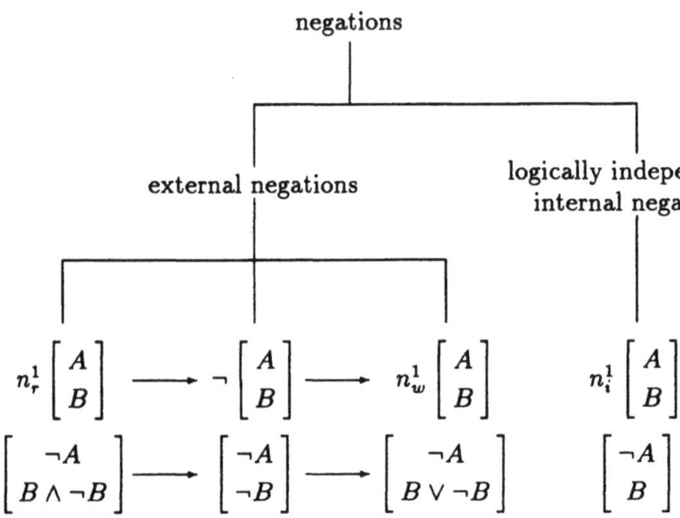

[5]Naturally the external negation is furthermore an n^2-product negation, which shows the uniform behavior of the classical negation in both dimensions.

2 A Reconstruction of Steinacker's Internal Negation as Restricted External Negation

Steinacker[6] formulates a system **Ni** taken as a basic calculus of *internal negation*. After going over the fundamentals of this system I will construct a biunique translation of its language into a sublanguage of $\mathcal{N}$. Steinacker's "internal" negation is represented by a variable functor which can be characterized as a restricted external negation.

2.1 Steinacker's calculus Ni

Let $\neg$ be the *external* (classical) negation, and $\sim$ be Steinacker's "*internal*" negation. We add to any given axiomatic formalization of classical logic the axiom

N0. $H \supset \neg \sim H$

Theorems of this system **Ni** are:

(1+) $\sim p \supset \neg p$. This indicates "$\sim$" as a candidate for a restricted negation.

(2+) $\sim \neg p \supset p$

(3+) $\sim \neg p \supset \neg \sim p$

(4+) $\sim p \supset \neg \sim \neg p$

(5+) $\sim (p \wedge q) \supset \neg p \vee \neg q$

(6+) $\sim p \vee \sim q \supset \neg(p \wedge q)$

(7+) $\sim (p \wedge q) \supset \neg \sim \neg p \vee \neg \sim \neg q$

(8+) $\sim p \vee \sim q \supset \neg \sim (\neg p \vee \neg q)$

(9+) $\neg(A \wedge \sim A)$

The following formulae are not theorems of **Ni**:

(1–) $p \supset \sim\sim p$

(2–) $\sim\sim p \supset p$

(3–) $\neg p \supset \sim p$

(4–) $p \supset \sim \neg p$

(5–) $\neg \sim p \supset \sim \neg p$

(6–) $\sim (p \wedge q) \supset \sim p \vee \sim q$

(7–) $\sim p \vee \sim q \supset \sim (p \wedge q)$.

2.2 The representation of Ni in $\mathcal{N}$

I postulate the following biunique mappings between the elements of the **Ni**–language and the elements of the $\mathcal{N}$–language. The symbol "$\Longleftrightarrow$" is to indicate the biunique mappings here:

(1) $p \Longleftrightarrow \begin{bmatrix} p_1 \\ p_2 \end{bmatrix}$ $\quad q \Longleftrightarrow \begin{bmatrix} q_1 \\ q_2 \end{bmatrix}$ $\quad$ in general: $p^i \Longleftrightarrow \begin{bmatrix} p_1^i \\ p_2^i \end{bmatrix}$

(2) Classical functors with the same shape correspond to each other.

[6] Cp. (Steinacker 1993, 334 f.).

(3) The "*internal*" negation is interpreted as a 1-place variable functor: $\sim \iff \sim$ with the following reduction property:[7]

$$\sim \begin{bmatrix} A \\ B \end{bmatrix} \Longrightarrow \begin{bmatrix} \neg A \wedge B \\ \neg A \wedge \neg B \end{bmatrix}$$

2.3 E-validity of corresponding $\mathcal{N}$-formulae

It is easy to verify that the following formulae corresponding to N0, - (9+) are E-valid:

(N0') $\models \begin{bmatrix} p_1 \\ p_2 \end{bmatrix} \supset \neg \sim \begin{bmatrix} p_1 \\ p_2 \end{bmatrix}$, because of $\begin{array}{l} \vdash p_1 \supset \neg(\neg p_1 \wedge p_2) \\ \vdash p_2 \supset \neg(\neg p_1 \wedge \neg p_2) \end{array}$

(1'+) $\models \sim \begin{bmatrix} p_1 \\ p_2 \end{bmatrix} \supset \neg \begin{bmatrix} p_1 \\ p_2 \end{bmatrix}$, because of $\begin{array}{l} \vdash \neg p_1 \wedge p_2 \supset \neg p_1 \\ \vdash \neg p_1 \wedge \neg p_2 \supset \neg p_2 \end{array}$

(2'+) $\models \sim \neg \begin{bmatrix} p_1 \\ p_2 \end{bmatrix} \supset \begin{bmatrix} p_1 \\ p_2 \end{bmatrix}$ because of $\begin{array}{l} \vdash \neg\neg p_1 \wedge \neg p_2 \supset p_1 \\ \vdash \neg\neg p_1 \wedge \neg\neg p_2 \supset p_2 \end{array}$

(3'+) $\models \sim \neg \begin{bmatrix} p_1 \\ p_2 \end{bmatrix} \supset \neg \sim \begin{bmatrix} p_1 \\ p_2 \end{bmatrix}$ because of $\begin{array}{l} \vdash \neg\neg p_1 \wedge \neg p_2 \supset \neg(\neg p_1 \wedge p_2) \\ \vdash \neg\neg p_1 \wedge \neg\neg p_2 \supset \neg(\neg p_1 \wedge \neg p_2) \end{array}$

(4'+) $\models \sim \begin{bmatrix} p_1 \\ p_2 \end{bmatrix} \supset \neg \sim \neg \begin{bmatrix} p_1 \\ p_2 \end{bmatrix}$ because of $\begin{array}{l} \vdash \neg p_1 \wedge p_2 \supset \neg(\neg\neg p_1 \wedge \neg p_2) \\ \vdash \neg p_1 \wedge \neg p_2 \supset \neg(\neg\neg p_1 \wedge \neg\neg p_2) \end{array}$

(5'+) $\models \sim \left(\begin{bmatrix} p_1 \\ p_2 \end{bmatrix} \wedge \begin{bmatrix} q_1 \\ q_2 \end{bmatrix} \right) \supset \left(\neg \begin{bmatrix} p_1 \\ p_2 \end{bmatrix} \vee \neg \begin{bmatrix} q_1 \\ q_2 \end{bmatrix} \right)$

(6'+) $\models \sim \left(\begin{bmatrix} p_1 \\ p_2 \end{bmatrix} \vee \sim \begin{bmatrix} q_1 \\ q_2 \end{bmatrix} \right) \supset \neg \left(\begin{bmatrix} p_1 \\ p_2 \end{bmatrix} \wedge \begin{bmatrix} q_1 \\ q_2 \end{bmatrix} \right)$

(7'+) $\models \sim \left(\begin{bmatrix} p_1 \\ p_2 \end{bmatrix} \wedge \begin{bmatrix} q_1 \\ q_2 \end{bmatrix} \right) \supset \left(\neg \sim \neg \begin{bmatrix} p_1 \\ p_2 \end{bmatrix} \vee \neg \sim \neg \begin{bmatrix} q_1 \\ q_2 \end{bmatrix} \right)$

(8'+) $\models \left(\sim \begin{bmatrix} p_1 \\ p_2 \end{bmatrix} \vee \sim \begin{bmatrix} q_1 \\ q_2 \end{bmatrix} \right) \supset \neg \sim \left(\neg \begin{bmatrix} p_1 \\ p_2 \end{bmatrix} \vee \neg \begin{bmatrix} q_1 \\ q_2 \end{bmatrix} \right)$

(9'+) $\models \neg \left(\begin{bmatrix} p_1 \\ p_2 \end{bmatrix} \wedge \sim \begin{bmatrix} p_1 \\ p_2 \end{bmatrix} \right)$, because of $\begin{array}{l} \vdash \neg(p_1 \wedge (\neg p_1 \wedge p_2)) \\ \vdash \neg(p_2 \wedge (\neg p_1 \wedge \neg p_2)) \end{array}$

[7] It should be mentioned that there is at least one other candidate for Steinacker's "internal" negation: $\sim \iff \sim_*$ with the following reduction property:

$$\sim_* \begin{bmatrix} A \\ B \end{bmatrix} \Longrightarrow \begin{bmatrix} \neg A \wedge \neg B \\ A \wedge \neg B \end{bmatrix}$$

It is up to the reader to show that the formulae (1'–) – (7'–) corresponding to (1–) – (7–) are not E–valid.

There are some other interesting properties:

(a) $\not\models \left(\begin{bmatrix} p_1 \\ p_2 \end{bmatrix} \equiv \begin{bmatrix} q_1 \\ q_2 \end{bmatrix}\right) \supset \left(\sim \begin{bmatrix} p_1 \\ p_2 \end{bmatrix} \equiv \sim \begin{bmatrix} q_1 \\ q_2 \end{bmatrix}\right)$

(b) $\models \sim\sim\sim\sim \begin{bmatrix} p_1 \\ p_2 \end{bmatrix} \equiv \sim \begin{bmatrix} p_1 \\ p_2 \end{bmatrix}$.

(c) There is no formula of the form $\sim X$ which is E–valid.

2.4 The variable functor "∼" as restricted external negation

The above mentioned formula (1'+) shows that negation "∼" is stronger than the external (classical) negation "¬". Therefore, we should presume that "∼" is a *restricted external* negation rather than an internal one. But how can we prove this assumption? Let us begin with the following E–valid expression:

$$\models \sim \begin{bmatrix} A \\ B \end{bmatrix} \equiv \neg \begin{bmatrix} A \\ B \end{bmatrix} \wedge \begin{bmatrix} A \vee B \\ A \supset B \end{bmatrix}, \text{ because of}$$

$$\begin{matrix} \vdash \neg A \wedge B \equiv \neg A \wedge (A \vee B) \\ \vdash A \neg \wedge \neg B \equiv \neg B \wedge (A \supset B) \end{matrix}. \text{ Then}$$

(1) $\models \begin{bmatrix} A \\ B \end{bmatrix} \supset \begin{bmatrix} A \vee B \\ A \supset B \end{bmatrix}$, because of $\begin{matrix} \vdash A \supset (A \vee B) \\ \vdash B \supset (A \supset B) \end{matrix}$ (logical connection)

(2) $\not\models \begin{bmatrix} A \vee B \\ A \supset B \end{bmatrix} \supset \begin{bmatrix} A \\ B \end{bmatrix}$ (consistency)

(3) $\not\models \begin{bmatrix} A \vee B \\ A \supset B \end{bmatrix}$ (proper restriction).

Therefore, it is shown that the variable functor "∼" can be characterized as a *restricted external* negation. Furthermore, because it satisfies all the properties supposed by Steinacker the negation "∼" of **Ni** should not be described as an internal one.

2.5 Truth operator as variable functor

There is a thesis in Steinacker's paper that the internal negation is a mixture of an operator (e.g. truth operator) and an internal usage of the external negation. Let "T" be another variable functor with the following reduction property:

$$T\begin{bmatrix} A \\ B \end{bmatrix} \Longrightarrow \begin{bmatrix} B \supset A \\ A \vee B \end{bmatrix}. \text{ Then we get}$$

$$\sim\begin{bmatrix} A \\ B \end{bmatrix} \equiv \neg T\begin{bmatrix} A \\ B \end{bmatrix}, \text{ because of } \begin{matrix} \vdash \neg A \wedge B \equiv \neg(B \supset A) \\ \vdash \neg A \wedge \neg B \equiv \neg(A \vee B) \end{matrix}.$$

Finally, with respect to

$$(\mathrm{T1})\ \not\models \neg T\begin{bmatrix} A \\ B \end{bmatrix} \equiv T\neg\begin{bmatrix} A \\ B \end{bmatrix}$$

$$(\mathrm{T2})\ \not\models T\begin{bmatrix} A \\ B \end{bmatrix} \wedge T\begin{bmatrix} C \\ D \end{bmatrix} \equiv T\left(\begin{bmatrix} A \\ B \end{bmatrix} \wedge \begin{bmatrix} C \\ D \end{bmatrix}\right) \text{ etc.}$$

we get the natural result that "T" differs essentially from the classical truth operator.[8]

2.6 Variable functors as modal operators

In an earlier paper of mine[9] it was shown that special variable functors can be interpreted as S5–like modal operators:

Theorem 2 *If H^{S5} is an S5–valid formula, then X_{HS5} is valid in the sense of "$\models$", where the biunique mappings concerning the translation of formulae are:*

$$p^i \Longleftrightarrow \begin{bmatrix} p_1^i \\ p_2^i \end{bmatrix} \quad \textit{(propositional variables vs. E–expressions containing propositional variables with the same index)}$$

$$\Diamond_{S5} \Longleftrightarrow \Diamond_{S5'}, \textit{ where } \Diamond_{S5'}\begin{bmatrix} A \\ B \end{bmatrix} \Longrightarrow \begin{bmatrix} A \vee B \\ A \vee B \end{bmatrix} \quad \textit{(possibility)}$$

$$\Box_{S5} \Longleftrightarrow \Box_{S5'}, \textit{ where } \Box_{S5'}\begin{bmatrix} A \\ B \end{bmatrix} \Longrightarrow \begin{bmatrix} A \wedge B \\ A \wedge B \end{bmatrix} \quad \textit{(necessity)}$$

and the functors are the classical ones.

Example:

$$\models \Box_{S5'}\begin{bmatrix} A \\ B \end{bmatrix} \equiv \neg\Diamond_{S5'}\neg\begin{bmatrix} A \\ B \end{bmatrix}, \begin{matrix} \vdash A \wedge B \equiv \neg(\neg A \vee \neg B) \\ \vdash A \wedge B \equiv \neg(\neg A \vee \neg B) \end{matrix}.$$

---

[8] Cp. (Steinacker 1993, 331).

[9] (Max 1987).

Non–extensionality of modal variable functors:

$$\not\models \left(\begin{bmatrix} p_1 \\ p_2 \end{bmatrix} \equiv \begin{bmatrix} q_1 \\ q_2 \end{bmatrix}\right) \supset \left(\Diamond_{S5'}\begin{bmatrix} p_1 \\ p_2 \end{bmatrix} \equiv \Diamond_{S5'}\begin{bmatrix} q_1 \\ q_2 \end{bmatrix}\right).$$

There is another result[10] stating that a logically equivalent representation of any functionally complete (and also incomplete) 4–valued propositional logic can be constructed by means of variable functors.

Lewis/Langford[11] create 4– and 8–valued truth–tables showing the distinctness of the Lewis systems S1–S5. After presenting their 4–valued truth–tables for S1 I offer a two–dimensional reading of the same tables. Then I show how these 4–valued functions can be represented by variable functors. Truth–tables for the Lewis system S1[12]:

H	$\neg H$	$\Diamond_{S1} H$	$\Box_{S1}$
1	4	1	2
2	3	2	4
3	2	1	3
4	1	3	4

$H_1 \wedge H_2$	1 2 3 4
1	1 2 3 4
2	2 2 4 4
3	3 4 3 4
4	4 4 4 4

$H_1 \rightarrow H_2$	1 2 3 4
1	2 4 3 4
2	2 2 3 3
3	2 4 2 4
4	2 2 2 2

Designated values: **1, 2**

Two–dimensional representation of the same truth–tables:

Now we can read the four truth–values as ordered pairs of the classical values 1 and 0: <k,l> with k,l ∈ {1,0}. For the sake of brevity I write only "kl". Let us assume the following biunique mappings:

$1 \Longleftrightarrow 11$ $\quad 2 \Longleftrightarrow 10$ $\quad 3 \Longleftrightarrow 01$ $\quad 4 \Longleftrightarrow 00$

We then obtain the following reading of the Lewis/Langford truth–tables:

H	$\neg H$	$\Diamond_{S1} H$	$\Box_{S1}$
11	00	11	10
10	01	10	00
01	10	11	01
00	11	01	00

$H_1 \wedge H_2$	11 10 01 00
11	11 10 01 00
10	10 10 00 00
01	01 00 01 00
00	00 00 00 00

$H_1 \rightarrow H_2$	11 10 01 00
11	10 00 01 00
10	10 10 01 01
01	10 00 10 00
00	10 10 10 10

Designated values: **11, 10**

Definition 7 *H^{S1} is* **S1–4–valid** *[symb.: $\vdash_{1,2}$] iff the truth–value of H^{S1} is* 1 (11), 2 (10) *for all truth–values of the propositional variables.*

Now I define a corresponding notion of validity for E–expressions:

[10](Max 1988).

[11]Cp. (Lewis/Langford 1932, 492–498).

[12]"→" indicates S1–strict implication. It would be sufficient to present the truth–tables for the primitives, e.g. $\neg$, $\wedge$, $\Diamond_{S1}$.

Definition 8 *The E-formula* $\begin{bmatrix} A \\ B \end{bmatrix}$ *is* **S1'-valid** [*symb.*: $\models_1$] *iff* $\vdash A$.[13]

This definition says that the S1'-validity of a pair of formulae depends only on the validity of the first line. Now we are able to state a theorem about the equivalence between the sets of S1–4-valid and S1'-valid formulae:

Theorem 3 $\vdash_{1,2} H^{S1}$ *iff* $\models_1 X_{HS1}$,
where the biunique mappings concerning the translation of formulae are:

$$p^i \Longleftrightarrow \begin{bmatrix} p^i_1 \\ p^i_2 \end{bmatrix} \quad \textit{(propositional variables vs. E-expressions containing propositional variables with the same index)}$$

$$\Diamond_{S1} \Longleftrightarrow \Diamond_{S1'}, \text{ where } \Diamond_{S1'} \begin{bmatrix} A \\ B \end{bmatrix} \Longrightarrow \begin{bmatrix} A \vee B \\ A \supset B \end{bmatrix} \qquad \textit{(possibility)}$$

$$\Box_{S1} \Longleftrightarrow \Box_{S1'}, \text{ where } \Box_{S1'} \begin{bmatrix} A \\ B \end{bmatrix} \Longrightarrow \begin{bmatrix} A \wedge B \\ \neg A \wedge B \end{bmatrix} \qquad \textit{(necessity)}$$

$$H_1 \rightarrow H_2 \Longleftrightarrow \rightarrow, \text{ where } \begin{bmatrix} A \\ B \end{bmatrix} \rightarrow \begin{bmatrix} C \\ D \end{bmatrix} \Longrightarrow \begin{bmatrix} (A \supset C) \wedge (B \supset D) \\ \neg(A \supset C) \wedge (B \supset D) \end{bmatrix}$$

(S1-strict implication)

and the functors are the classical ones.

To prove this theorem it has to be shown that (1) there is a biunique connection between the 4-valued truth-tables and the truth-tables of the two-dimensional expressions, and that (2) the reduction rules of variable operators act in a syntactic manner like the 4-valued truth-operation in a semantic way.

Only to sketch this proof I show the pseudo-two-dimensional value-tables of the above mentioned variable functors:

$\begin{bmatrix} A \\ B \end{bmatrix}$	$\Diamond_{S1'} \begin{bmatrix} A \\ B \end{bmatrix}$	$\Box_{S1'} \begin{bmatrix} A \\ B \end{bmatrix}$
1/1	1/1	1/0
1/0	1/0	0/0
0/1	1/1	0/1
0/0	0/1	0/0

$\begin{bmatrix} A \\ B \end{bmatrix} \rightarrow \begin{bmatrix} C \\ D \end{bmatrix}$	1/1	1/0	0/1	0/0
1/1	1/0	0/0	0/1	0/0
1/0	1/0	1/0	0/1	0/1
0/1	1/0	0/0	1/0	0/0
0/0	1/0	1/0	1/0	1/0

The use of the notation "k/l" is to indicate that we do not have real two-dimensional values here, but only *pseudo*-values. It would be much better

[13]Cp. definition 3 for defining $\models_1 X$.

to have a vertical dash showing that the left part of each column represents the truth–table of the first dimension of any expression and the right part of each column the second one.

If we neglect product operations, reduction rules of variable functors act in such a way that, informally speaking, they put both lines together in each reduction step. This corresponds to the functional dependency of both dimensions of the 4 truth–values read as ordered pairs of classical values.

It is well–known that every *S1–provable* formula is S1–4–valid but not vice versa.[14] Therefore, we get a corresponding theorem in this weaker form:

Theorem 4 *If H^{S1} is an S1–provable formula, then* $\models_1 X_{H^{S1}}$.

This can easily be shown by demonstrating that each formula which bi-uniquely corresponds to an axiom of S1 is S1'–valid. And that each analogon of the inference rules of S1 preserves S1'–validity.

Regarding these results I call the variable functors $\Diamond_{S1'}$, $\Box_{S1'}$, and $\rightarrow$ *S1–like* modal operators.

2.7 Steinacker's internal negation as modal–restricted external negation

The reader will surely have noticed that the S1–like possibility $\Diamond_{S1'}$ looks like the restriction N of the external negation leading to Steinacker's "internal" negation $\sim$.[15] Let us recall the characteristic equivalence using now $\Diamond_{S1'}$:

$$\models \sim \begin{bmatrix} A \\ B \end{bmatrix} \equiv \neg \begin{bmatrix} A \\ B \end{bmatrix} \wedge \Diamond_{S1'} \begin{bmatrix} A \\ B \end{bmatrix}$$

There is a remarkable connection between the fact that in our representation S1–like possibility figures as restriction and Steinacker's argumentation that the semantic representation of such a negation necessitates the extension of the set of possible worlds by adding non–normal worlds. It is possible to read the expression $\sim \begin{bmatrix} A \\ B \end{bmatrix}$, i.e. $\begin{bmatrix} \neg A \wedge B \\ \neg A \wedge \neg B \end{bmatrix}$ as "there is a negation in the A–worlds together with contradictory (non–normal) B–worlds".

One nice advantage of the two–dimensionally syntactical approach is to make the internal relation directly explicit:

$$\models \sim \begin{bmatrix} A \\ B \end{bmatrix} \supset \neg A$$

[14]There is no characteristic finite matrix for any of the systems S1–S5. Cp. (Dugundji 1940).

[15]Cp. section 2.4.

Therefore, $\sim \begin{bmatrix} A \\ B \end{bmatrix}$ is even stronger than $\neg A$. The formula containing $\neg A$ does not correspond to any formula of a 4-valued logic. Furthermore, in our S1'-fragment there is no valid expression of the form

$\models A \equiv X$, where X is not a classical formula.

Definition 9 *The classical formula A is* **classically inconsistent** *(is a* **contradiction***) [symb.: $\dashv A$] iff the truth-value of A is 0 for all truth-values of the propositional variables.*

Definition 10 *The E-formula* $\begin{bmatrix} A \\ B \end{bmatrix}$ *is* **explicitly inconsistent**

[symb.: ${}_{e}\dashv \begin{bmatrix} A \\ B \end{bmatrix}$*] iff* $\dashv A$ *and* $\dashv B$*.*

Definition 11 *The E-formula* $\begin{bmatrix} A \\ B \end{bmatrix}$ *is* **implicitly inconsistent**

[symb.: ${}_{i}\dashv \begin{bmatrix} A \\ B \end{bmatrix}$*] iff (i)* $\dashv A \wedge B$*, (ii)* $\not\dashv A$*, and (iii)* $\not\dashv B$*.*

Definition 12 *Let Z be any NC-formula and* $\begin{bmatrix} A_Z \\ B_Z \end{bmatrix}$ *that E-formula which is the result of complete reduction of Z, i.e. that all occurrences of variable functors and all occurrences of classical functors outside the scope of brackets are eliminated:*

Z is inconsistent *in any sense iff* $\begin{bmatrix} A_Z \\ B_Z \end{bmatrix}$ *is inconsistent in the same sense.*

Now we are able to show some other properties with respect to explicit and implicit inconsistency:

$${}_{e}\not\dashv \sim \begin{bmatrix} A \\ B \end{bmatrix} \quad {}_{e}\not\dashv \Box_{S1'} \begin{bmatrix} A \\ B \end{bmatrix} \quad {}_{e}\not\dashv \begin{bmatrix} A \\ B \end{bmatrix} \rightarrow \begin{bmatrix} C \\ D \end{bmatrix}, \text{ but}$$

$${}_{i}\dashv \sim \begin{bmatrix} A \\ B \end{bmatrix} \quad {}_{i}\dashv \Box_{S1'} \begin{bmatrix} A \\ B \end{bmatrix} \quad {}_{i}\dashv \begin{bmatrix} A \\ B \end{bmatrix} \rightarrow \begin{bmatrix} C \\ D \end{bmatrix}$$

$${}_{e}\not\dashv \Diamond_{S1'} \begin{bmatrix} A \\ B \end{bmatrix} \quad {}_{i}\not\dashv \Diamond_{S1'} \begin{bmatrix} A \\ B \end{bmatrix}$$

With respect to the *law of excluded middle* the situation is much more restricted. The formulation of the law of excluded middle using our negation "$\sim$" is *logically equivalent* to S1-like possibility. I.e., the validity of this central law is equal to a specific empirical situation.

$$\models (X \vee \sim X) \equiv \Diamond_{S1'} X.$$

Another interesting point is that the formulation of necessity using negation "$\sim$" leads to the 4–valued notion of necessity formulated by Łukasiewicz:[16]

$$\Box_{Ł} \begin{bmatrix} A \\ B \end{bmatrix} \Longrightarrow \begin{bmatrix} A \\ B \wedge \neg B \end{bmatrix} \qquad \models \sim \Diamond_{S1'} \sim \begin{bmatrix} A \\ B \end{bmatrix} \equiv \Box_{Ł} \begin{bmatrix} A \\ B \end{bmatrix}.$$

With respect to S1–like possibility there is no difference between $\sim$ and our product negation n_r^1:

$$\models \sim \Diamond_{S1'} X \equiv n_r^1 \Diamond_{S1'} X.$$

It is also possible to combine S*–like modal operators to get some interesting connections:

$$\models \Box_{S5'} \Diamond_{S1'} \begin{bmatrix} A \\ B \end{bmatrix} \equiv B \qquad \models \Diamond_{S5'} \Diamond_{S1'} X.$$

The reader could raise the following objection: Why do we use two different notions of validity: $\models$ with regard to Steinacker's negation and $\models_1$ in the case of the S1'–fragment.[17] Our first observation is that except (9+) all of Steinacker's formulae have the form $(H_1 \supset H_2)$. If we take the *strict* versions of these formulae, i.e. the form $(H_1 \rightarrow H_1)$ or $\Box_{S1}(H_1 \supset H_2)$, we get an equivalence expressed by the following theorem:

Theorem 5 $\models X$ *iff* $\models_1 \Box_{S1'} X$.

3 Negation and Tautological Entailment

Dealing with many–valued logics it is a well–known semantic tool to establish one or more *designated values*. A possible *metaphysical* misunderstanding consists in the identification of *designated values* with "real" truth–values. In a pure technical respect we need designated values only to define a notion of *validity* within many–valued logics. But there are a lot of *intuitive* readings of values depending on the intended field of application of a chosen many–valued logic. 3– and 4–valued logics, e.g., can be used to discuss the "logic" of quantum physics, presuppositions of natural languages, time modelling, vagueness of notions etc.

Usually it is presumed that expressing different intuitions with regard to the values of a many–valued logic necessitates different systems or formulations (e.g. functionally incomplete ones) of such logics. Then we can observe

[16]Cp. (Łukasiewicz 1953). By the way, this necessity is a product operation.

[17]Additionally, there is the $\models_1$–validity of (6'–). But if we take the strict version we get $\not\models_1 \Box_{S1'}$ (6'–).

a highly emotional discussion about the "right" or "intuitive acceptable" negation, conjunction, disjunction etc.

The discussion about the "right" semantics for the system E_{fde}[18] of tautological entailment shows the opposite case. In accordance with Anderson/Belnap we call an n–valued matrix "*characteristic* for a calculus when a formula A is provable just in case it assumes designated values for every assignment of its variables."[19] They also mention that there is a proof by Smiley (in correspondence) showing that the 4–valued matrix discussed below is characteristic for E_{fde}.

Nevertheless there are several variants of reading these 4 values and alternative semantic results using, e.g., algebraic or lattice theories. Given the Smiley matrix I am going to show that it allows several *two–dimensional intuitive* readings of the 4 values and that these differences can be made syntactically explicit in my two–dimensional approach. With respect to our topic it is very important that we get several types of negations, one of them comes out as *restricted external negation* in the context of pseudo–3–valued arguments.

3.1 Language of E_{fde}

Primitive symbols

(1) p^i form of propositional variables

(2) $\sim_e, \wedge_e, \vee_e, \rightarrow_e$ propositional functors

Formation rules

(1) A propositional variable standing alone is a formula.

(2) If H_1, H_2 are formulae in which "$\rightarrow_e$" does not occur, then $\sim_e H_1$, $(H_1 \wedge_e H_2)$, $(H_1 \vee_e H_2)$, and $(H_1 \rightarrow_e H_2)$ are formulae.

3.2 Semantics of E_{fde}

H	$\sim_e H$
1	4
2	2
3	3
4	1

$H_1 \wedge_e H_2$	1	2	3	4
1	1	2	3	4
2	2	2	4	4
3	3	4	3	4
4	4	4	4	4

$H_1 \vee_e H_2$	1	2	3	4
1	1	1	1	1
2	1	2	1	2
3	1	1	3	3
4	1	2	3	4

$H_1 \rightarrow_e H_2$	1	2	3	4
1	1	2	3	4
2	1	1	3	3
3	1	2	1	2
4	1	1	1	1

[18]My index $_{fde}$ is to indicate that this system does not contain expressions with nested arrows, i.e. all entailment expressions are first degree entailments.

[19](Anderson/Belnap 1975, 161).

Designated value: **1**.[20]

It is a matter of fact that we have several *intuitive* readings of this 4–valued truth–tables. But it is also a fact that such *one–dimensional* truth–tables cannot be used for expressing a variety of readings *explicitly*. Let us assume that "1" is the candidate for the value which I want to name "*intuitively designated value*". There are then some *structural restrictions* with respect to the value "1" in these value–tables which I am going to use below for identifying the *intuitively designated values*:

(1) "1" occurs exactly once both in the negation– and in the conjunction–table.
(2) "1" occurs exactly 9 times both in the disjunction– and in the entailment–table.
(3) "1" and "4" change their positions in the negation–table.

3.3 Some observations regarding negation in E_{fde}

Up to now we have no more than a 4–valued table for $\sim_e$. What can we say about its properties? First, there is a "classical" part consisting in "1" and "4". These two values change their positions like in the familiar classical case. But there is also a "strange" part consisting in "2" and "3". Nothing happens. It looks like an invariant part: here is no difference between unnegated and negated expression. The whole expression seems to be a "mixture" of *negating* (exchanging the place) and *asserting* (keeping the place).

Putting this in a picture we can get the following diagram:

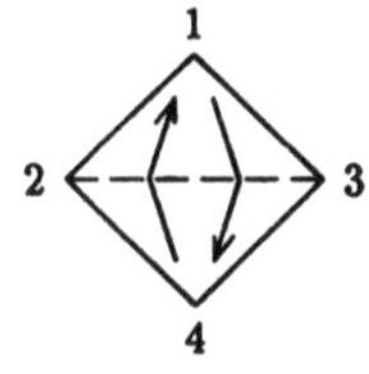

Negation is acting like turning a diamond along the axis 2 — 3 (indicated by the broken line between 2 and 3). E.g. 2 and 3 are fixed points allowing 1 and 4 to change their places.

3.4 Dunn's version

Dunn offers an interesting way of developing a two–dimensional semantics for E_{fde}, and he tries it in a *unifying* manner:

> The motivating idea is that *A* should relevantly imply *B* iff, in some appropriate sense, whatever *B* is about, *A* is about as well. This is

[20]There is another table for $\rightarrow_e$, first established by Smiley. The only difference is that in his table "2" and "3" have been replaced by "4". But it does not make any difference with respect to the designated value "1".

> meant as an explication of the classical metaphor of the containment of the predicate in the subject of an analytic truth (cf. *Kant's Critique of Pure Reason*, B10).[21]

Let us first look at the logical connectives:

Conjunction: A conjunction gives us definite information about something iff either conjunct gives us definite information about that thing.

Disjunction: A disjunction gives us definite information about something iff each disjunct gives us definite information about that thing.

Negation: Given what A gives definite information about, we cannot tell what $\sim_e A$ gives definite information about, and conversely.[22] Understanding the meaning of a sentence involves understanding the meaning of its negation as well.

> This informal discussion was intended to motivate the notion of an interpretation of a formula in terms of an "aboutness" valuation, which assigns in some inductive fashion to each formula a pair of classes, the first class to be thought of as the things that the formula gives definite information about, and the second class to be thought of as the things that its negation gives information about.[23]

Dunn calls each pair of subsets of X (X_i, X_j) a ***proposition surrogate***, "as seems appropriate since such a pair when assigned to a formula gives a partial representation of the meaning of the formula."[24]

Dunn defines the following operations upon the set of all propositional surrogates of a given universe of discouse X.[25]

$$\sim_e (X_1, X_2) = (X_2, X_1)$$
$$(X_1, X_2) \wedge_e (Y_1, Y_2) = (X_1 \cup Y_1, X_2 \cap Y_2)$$
$$(X_1, X_2) \vee_e (Y_1, Y_2) = (X_1 \cap Y_1, X_2 \cup Y_2)$$
$$(X_1, X_2) \rightarrow_e (Y_1, Y_2) = (Y_1 \subseteq X_1, X_2 \subseteq Y_2)$$

---

[21](Dunn 1966, 121).

[22]Cp. (Dunn 1966, 122–124). This formulation indicates that negation — in this reading — cannot be a ***product operation***, and that this interpretation is not homogeneous with regard to 1– and 2–place connectives. See my discussion below.

[23](Dunn 1966, 124 f.).

[24](Dunn 1966, 126).

[25]Cp. (Dunn 1966, 126). It is one of Dunn's motives for choosing definitions in such a way to get a ***de Morgan lattice***. I use already introduced symbols for these operations.

3.5 Corresponding reduction rules

It is very natural to take variable functors chracterized by the following reduction rules.[26]

$$\sim_E \begin{bmatrix} A \\ B \end{bmatrix} \Longrightarrow \begin{bmatrix} B \\ A \end{bmatrix} \qquad \begin{bmatrix} A \\ B \end{bmatrix} \vee_E \begin{bmatrix} C \\ D \end{bmatrix} \Longrightarrow \begin{bmatrix} A \wedge C \\ B \vee D \end{bmatrix}$$

$$\begin{bmatrix} A \\ B \end{bmatrix} \wedge_E \begin{bmatrix} C \\ D \end{bmatrix} \Longrightarrow \begin{bmatrix} A \vee C \\ B \wedge D \end{bmatrix} \qquad \begin{bmatrix} A \\ B \end{bmatrix} \rightarrow_E \begin{bmatrix} C \\ D \end{bmatrix} \Longrightarrow \begin{bmatrix} C \supset A \\ B \supset D \end{bmatrix}$$

The structural correspondences should be immediately obvious.

Here are some statements regarding these expressions:

- $\wedge_E$, $\vee_E$, and $\rightarrow_E$ are product–operations (cp. definition 4).

Definition 13 *A variable functor of the form V^n is an* **internally dual operation** *iff its reduction rule has the following form:*

$$V^n \begin{bmatrix} A_1 \\ B_1 \end{bmatrix} \cdots \begin{bmatrix} A_n \\ B_n \end{bmatrix} \Longrightarrow \begin{bmatrix} \Phi^{2n} A_1 \dots A_n, B_1 \dots B_n \\ \Psi^{2n} A_1 \dots A_n, B_1 \dots B_n \end{bmatrix}, \textit{ and}$$

$$\vdash \Phi^{2n} C_1 \dots C_{2n} \equiv \neg \Psi^{2n} \neg C_1 \dots \neg C_{2n}.$$

- $\wedge_E$ and $\vee_E$ are internally dual operations, but not $\rightarrow_E$ and of course not $\sim_E$.

Definition 14 *A variable functor of the form V^n is an* **internal switch operation** *iff its reduction rule has the following form:*

$$V^n \begin{bmatrix} A_1 \\ B_1 \end{bmatrix} \cdots \begin{bmatrix} A_n \\ B_n \end{bmatrix} \Longrightarrow \begin{bmatrix} \Phi^n B_1 \dots B_n \\ \Psi^n A_1 \dots A_n \end{bmatrix}.$$

- In our list in this section $\sim_E$ is the only internal switch operation.

This indicates that we have to consider three types of operations in our representation (and in Dunn's as well):
(1) both product– and internally dual operations: $\wedge_E$, $\vee_E$,
(2) product but not internally dual operation: $\rightarrow_E$, and
(3) internal switch operation: $\sim_E$

3.6 Formal vs. intuitive semantics

Regarding our new variable functors we are able to compute the corresponding pseudo–two–dimensional value–tables:

[26] See also Dunn's remark about the possibility of representing a proposition surrogate (X_1, X_2) of a topic set X by a pair $(\{a_{1_x}\}_{x \in X}, \{a_{2_x}\}_{x \in X})$ where each $a_{i_x} \in \mathbf{2}$. "Thus by coding of 'in' and 'out' with 1 and 0 we can represent the surrogate. We can let $a_{1_x} = 1$ mean that $x \in X_1$ and a_{2_x} that $x \in X_2$." (Dunn 1966, 131).

$\begin{bmatrix} A \\ B \end{bmatrix}$	$\neg_E \begin{bmatrix} A \\ B \end{bmatrix}$
1/1	1/1
1/0	0/1
0/1	1/0
0/0	0/0

$\begin{bmatrix} A \\ B \end{bmatrix} \wedge_E \begin{bmatrix} C \\ D \end{bmatrix}$	1/1	1/0	0/1	0/0
1/1	1/1	1/0	1/1	1/0
1/0	1/0	1/0	1/0	1/0
0/1	1/1	1/0	0/1	0/0
0/0	1/0	1/0	0/0	0/0

$\begin{bmatrix} A \\ B \end{bmatrix} \vee_E \begin{bmatrix} C \\ D \end{bmatrix}$	1/1	1/0	0/1	0/0
1/1	1/1	1/1	0/1	0/1
1/0	1/1	1/0	0/1	0/0
0/1	0/1	0/1	0/1	0/1
0/0	0/1	0/0	0/1	0/0

$\begin{bmatrix} A \\ B \end{bmatrix} \rightarrow_E \begin{bmatrix} C \\ D \end{bmatrix}$	1/1	1/0	0/1	0/0
1/1	1/1	1/0	1/1	1/0
1/0	1/1	1/1	1/1	1/1
0/1	0/1	0/0	1/1	1/0
0/0	0/1	0/1	1/1	1/1

Our observation is that the only candidate for an *intuitively designated value* is the combination 0/1. It occurs (1) exactly once both in the negation– and in the conjunction–table and (2) exactly 9 times both in the disjunction– and in the entailment–table. (3) it exchanges its position with the combination 1/0 with regard to the negation–table (cp. section 3.2). I.e. we can read the Smiley–tables in the following way two–dimensionally:

$$1 \Longleftrightarrow 01 \qquad 2 \Longleftrightarrow 00 \qquad 3 \Longleftrightarrow 11 \qquad 4 \Longleftrightarrow 10$$

But now we are faced with a problem concerning validity in E_{fde}. Dunn's definition of validity runs as follows:[27]

> A first degree entailment of the form $A \rightarrow B$ is *true* in its aboutness model iff when the aboutness valuation of A is (X_1, X_2) and of B is (Y_1, Y_2), then $Y_1 \subseteq X_1$ and $X_2 \subseteq Y_2$. It is *valid* iff it is true in every aboutness model.

If we cross the bridge over the 4–valued interpretation the corresponding notion of validity is our familiar $\models$–validity. But this type of validity corresponds to the combination 1/1, therefore to 11 and finally to 1.

Definition 15 *H^{fde} is* **fde–valid** [*symb.:* $\vdash_{fde} H^{fde}$] *iff the value of H^{fde} is 1 with regard to the Smiley–matrix for all values of the propositional variables*

Theorem 6 $\vdash_{fde} H^{fde}$ *iff* $\models X_{H^{fde}}$,
where the biunique mappings concerning the translation of formulae are:

$$p^i \Longleftrightarrow \begin{bmatrix} p_1^i \\ p_2^i \end{bmatrix} \quad \sim_e \Longleftrightarrow \sim_E \quad \wedge_e \Longleftrightarrow \wedge_E \quad \vee_e \Longleftrightarrow \vee_E \quad \rightarrow_e \Longleftrightarrow \rightarrow_E .$$

[27](Dunn 1966, 128 f.).

Taking this result for granted it is clear that, at least with respect to the system E_{fde}, we are relatively free in choosing a tautological entailment relation. One main point in this respect is the possible distinctness of the *designated value* (11 vs. $\models$) and the *intuitively designated value* (01).

If it is desirable to identify designated and intuitively designated values, it forces some unifying steps: (1) modifying the reduction properties of $\rightarrow_E$ and (2) defining another notion of validity.

$$\begin{bmatrix} A \\ B \end{bmatrix} \rightarrow_{E'} \begin{bmatrix} C \\ D \end{bmatrix} \Longrightarrow \begin{bmatrix} C \wedge \neg A \\ B \supset D \end{bmatrix}$$

It should be noticed that $\rightarrow_{E'}$ is not only a product operation but also an internally dual operation.

Definition 16 *The E-formula* $\begin{bmatrix} A \\ B \end{bmatrix}$ *is* E'_{fde}-valid [*symb.:* $\models_{01} \begin{bmatrix} A \\ B \end{bmatrix}$] *iff*

(*i*) $\dashv A$ *and* (*ii*) $\vdash B$

It is obvious that $\models_{01}$-validity corresponds to the designated value 01 because it refers to a first classically inconsistent column and a second classically valid column.

Theorem 7 (Classical style representation)

$$\models_{01} \begin{bmatrix} A \\ B \end{bmatrix} \textit{ iff} \vdash (\neg A \wedge B).$$

Definition 17 *Let* $X_{E'}$ *be any NC-formula containing only* $\begin{bmatrix} p_1^i \\ p_2^i \end{bmatrix}$, $\sim_E$, $\wedge_E$, $\vee_E$, *and* $\rightarrow_{E'}$ *as primitives, let* $\begin{bmatrix} A_{X_{E'}} \\ B_{X_{E'}} \end{bmatrix}$ *be that E-formula which is the result of the complete reduction of* $X_{E'}$, *i.e. that all occurrences of variable functors and all occurrences of classical functors outside the scope of brackets have been eliminated:*

$$\models_{01} X_{E'} \textit{ iff} \models_{01} \begin{bmatrix} A_{X_{E'}} \\ B_{X_{E'}} \end{bmatrix}.$$

Theorem 8 $\vdash_{fde} H^{fde}$ *iff* $\models_{01} X_{H^{fde}}$,
where the biunique mappings concerning the translation of formulae are:

$$p^i \Longleftrightarrow \begin{bmatrix} p_1^i \\ p_2^i \end{bmatrix} \qquad \sim_e \Longleftrightarrow \sim_E \qquad \wedge_e \Longleftrightarrow \wedge_E \qquad \vee_e \Longleftrightarrow \vee_E \qquad \rightarrow_e \Longleftrightarrow \rightarrow_{E'}.$$

3.7 Another intuitive interpretation

We have already seen that the basic interpretation of the 2 dimensions is meant to state that one dimension expresses what the expression itself gives information about while the second one what the negation of that expression gives information about.

Dunn pointed out that in the cases of conjunction and disjunction the first dimension is analogous in its behavior to the treatment of *information content* in Carnap/Bar-Hillel.[28] In these cases and with respect to $\rightarrow_E$, $\rightarrow_{E'}$ the second dimension is analogous to the classical case.

Dropping the assumption that both components tell us something about *information* or *content* we are free to give each component its own interpretation. It is therefore possible to combine the content–reading with the truth–reading:

(1) I postulate that the first dimension/component A of an E–expression $\begin{bmatrix} A \\ B \end{bmatrix}$ tells us something about the information content of the whole expression.

(2) I postulate that the second dimension/component B of this expression tells us something about the truth of the whole expression.

In this sense *information content* has a lot in common with *relevance*. In the case of conjunction and disjunction the content/relevance dimension is the *dual* of the truth dimension. In the case of entailment $\rightarrow_E$, the first line is a replication indicating that, with respect to inference, content is the *opposite* direction of classical inference. The negation $\sim_E$ is the only connective which brakes through this distinctness of both dimensions. It produces a change of relevance and truth dimension.

3.8 Internal vs. external negation

Going back to the original *expression/its negation expression* distiction it is clear that the classical–like truth–values are 01 and 10. Non–classical–like values are 11 and 00. Within our two–dimensional approach we can explicitly analyze this different behavior of negation $\sim_E$ with respect to both of these pairs:

(1) 10 and 01 can be represented by expressions of the form $\begin{bmatrix} p_1^i \\ \neg p_2^i \end{bmatrix}$.

(2) 11 and 00 can be represented by expressions of the form $\begin{bmatrix} p^i \\ p^i \end{bmatrix}$.

[28]Cp. (Carnap/Bar-Hillel 1952, 16).

$$\models \sim_E \begin{bmatrix} p \\ \neg p \end{bmatrix} \equiv \neg \begin{bmatrix} p \\ \neg p \end{bmatrix} \qquad \models \sim_E \begin{bmatrix} p \\ p \end{bmatrix} \equiv \begin{bmatrix} p \\ p \end{bmatrix} \qquad \models \begin{bmatrix} p \\ p \end{bmatrix} \equiv p.$$ [29]

If the argument of $\sim_E$ is a classical-like expression $\sim_E$ acts like the classical negation. But if the argument is a non-classical-like expression it does nothing.

This indicates that a reduction of our analogous expression of the form $\begin{bmatrix} p_1^i \\ p_2^i \end{bmatrix}$ to expressions of the form $\begin{bmatrix} p^i \\ \neg p^i \end{bmatrix}$ would give us the opportunity to eliminate negation $\sim_E$ and to use our familiar classical negation.

This seems to be a surprising observation since $\sim_E$ as negation of an unrestricted argument is obviously an *internal negation*:

$$\not\models \sim_E \begin{bmatrix} A \\ B \end{bmatrix} \supset \neg \begin{bmatrix} A \\ B \end{bmatrix} \qquad \not\models \neg \begin{bmatrix} A \\ B \end{bmatrix} \supset \sim_E \begin{bmatrix} A \\ B \end{bmatrix}.$$

What happens if we cancel one of the non-classical values, say 11? In our representation we are forced to adopt at least one *internal* functional dependency within expressions simulating propositional variables of any 3-valued logic.[30] We can choose, e.g.:

$\begin{bmatrix} p_1^i \\ \neg p_1^i \wedge p_2^i \end{bmatrix}$	$\sim_E \begin{bmatrix} p_1^i \\ \neg p_1^i \wedge p_2^i \end{bmatrix}$	$\neg \begin{bmatrix} p_1^i \\ \neg p_1^i \wedge p_2^i \end{bmatrix}$	$\Diamond_{S5'} \begin{bmatrix} p_1^i \\ \neg p_1^i \wedge p_2^i \end{bmatrix}$
1/0	0/1	0/1	1/1
1/0	0/1	0/1	1/1
0/1	1/0	1/0	1/1
0/0	0/0	1/1	0/0

With respect to such a pseudo-3-valued elementary expression we are able to offer a reading of $\sim_E$ as *restricted external negation*:

$$\models \sim_E \begin{bmatrix} p_1^i \\ \neg p_1^i \wedge p_2^i \end{bmatrix} \equiv \left(\neg \begin{bmatrix} p_1^i \\ \neg p_1^i \wedge p_2^i \end{bmatrix} \wedge \Diamond_{S5'} \begin{bmatrix} p_1^i \\ \neg p_1^i \wedge p_2^i \end{bmatrix} \right)$$

$$\models \begin{bmatrix} p_1^i \\ \neg p_1^i \wedge p_2^i \end{bmatrix} \supset \Diamond_{S5'} \begin{bmatrix} p_1^i \\ \neg p_1^i \wedge p_2^i \end{bmatrix}$$

[29] It is worth notifying that the arguments in each expression are pseudo-2-valued expressions, and that there is no analogous expression in a 4-valued system. Further, it is not possible to define an operator within the functionally incomplete 4-valued matrix of E_{fde} which allows the logically equivalent representation of p.

[30] Within our two-dimensional representation it forces us to double one pseudo-value combination k/l.

$$\not\models \Diamond_{S5'} \begin{bmatrix} p_1^i \\ \neg p_1^i \wedge p_2^i \end{bmatrix} \supset \begin{bmatrix} p_1^i \\ \neg p_1^i \wedge p_2^i \end{bmatrix}$$

$$\not\models \begin{bmatrix} p_1^i \\ \neg p_1^i \wedge p_2^i \end{bmatrix} \qquad \models \sim_E \begin{bmatrix} p_1^i \\ \neg p_1^i \wedge p_2^i \end{bmatrix} \supset \neg \begin{bmatrix} p_1^i \\ \neg p_1^i \wedge p_2^i \end{bmatrix}$$

There are some interesting connections between negation $\sim_E$, S5–like–modality, and classical functors:

$$\models \left(\begin{bmatrix} p_1 \\ p_2 \end{bmatrix} \vee \sim_E \begin{bmatrix} p_1 \\ p_2 \end{bmatrix} \right) \equiv \Diamond_{S5'} \begin{bmatrix} p_1 \\ p_2 \end{bmatrix}$$

i.e. with respect to a formulation of the law of exluded middle using classical disjunction we obtain that the law is logically equivalent with an empirical situation, namely the S5–like possibility of the expression under consideration. We get the dual case regarding the law of contradiction:

$$\models \left(\begin{bmatrix} p_1 \\ p_2 \end{bmatrix} \wedge \sim_E \begin{bmatrix} p_1 \\ p_2 \end{bmatrix} \right) \equiv \Box_{S5'} \begin{bmatrix} p_1 \\ p_2 \end{bmatrix}$$

$$\models \Diamond_{S5'} \sim_E \begin{bmatrix} p_1 \\ p_2 \end{bmatrix} \equiv \sim_E \Diamond_{S5'} \begin{bmatrix} p_1 \\ p_2 \end{bmatrix} \qquad \models \Diamond_{S5'} \sim_E \begin{bmatrix} p_1 \\ p_2 \end{bmatrix} \equiv \Diamond_{S5'} \begin{bmatrix} p_1 \\ p_2 \end{bmatrix}.$$

$$\models \Box_{S5'} \sim_E \begin{bmatrix} p_1 \\ p_2 \end{bmatrix} \equiv \sim_E \Box_{S5'} \begin{bmatrix} p_1 \\ p_2 \end{bmatrix} \qquad \models \Box_{S5'} \sim_E \begin{bmatrix} p_1 \\ p_2 \end{bmatrix} \equiv \Box_{S5'} \begin{bmatrix} p_1 \\ p_2 \end{bmatrix}.$$

3.9 Alternative two–dimensional interpretations

It has already been indicated by Dunn that "we could have just as well assigned to each formula A a pair $(\overline{X}_1, X_2)$, where X_1 is the set of topics that A is about and X_2 is the set of topics that $\overline{A}$ is about," to get an "algebraically neater" representation. Dunn argues that "this mixing of <u>un</u>aboutness with aboutness gets conceptually confusing".[31] But in my syntactic and two–dimensional account such a confusion does not arise. Let us first mention the new two–dimensional reading of the 4 values which looks much more convenient:

$$1 \Longleftrightarrow 11 \qquad 2 \Longleftrightarrow 10 \qquad 3 \Longleftrightarrow 01 \qquad 4 \Longleftrightarrow 00$$

Theorem 9 $\vdash_{fde} H^{fde}$ *iff* $\models X_{H^{fde}}$,
where the biunique mappings concerning the translation of formulae are:

$$p^i \Longleftrightarrow \begin{bmatrix} p_1^i \\ p_2^i \end{bmatrix} \qquad \sim_e \Longleftrightarrow \sim_{E*}, \textit{ where } \sim_{E*} \begin{bmatrix} A \\ B \end{bmatrix} \Longrightarrow \begin{bmatrix} \neg B \\ \neg A \end{bmatrix}.$$

$$\wedge_e \Longleftrightarrow \wedge \quad \vee_e \Longleftrightarrow \vee \quad \rightarrow_e \Longleftrightarrow \supset.$$

[31](Dunn 1966, 130 f.). Cp. also (Stelzner, this volume).

Now we can use classical conjunction, disjunction, even material implication. There is only one difference between classical logic and the system E_{fde} — our new variable functor $\sim_{E*}$.[32] This type of negation is a combination of classical negation and our internal switch operation $\sim_E$:

$$\models \sim_{E*} X \equiv \neg \sim_E X \qquad \models \sim_{E*} X \equiv \sim_E \neg X.$$

The third interesting reading of the system E_{fde} is the somewhat unusual attempt to read this system as a representation of natural language presuppositions. Let us try the following reading:[33]

- The first dimension is to represent the *truth* of an expression.
- The second dimension is to represent the *correctness* of the same expression. An expression is *correct* if all of its presuppositions are true.

Let X be any expression of the form $\begin{bmatrix} A \\ B \end{bmatrix}$.

$A = 1$ means X is true. $B = 1$ means X is correct.
$A = 0$ means X is false. $B = 0$ means X is incorrect.

In classical logic it is generally presupposed that each expression is a correct one. Therefore, the classical value "true" corresponds to "11", and "false" to "01". This leads us to the following biunique mappings:

$$1 \Longleftrightarrow 11 \qquad 2 \Longleftrightarrow 10 \qquad 3 \Longleftrightarrow 00 \qquad 4 \Longleftrightarrow 01$$

To get the negation working in such a case (exchanging 11 and 01) we need another type of operation. It is necessary to use a *2-place* classical functor within the formulation of the reduction rule needed here. To be more specific, the first dimension has to postulate an *internal distinctness*:

$$\sim_{Ep} \begin{bmatrix} A \\ B \end{bmatrix} \Longrightarrow \begin{bmatrix} A \not\equiv B \\ B \end{bmatrix}$$

The other 2-place functors are classical in the first dimension but very complex in the second one:

$$\begin{bmatrix} A \\ B \end{bmatrix} \wedge_{Ep} \begin{bmatrix} C \\ D \end{bmatrix} \Longrightarrow \begin{bmatrix} A \wedge C \\ ((A \equiv B) \wedge (C \equiv D)) \vee (\neg A \wedge \neg D) \vee (\neg B \wedge \neg C) \end{bmatrix}$$

$$\begin{bmatrix} A \\ B \end{bmatrix} \vee_{Ep} \begin{bmatrix} C \\ D \end{bmatrix} \Longrightarrow \begin{bmatrix} A \vee C \\ ((A \equiv B) \wedge (C \equiv D)) \vee (A \wedge D) \vee (B \wedge C) \end{bmatrix}$$

$$\begin{bmatrix} A \\ B \end{bmatrix} \rightarrow_{Ep} \begin{bmatrix} C \\ D \end{bmatrix} \Longrightarrow \begin{bmatrix} A \supset C \\ ((A \equiv B) \wedge (C \equiv D)) \vee (\neg A \wedge D) \vee (\neg B \wedge C) \end{bmatrix}$$

[32] If we consider this sublanguage with $\neg$ instead of $\sim_{E*}$ it can easily be shown that this system is equivalent to the classical logic, i.e. that the use of two-dimensional expressions instead of ordinary propositional variables does not add anything in this sublanguage.

[33] This reading of two-dimensional values has been introduced by Bergmann (1981).

Theorem 10 $\vdash_{fde} H^{fde}$ *iff* $\models X_{Hfde}$,
where the biunique mappings concerning the translation of formulae are:

$$p^i \Longleftrightarrow \begin{bmatrix} p_1^i \\ p_2^i \end{bmatrix} \quad \sim_e \Longleftrightarrow \sim_{Ep} \quad \wedge_e \Longleftrightarrow \wedge_{Ep} \quad \vee_e \Longleftrightarrow \vee_{Ep} \quad \rightarrow_e \Longleftrightarrow \rightarrow_{Ep}.$$

References

Anderson, A.R., & Belnap Jr., N.D.: *Entailment. The Logic of Relevance and Necessity*, Vol. 1. Princeton 1975.

Bergmann, M.: Presupposition and two-dimensional logic. *Journal of Philosophical Logic* 10 (1981), 27–53.

Carnap, R., & Bar-Hillel, Y.: An outline of a theory of semantic information. *M.I.T. Research Lab. of Electronics Technical Report*, No. 247, 1952.

Dugundji, J.: Note on a property of matrices for Lewis and Langford's calculi of propositions, *Journal of Symbolic Logic* 5 (1940), 150–151.

Dunn, J.M. 1966, *The algebra of intensional logics*. Doctoral dissertation, University of Pittsburgh. Ann Arbor (University Microfilms).

Kleene, S.C.: *Introduction to Metamathematics*. Amsterdam 1952.

Lewis, C.I., & Langford, C.H.: *Symbolic Logic*. New York 1932, 1959^2.

Łukasiewicz, J.: A system of modal logic, *Journal of Computing Systems* 1 (1953), 111–149.

Max, I.: Zur Explikation der Modaloperatoren von S5 mittels G–Funktorenvariablen. In: *Logik in der Semantik — Semantik in der Logik*, ed. by E. Dölling, Berlin, 1987, 123–133.

Max, I.: Vorschläge zur logischen Explikation von Negationen mittels Funktorenvariablen, *Linguistische Studien*, Reihe A, 182 (1988), 105–161.

Max, I.: The equivalence between n–dimensional classical logics and 2^n–valued logics, *Journal of Symbolic Logic* 59 (1994), 709.

Steinacker, P.: Die epistemische Komponente einer nichtklassischen Negation. In: *Philosophie und Logik. Frege-Kolloquien 1989/91*, ed. by W. Stelzner, Berlin 1993, 329–337.

Stelzner, W.: *Epistemische Logik. Zur logischen Analyse von Akzeptationsformen*. Berlin 1984.

Stelzner, W.: Negation and relevance. This volume.

Negation and Relevance

Werner Stelzner

One problem with negations is that we have so many of them. A task connected with this problem is to look for adequate classificational and unifying principles in this universe of negations. As the title of my paper indicates, I will confine my considerations to negations in some prominent logical systems intended to capture different aspects of relevance by explicating different relevant entailment relations: Tautological Entailments[1], logical entailment relations explicated by Sinowjew (among them first–degree analytic implication)[2], some systems based on fulfillment condition semantics[3], and mixed systems of this kind.

1 Value semantics for first–degree entailment systems

For all these systems we have an adequate value matrix semantics: a 4–valued semantics for Tautological Entailments and the conceptualist systems of Sinowjew, an 8–valued classical fulfillment condition semantics, an 8–valued semantics for tautological variants of the Sinowjew systems, and a 16–valued semantics for the tautological fulfillment condition system.

1.1 Tautological Entailments

The main effect of Tautological Entailment is that it excludes all such classically sound entailments as tautologically unsound, which take their classical soundness not only from the definitions of the truth–functions negation, conjunction and disjunction, but from the classical impossibility to have both a sentence and its negation in the same valuation as true or neither of them as true. Negation and the other truth–functions are therefore seen as entirely classical, valuations however can turn out to be non–classical.

Negation, Conjunction, Disjunction, and Tautological Entailments:[4]

$\sim$	$\wedge$	1	2	3	4
4	1	1	2	3	4
2	2	2	2	4	4
3	3	3	4	3	4
1	4	4	4	4	4

$\vee$	1	2	3	4
1	1	1	1	1
2	1	2	1	2
3	1	1	3	3
4	1	2	3	4

$\rightarrow$	1	2	3	4
1	1	2	3	4
2	1	1	3	3
3	1	2	1	2
4	1	1	1	1

[1] Cf. (Anderson/Belnap 1975).
[2] Cf. (Zinov'ev 1982)
[3] Cf. (Stelzner 1992a)
[4] The designated value here and in the following semantics is the value 1.

1.2 Conceptualist systems

Conceptualist systems like those of Sinowjew[5] and Parry's analytic implication[6] presuppose classical valuations and combine classical entailment relations with relevance filters that are defined by special containment conditions. These state:
a) the conceptual material of the conclusion of a relevant entailment should not exceed the conceptual material of the premises (conclusion relevant conceptualist entailment)
b) the conceptual material of the premises of a relevant entailment should not exceed the conceptual material of the conclusion (premise relevant conceptualist entailment)
c) the conceptual material of the conclusion of a relevant entailment and the conceptual material of the premises should be the same (maximal (conclusion and premise) relevant conceptualist entailment).

In the conceptualist systems "conceptual material" ist codified as "set of variables". While Parry's analytic implication is confined to conceptualist conclusion relevance, Sinowjew's systems of strong logical entailment ($\vdash_s$), converse logical entailment ($\vdash_k$), and maximal logical entailment ($\vdash_m$) capture conclusion, premise, and maximal conceptualist relevance.

Negation, conjunction, disjunction, and conceptualist conclusion relevant, premise relevant and maximal relevant entailment[7] relations:

$\sim$	$\wedge$	1	2	3	4
3	1	1	2	3	4
4	2	2	2	4	4
1	3	3	4	3	4
2	4	4	4	4	4

$\vee$	1	2	3	4
1	1	2	1	2
2	2	2	2	2
3	1	2	3	4
4	2	2	4	4

$\vdash_s$	1	2	3	4
1	1	2	3	4
2	1	1	3	3
3	1	2	1	2
4	1	1	1	1

$\vdash_k$	1	2	3	4
1	1	1	3	3
2	2	1	4	3
3	1	1	1	1
4	2	1	2	1

$\vdash_m$	1	2	3	4
1	1	2	3	4
2	2	1	4	3
3	1	2	1	2
4	2	1	2	1

The differences between these conceptualist types of entailment don't arise from different negations or other connectives in these systems but from the

[5]Cf. (Sinowjew 1962, 1967,1970a, 1970b), (Sinowjew/Wessel 1975), (Stelzner 1992b).
[6]Cf. (Parry 1933, 1989), (Dunn 1972).
[7]The semantic presented here is adequate for the axiomatizations that were first given by Sinowjew (cf. (Sinowjew 1962, 1967,1970a, 1970b), (Sinowjew/Wessel 1975), (Stelzner 1992b)).

differences in conceptualist entailment conditions:
So we have as sound

$$\sim p \vdash_k \sim (p \wedge q)\ , \qquad \sim (p \vee q) \vdash_s \sim p\ ,$$

but as unsound

$$\sim p \vdash_s \sim (p \wedge q)\ , \qquad \sim (p \vee q) \vdash_k \sim p\ .$$

This is not due to different negations in these formulas, but rather to different containment conditions to be fulfilled by conclusion and by premise relevant conceptualist entailments. If the set of variables occurring in the premises equals the set of variables occurring in the conclusion, then we have no difference between premise relevance and conclusion relevance. Both conditions are therefore fulfilled, and by this condition the set of sound maximal logical entailments is the intersection of the sets of sound premise relevant entailments and the set of sound conclusion relevant entailments. Maximal logical entailment therefore has as sound formulas not only (and of course not all) classically sound equivalents but all those classically sound entailments whose premises and conclusion contain identical sets of variables (e.g., we have $p \wedge q \vdash_m p \vee q$).

1.3 Negation, conjunction, disjunction, and classical fulfillment conditions entailment

Turning away from the purely syntactic relevance criteria of conceptualism, in analytic containment[8] fulfillment conditions semantics[9] we have a merely semantic relevance filter. This filter is given by comparing the containments between the sets of variables occurring positively and negatively in the premises with those of the conclusion or by comparing the partial fulfillments of the premises with those of the conclusion. For premise relevance ($_R\!\rightarrow$), every partial fulfillment of the premises is a partial fulfillment of the conclusion. For conclusion relevance, any partial fulfillment of the conclusion is a partial fulfillment of the premises.

~	∧	1	2	3	4	5	6	7	8
5	1	1	2	3	4	5	6	7	8
7	2	2	2	4	4	6	6	8	8
6	3	3	4	3	4	7	8	7	8
8	4	4	4	4	4	8	8	8	8
1	5	5	6	7	8	5	6	7	8
3	6	6	6	8	8	6	6	8	8
2	7	7	8	7	8	7	8	7	8
4	8	8	8	8	8	8	8	8	8

∨	1	2	3	4	5	6	7	8
1	1	2	3	4	1	2	3	4
2	2	2	4	4	2	2	4	4
3	3	4	3	4	3	4	3	4
4	4	4	4	4	4	4	4	4
5	1	2	3	4	5	6	7	8
6	2	2	4	4	6	6	8	8
7	3	4	3	4	7	8	7	8
8	4	4	4	4	8	8	8	8

[8] Cf. (Angell 1977, 1989).
[9] Cf. (Stelzner 1992a, 1994).

$R\rightarrow$	1	2	3	4	5	6	7	8
1	1	2	3	4	5	6	7	8
2	1	1	3	3	5	5	7	7
3	1	2	1	2	5	6	5	6
4	1	1	1	1	5	5	5	5
5	1	2	3	4	1	2	3	4
6	1	1	3	3	1	1	3	3
7	1	2	1	2	1	2	1	2
8	1	1	1	1	1	1	1	1

$P\rightarrow$	1	2	3	4	5	6	7	8
1	1	1	1	1	5	5	5	5
2	2	1	2	1	6	5	6	5
3	3	3	1	1	6	7	5	5
4	4	3	2	1	8	7	6	5
5	1	1	1	1	1	1	1	1
6	2	1	2	1	2	1	2	1
7	3	3	1	1	3	3	1	1
8	4	3	2	1	4	3	2	1

Because we are dealing with first–degree entailment systems, we could replace all values for entailments that differ from 1 by one of the undesignated values or by a negative designated value in the tables for the different entailments. This stresses the relational character of these entailments. Another way to stress the relational character of the entailment relations consists in defining them by their characteristic diagramms:

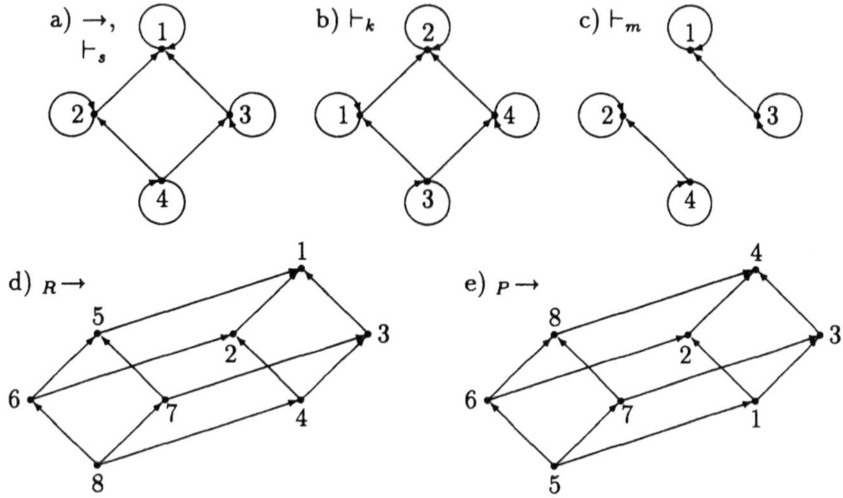

To get a readable picture in Diagramms d) and e) we didn't indicate the reflexivity of the arrow-relation (which of course is presumed). For all diagramms the transitivity of the arrow-relation is given.

2 Value composition by components

For a more detailed exposition of first–degree entailment relations it is useful to connect the sentences in question in a k–valued semantics with values represented by ordered n–tupels $< v_1, v_2, ..., v_n >$ of basic values

$(v_1, v_2, ..., v_n \in \{t,f\})$, where $n = \log_2 k$. In the Anderson/Belnap system of Tautological Entailments[10] and in the different conceptualist entailment systems of Sinowjew (among them the first–degree analytic implication) we have as assigned n–tupels of values pairs of values $< v_1, v_2 >$, with $v_1, v_2 \in \{t,f\}$. The above adequate four–valued semantics for these systems are based on these pairs of values in specific ways.

The four values can then be composed as:
1 =<t,t>, 2 =<t,f>, 3 =<f,t>, 4 =<f,f>

Another way is to follow Belnap's[11] suggestion to treat the four values as:
1 =<t,f> (True) 2 =<t,t> (Both) 3 =<f,f> (None) 4 =<f,t> (False).

For reasons of alphabetic ordering and the more homogeneous composition of values of compound sentences from the value components of the constituent sentences we won't follow Belnap's suggestion in this context. For negation this gives:

Tautological Entailment	
H	$\sim H$
1 t t	4 f f
2 t f	2 t f
3 f t	3 f t
4 f f	1 t t

Conceptualist Entailment	
H	$\sim H$
1 t t	3 f t
2 t f	4 f f
3 f t	1 t t
4 f f	2 t f

Fulfillment Entailment	
H	$\sim H$
1 t t t	5 f t t
2 t t f	7 f f t
3 t f t	6 f t f
4 t f f	8 f f f
5 f t t	1 t t t
6 f t f	3 t f t
7 f f t	2 t t f
8 f f f	4 t f f

When looking at the treatment of negation in this framework, it is not surprising that we get different definitions for the four–valued negation in Tautological Entailments, for the negation in the four–valued conceptualist systems and for negation in the 8–valued fulfillment condition semantics. It is easy to see that the component values of these different negations are generated in different ways from the components of the unnegated sentences. For the conceptualist negation this works entirely componentwise (first component with negation, second component with assertion: $\sim< v_1, v_2 >=<\sim v_1, v_2 >$). In the fulfillment semantics it works for the first component componentwise (negation); the second and third component change their position: $\sim< v_1, v_2, v_3 >=<\sim v_1, v_3, v_2 >$. For Tautological

[10]Cf. (Dunn 1966), (Anderson/Belnap 1975), (Anderson/Belnap/Dunn 1992).
[11](Belnap 1977).

Entailment to receive the value of the negated sentence from the components of the unnegated sentence, in the component notation used here, these components have to change their position in the table of their alphabetic listing diagonally. So, unlike negation in the conceptualist systems, the negation of four–valued Tautological Entailments is not composed componentwise, i.e., the single component in the negated sentence is not a function of the single component in the unnagated sentence.

The values for conjunction, alternative, and entailment are generated componentwise in all the different entailment systems. For conjunction and conlusion relevant entailment this applies in the same manner for all the systems given above:

$$< v_1, ..., v_n > \wedge < v_{n+1}, ..., v_{2n} > = < v_1 \wedge v_{n+1}, ..., v_n \wedge v_{2n} >$$
$$< v_1, ..., v_n > \vdash < v_{n+1}, ..., v_{2n} > = < v_1 \supset v_{n+1}, ..., v_n \supset v_{2n} >$$

Differences occur in connection with disjunction. Here we have for:

tautological entailment: $< v_1, v_2 > \vee < v_3, v_4 > = < v_1 \vee v_3, v_2 \vee v_4 >$
conceptualist systems: $< v_1, v_2 > \vee < v_3, v_4 > = < v_1 \vee v_3, v_2 \wedge v_4 >$
fulfillment semantics: $< v_1, v_2, v_3 > \vee < v_4, v_5, v_6 > =$
$= < v_1 \vee v_4, v_2 \wedge v_5, v_3 \wedge v_6 >$

3 Tautological Conceptualism and Tautological Fulfillment Semantics

3.1 Conceptualist systems

There are well–known problems of handling contradictions and tautologies in conceptualist systems.
E.g., we have:

(1) $p \wedge \sim p \wedge q \vdash_s \sim q$
(2) $p \vee (\sim q \vee q) \vdash_s \sim p \vee (\sim q \vee q)$
(3) $p \vee (\sim q \vee q) \vdash_s q \vee (\sim p \vee p)$
(4) $\sim p \wedge p \vee (\sim q \vee q) \vdash_s \sim q \wedge q$

Though these entailments are conceptualistically perfectly acceptable, even hard–line defenders of the conceptualist attitude feel in some way unhappy about their soundness. To block the soundness of these entailments in the case of conceptualist thinking, Wessel[12] confines the system of strong logical entailment in such a way that inferring anything from contradictions is excluded and that there are no sound entailments having a tautology as their consequent. This solution is reminiscent of Wittgenstein's suggestion about

[12](Wessel 1979).

how to avoid the explosion of classical logic in the case of contradictory premises: Don't draw any conclusion from such premises.[13]

Wessel presents his solution with the desired side–effect that the disjunctive syllogism is sound in the revised system, while blaming Tautological Entailment for not having disjunctive syllogism as a sound principle of logic. In my opinion, to be bothered by "bad" entailments with contradictory premises and nevertheless to have disjunctive syllogism is inconsistent from an intuitive point of view. To prevent bad inferences from inconsistent premises is pragmatically useful only in case that we take contradictions to be possibly true, because only in this case do we come from the contradictory premises to the explosive conclusion. Classical logic should therefore not bother itself with its explosiveness in the case of contradictory premises, as such premises never come true in classical valuations. If we bother, we consider true contradictions as a possibility. But then we cannot have disjunctive syllogism either, because there we obtain a local classical contradiction leading to the explosion of the system in the sense that under this condition everything in the system is sound. Forbidding contradictory premises is therefore not a coherent solution to the problem. Moreover, we have to face formal problems with such a solution (e.g., substitution doesn't hold any longer). There are more substantial arguments not to go this way.

On the other hand we can put things together and connect the conceptualist attitude with the reasoning in the non–limited logical space of epistemic situations, a characteristic intuitive background for Tautological Entailments. We then obtain a new (Tautological conceptualist) entailment with appropriate negation for the synthesized system, for which we have an adequate 8–valued semantics with values represented by triples of values consisting of the two elements of Tautological Entailment valuations and the second component of the conceptualist valuation.

The valuation of compound sentences now follows in the first two elements regulations known from 4–valued Tautological Entailment semantics. In the third component the value triple is treated in the same way as the second element of value pairs for conceptualist systems.

[13]Wittgenstein defended this view in the early 30s in his conversations with Waismann and didn't give it up even in the face of very substantial attacks by Turing. So in his 1939 *Lectures on the Foundation of Mathematics*, Wittgenstein pointed out: "You might get $p \cdot \overline{p}$ by means of Frege's System. If you can draw any conclusion you like from it, then that, as far as I can see, is all the trouble you can get into. And I would say, 'Well then, just don't draw any conclusions from a contradiction'" (Wittgenstein 1976, 220).

Negation, conjunction, disjunction and Tautological analytic implication:

$\sim$	$\wedge$	1	2	3	4	5	6	7	8
7 f f t	1 t t t	1	2	3	4	5	6	7	8
8 f f f	2 t t f	2	2	4	4	6	6	8	8
3 t f t	3 t f t	3	4	3	4	7	8	7	8
4 t f f	4 t f f	4	4	4	4	8	8	8	8
5 f t t	5 f t t	5	6	7	8	5	6	7	8
6 f t f	6 f t f	6	6	8	8	6	6	8	8
1 t t t	7 f f t	7	8	7	8	7	8	7	8
2 t t f	8 f f f	8	8	8	8	8	8	8	8

$\vee$	1	2	3	4	5	6	7	8
1	1	2	1	2	1	2	1	2
2	2	2	2	2	2	2	2	2
3	1	2	3	4	1	2	3	4
4	2	2	4	4	2	2	4	4
5	1	2	1	2	5	6	5	6
6	2	2	2	2	6	6	6	6
7	1	2	3	4	5	6	7	8
8	2	2	4	4	6	6	8	8

$\vdash_{ts}$	1	2	3	4	5	6	7	8
1	1	2	3	4	5	6	7	8
2	1	1	3	3	5	5	7	7
3	1	2	1	2	5	6	5	6
4	1	1	1	1	5	5	5	5
5	1	2	3	4	1	2	3	4
6	1	1	3	3	1	1	3	3
7	1	2	1	2	1	2	1	2
8	1	1	1	1	1	1	1	1

$\vdash_{tk}$	1	2	3	4	5	6	7	8
1	1	1	3	3	5	5	7	7
2	2	1	4	3	6	5	8	7
3	1	1	1	1	5	5	5	5
4	2	1	2	1	6	5	6	5
5	1	1	3	3	1	1	3	3
6	2	1	4	3	2	1	4	3
7	1	1	1	1	1	1	1	1
8	2	1	2	1	2	1	2	1

From the standpoint of this 8–valued evaluation system, both the system of Tautological Entailments and the conceptualist systems are specifications with different designated values. Sound tautological entailments take 1 and 2 as designated values and sound strong logical entailments (first–degree analytic implication) take 1 and 3 as designated values. The designated values for classical entailment are 1 to 6, and here of course we have formulas showing in their course of values only these values: classical tautologies. In this context, the different undesignated values in the eight–valued truth table obtain a decisive sense.

Neither in the system of Tautological Entailments nor in the conceptualist systems do we have any tautologies, in Tautological Entailment we don't have them due to the possiblity of empty valuations and contradictory valuations. In conceptualist systems we don't have tautologies, reflecting the fact that the fulfillment of the conceptualist condition cannot be stated as a logical truth. And for the same reasons we don't have tautologies in the mixed systems.

The simplest way to axiomatize the synthesized system we get is to put an adequate axiomatization of Tautological Entailments together with an adequate axiomatization of a conceptualist system and then add the following rule:

$$\frac{\begin{array}{l} H_1 \rightarrow_t H_2 \\ H_1 \vdash_s H_2 \end{array}}{H_1 \vdash_{ts} H_2}$$

Confronted with the simple semantic treatment, building up this or another kind of axiomatization for the given entailment is rather pointless.

3.2 Fulfillment condition semantics based systems

Even though tautological conceptualism avoides irrelevancies connected with classically impossible valuations (true (classical) contradictions, false (classical) tautologies), truth irrelevancies of a stronger type remain sound formulas, because in conceptualism a formula and the negation of this formula are pragmatically (i.e., by their conceptual content) rated as identical. So we have as sound tautological conceptualist entailment

$p \wedge q \vdash_{ts,tk} p \vee \sim q.$

This gives rise to the construction of the Gettier–Paradox[14] in conceptualist systems, even though we don't have the disjunction introduction rule used by Gettier in his original paper. This type of irrelevancy doesn't occur in the fulfillment condition based systems. There we don't have that a formula and its negation have the same fulfillment constituents in every case. Because of this, in these systems formulas of type (1)–(3) are not sound. They are irrelevant in the fulfillment condition sense.

But nevertheless contradictory premises can have unliked consequences in fulfillment condition semantics as well. Formulas of type

$(4^*) \quad (\sim p \wedge p) \vee (\sim q \vee q) \;{}_R\!\!\rightarrow \sim q \wedge q$

are relevantly sound in the conclusion relevance sense. In the case of a local contradiction in the premises, we can therefore relevantly entail from a classical tautology a classical contradiction.

To get formally rid of this we could apply the Wittgenstein-strategy and refuse to use contradictions in premises. Or we could apply some update operation for fulfillment conditions, limiting in this way the explosive deductive power of contradictions in the premises. All these updates, on the other hand, have undesired formal consequences: substitution does not hold as a rule and non–monotonity arises.[15]

If we go this way, that is, if we go with the conceptualist systems and build up Tautological fulfillment semantics, these formal problems don't arise and pragmatic incoherencies as described above don't arise either. Going

[14] Cf. (Gettier 1963).
[15] Different such update operations are developed in (Stelzner 1992a, 1994).

this line in the construction of adequate truth tables we can stipulate for the first value component in the eight values of the fulfillment condition semantics the two components of tautological valuations, coming now to a 16–valued semantics, in which again at least one interesting new negation can be introduced: a negation for Tautological fulfillment relevant negation:

Negation and conjunction for Tautological fulfillment entailment:[16]

~	∧	1	2	3	4	5	6	7	8	9	10	11	12	13	14	15	16
13 f f t t	1 t t t t	1	2	3	4	5	6	7	8	9	10	11	12	13	14	15	16
15 f f f t	2 t t t f	2	2	4	4	6	6	8	8	10	10	12	12	14	14	16	16
14 f f t f	3 t t f t	3	4	3	4	7	8	7	8	11	12	11	12	15	16	15	16
16 f f f f	4 t t f f	4	4	4	4	8	8	8	8	12	12	12	12	16	16	16	16
5 t f t t	5 t f t t	5	6	7	8	5	6	7	8	13	14	15	16	13	14	15	16
7 t f f t	6 t f t f	6	6	8	8	6	6	8	8	14	14	16	16	14	14	16	16
6 t f t f	7 t f f t	7	8	7	8	7	8	7	8	15	16	15	16	15	16	15	16
8 t f f f	8 t f f f	8	8	8	8	8	8	8	8	16	16	16	16	16	16	16	16
9 f t t t	9 f t t t	9	10	11	12	13	14	15	16	9	10	11	12	13	14	15	16
11 f t f t	10 f t t f	10	10	12	12	14	14	16	16	10	10	12	12	14	14	16	16
10 f t t f	11 f t f t	11	12	11	12	15	16	15	16	11	12	11	12	15	16	15	16
12 f t f f	12 f t f f	12	12	12	12	16	16	16	16	12	12	12	12	16	16	16	16
1 t t t t	13 f f t t	13	14	15	16	13	14	15	16	13	14	15	16	13	14	15	16
3 t t f t	14 f f t f	14	14	16	16	14	14	16	16	14	14	16	16	14	14	16	16
2 t t t f	15 f f f t	15	16	15	16	15	16	15	16	15	16	15	16	15	16	15	16
4 t t f f	16 f f f f	16	16	16	16	16	16	16	16	16	16	16	16	16	16	16	16

Tautological fulfillment conditions entailment:

$T\rightarrow$	1	2	3	4	5	6	7	8	9	10	11	12	13	14	15	16
1	1	2	3	4	5	6	7	8	9	10	11	12	13	14	15	16
2	1	1	3	3	5	5	7	7	9	9	11	11	13	13	15	15
3	1	2	1	2	5	6	5	6	9	10	9	10	13	14	13	14
4	1	1	1	1	5	5	5	5	9	9	9	9	13	13	13	13
5	1	2	3	4	1	2	3	4	9	10	11	12	9	10	11	12
6	1	1	3	3	1	1	3	3	9	9	11	11	9	9	11	11
7	1	2	1	2	1	2	1	2	9	10	9	10	9	10	9	10
8	1	1	1	1	1	1	1	1	9	9	9	9	9	9	9	9
9	1	2	3	4	5	6	7	8	1	2	3	4	5	6	7	8
10	1	1	3	3	5	5	7	7	1	1	3	3	5	5	7	7
11	1	2	1	2	5	6	5	6	1	2	1	2	5	6	5	6
12	1	1	1	1	5	5	5	5	1	1	1	1	5	5	5	5
13	1	2	3	4	1	2	3	4	1	2	3	4	1	2	3	4
14	1	1	3	3	1	1	3	3	1	1	3	3	1	1	3	3
15	1	2	1	2	1	2	1	2	1	2	1	2	1	2	1	2
16	1	1	1	1	1	1	1	1	1	1	1	1	1	1	1	1

[16]We don't write out the table for disjunction generated according to De Morgan laws.

Like in the case of Tautological first–degree analytic implication, in this 16–valued semantics we come to adequate semantics for the system of Tautological Entailments and classical fulfillment condition entailment if we add new designated values: For Tautological Entailments the designated values are 1, 2, 3, 4, for fulfillment condition entailment the designated values are 1, 5, and for first–degree analytic implication the designated values are 1, 2, 5, 6.

4 Integrative intuitive semantics

Using the component semantics we got more structure to explain some technical connections between the different relevant entailment systems and the negations of these systems. With negation the result was that the basic systems were different and for the mixed systems that the negations are constructed using components from the negations of the basic systems. This is not much surprising and adds almost nothing to our understanding as to what negation means in different systems. What we gave seems to be not more than a kind of technical explanation. But, extending Nuel Belnap's epistemic states interpretation of the value components in the four values T, B, N, F for Tautological Entailments, the value component treated so far can be the starting point for a technically adequate and intuitive (may be a little better) explanation of what, e.g., negations in the mentioned systems mean, how they differ, and why they look like different negations in the different systems.

Intuitively, we treat the Quadruple $< v_1, v_2, v_3, v_4 >$ of values connected with a sentenced H in the mentioned relevant systems as:

Epistemic values:

v_1 =t : H is assigned the value true
v_1 =f : H is not assigned the value true
v_2 =t : $\sim H$ is not assigned the value true
v_2 =f : $\sim H$ is assigned the value true

Pragmatic values:

v_3 =t : the fulfillment of H is pragmatically free
v_3 =f : the fulfillment of H is not pragmatically free
v_4 =t : the fulfillment of $\sim H$ is pragmatically free
v_4 =f : the fulfillment of $\sim H$ is not pragmatically free

So, in fact, the value quadruples are reports about situations which can be ascribed by means of the predicates A ("is assigned the value t") and F ("is free for fulfillment"):

$w(H) = 1$: $A(H) \wedge \sim A(\sim H) \wedge F(H) \wedge F(\sim H)$ or shorter: t t t t

$w(H) = 2$:	$A(H) \wedge \sim A(\sim H) \wedge F(H) \wedge \sim F(\sim H)$	t t t f
$w(H) = 3$:	$A(H) \wedge \sim A(\sim H) \wedge \sim F(H) \wedge F(\sim H)$	t t f t
$w(H) = 4$:	$A(H) \wedge \sim A(\sim H) \wedge \sim F(H) \wedge \sim F(\sim H)$	t t f f
$w(H) = 5$:	$A(H) \wedge A(\sim H) \wedge F(H) \wedge F(\sim H)$	t f t t
⋮	⋮	⋮
$w(H) = 16$:	$\sim A(H) \wedge A(\sim H) \wedge \sim F(H) \wedge \sim F(\sim H)$	f f f f

Let us write out the table for negation:

	H	$\sim H$	
1	$A(H) \wedge \overline{A(\overline{H})} \wedge F(H) \wedge F(\overline{H})$	$\overline{A(\overline{H})} \wedge A(\overline{\overline{H}}) \wedge F(\overline{H}) \wedge F(\overline{\overline{H}})$	13
2	$A(H) \wedge \overline{A(\overline{H})} \wedge F(H) \wedge \overline{F(\overline{H})}$	$\overline{A(\overline{H})} \wedge A(\overline{\overline{H}}) \wedge \overline{F(\overline{H})} \wedge F(\overline{\overline{H}})$	15
3	$A(H) \wedge \overline{A(\overline{H})} \wedge \overline{F(H)} \wedge F(\overline{H})$	$\overline{A(\overline{H})} \wedge A(\overline{\overline{H}}) \wedge F(\overline{H}) \wedge \overline{F(\overline{\overline{H}})}$	14
4	$A(H) \wedge \overline{A(\overline{H})} \wedge \overline{F(H)} \wedge \overline{F(\overline{H})}$	$\overline{A(\overline{H})} \wedge A(\overline{\overline{H}}) \wedge \overline{F(\overline{H})} \wedge \overline{F(\overline{\overline{H}})}$	16
5	$A(H) \wedge A(\overline{H}) \wedge F(H) \wedge F(\overline{H})$	$A(\overline{H}) \wedge A(\overline{\overline{H}}) \wedge F(\overline{H}) \wedge F(\overline{\overline{H}})$	5
6	$A(H) \wedge A(\overline{H}) \wedge F(H) \wedge \overline{F(\overline{H})}$	$A(\overline{H}) \wedge A(\overline{\overline{H}}) \wedge \overline{F(\overline{H})} \wedge F(\overline{\overline{H}})$	7
7	$A(H) \wedge A(\overline{H}) \wedge \overline{F(H)} \wedge F(\overline{H})$	$A(\overline{H}) \wedge A(\overline{\overline{H}}) \wedge F(\overline{H}) \wedge \overline{F(\overline{\overline{H}})}$	6
8	$A(H) \wedge A(\overline{H}) \wedge \overline{F(H)} \wedge \overline{F(\overline{H})}$	$A(\overline{H}) \wedge A(\overline{\overline{H}}) \wedge \overline{F(\overline{H})} \wedge \overline{F(\overline{\overline{H}})}$	8
9	$\overline{A(H)} \wedge \overline{A(\overline{H})} \wedge F(H) \wedge F(\overline{H})$	$\overline{A(\overline{H})} \wedge \overline{A(\overline{\overline{H}})} \wedge F(\overline{H}) \wedge F(\overline{\overline{H}})$	9
10	$\overline{A(H)} \wedge \overline{A(\overline{H})} \wedge F(H) \wedge \overline{F(\overline{H})}$	$\overline{A(\overline{H})} \wedge \overline{A(\overline{\overline{H}})} \wedge \overline{F(\overline{H})} \wedge F(\overline{\overline{H}})$	11
11	$\overline{A(H)} \wedge \overline{A(\overline{H})} \wedge \overline{F(H)} \wedge F(\overline{H})$	$\overline{A(\overline{H})} \wedge \overline{A(\overline{\overline{H}})} \wedge F(\overline{H}) \wedge \overline{F(\overline{\overline{H}})}$	10
12	$\overline{A(H)} \wedge \overline{A(\overline{H})} \wedge \overline{F(H)} \wedge \overline{F(\overline{H})}$	$\overline{A(\overline{H})} \wedge \overline{A(\overline{\overline{H}})} \wedge \overline{F(\overline{H})} \wedge \overline{F(\overline{\overline{H}})}$	12
13	$\overline{A(H)} \wedge A(\overline{H}) \wedge F(H) \wedge F(\overline{H})$	$A(\overline{H}) \wedge \overline{A(\overline{\overline{H}})} \wedge F(\overline{H}) \wedge F(\overline{\overline{H}})$	1
14	$\overline{A(H)} \wedge A(\overline{H}) \wedge F(H) \wedge \overline{F(\overline{H})}$	$A(\overline{H}) \wedge \overline{A(\overline{\overline{H}})} \wedge \overline{F(\overline{H})} \wedge F(\overline{\overline{H}})$	3
15	$\overline{A(H)} \wedge A(\overline{H}) \wedge \overline{F(H)} \wedge F(\overline{H})$	$A(\overline{H}) \wedge \overline{A(\overline{\overline{H}})} \wedge F(\overline{H}) \wedge \overline{F(\overline{\overline{H}})}$	2
16	$\overline{A(H)} \wedge A(\overline{H}) \wedge \overline{F(H)} \wedge \overline{F(\overline{H})}$	$A(\overline{H}) \wedge \overline{A(\overline{\overline{H}})} \wedge \overline{F(\overline{H})} \wedge \overline{F(\overline{\overline{H}})}$	4

The following principles are used for generating the truth tables (where R stands for A or F respectively):

(1) $R(p \wedge q) \Longleftrightarrow R(p) \wedge R(q)$

(2^A) $A(p \vee q) \Longleftrightarrow A(p) \vee A(q)$

(2^F) $F(p \vee q) \Longleftrightarrow F(p) \wedge F(q)$

(3) $R(\sim (p \wedge q)) \Longleftrightarrow R(\sim p \vee \sim q)$

(4) $R(\sim (p \vee q)) \Longleftrightarrow R(\sim p \wedge \sim q)$

(5) $R(\sim\sim p) \Longleftrightarrow R(p)$

(6^C) Conclusion relevance: $< v_1, v_2, v_3, v_4 > \rightarrow < v_5, v_6, v_7, v_8 >=$

$$=< v_1 \supset v_5, v_2 \supset v_6, v_3 \supset v_7, v_4 \supset v_8 >$$

(6[P]) Premise relevance: $< v_1, v_2, v_3, v_4 > \rightarrow < v_5, v_6, v_7, v_8 >=$
$=< v_1 \supset v_5, v_2 \supset v_6, v_7 \supset v_3, v_8 \supset v_4 >$

Using these principles, we get by purely classical logic the translation of all the 16–valued truth tables given above.

The so translated negation truth table reports the values for the negation of $H(\sim H)$ purely on the basis of the values assigned to H. No such value assignment is changed by this negation. In this sense the negation given here behaves like a De Morgan Negation: It says which value the negated sentence has, if a value assignment for the unnegated sentence is given. It doesn't behave like Boolean Negation, which says that the reported values for the components of the value of H are misreported and have to change for every component of the value report. Technically, Boolean negation then turns out to be generated by negating the values of the unnegated sentence componentwise.

5 A unifying treatment of De Morgan Negation

Looking back to the negations mentioned so far, we see that they all are covered by the 16–valued negation if we confine the space of relevant situations which are taken into account in different entailment systems by one or more of the following principles PC0, PC1, and PC2. And this, of course, holds for the other logical connectives and the entailment relations too:

Classical assignment condition:

PC0. $A(H) \Longleftrightarrow \sim A(\sim H)$

If this condition holds, situations 5 to 12 are excluded because there we have that $A(H) \Longleftrightarrow A(\sim H)$.

	H			$\sim H$	
1	$A(H) \wedge \overline{A(\overline{H})} \wedge F(H) \wedge F(\overline{H})$	1	5	$\overline{A(\overline{H})} \wedge A(\overline{\overline{H}}) \wedge F(\overline{H}) \wedge F(\overline{\overline{H}})$	13
2	$A(H) \wedge \overline{A(\overline{H})} \wedge F(H) \wedge \overline{F(\overline{H})}$	2	7	$\overline{A(\overline{H})} \wedge A(\overline{\overline{H}}) \wedge \overline{F(\overline{H})} \wedge F(\overline{\overline{H}})$	15
3	$A(H) \wedge \overline{A(\overline{H})} \wedge \overline{F(H)} \wedge F(\overline{H})$	3	6	$\overline{A(\overline{H})} \wedge A(\overline{\overline{H}}) \wedge F(\overline{H}) \wedge \overline{F(\overline{\overline{H}})}$	14
4	$A(H) \wedge \overline{A(\overline{H})} \wedge \overline{F(H)} \wedge \overline{F(\overline{H})}$	4	8	$\overline{A(\overline{H})} \wedge A(\overline{\overline{H}}) \wedge \overline{F(\overline{H})} \wedge \overline{F(\overline{\overline{H}})}$	16
13	$\overline{A(H)} \wedge A(\overline{H}) \wedge F(H) \wedge F(\overline{H})$	5	1	$A(\overline{H}) \wedge \overline{A(\overline{\overline{H}})} \wedge F(\overline{H}) \wedge F(\overline{\overline{H}})$	1
14	$\overline{A(H)} \wedge A(\overline{H}) \wedge F(H) \wedge \overline{F(\overline{H})}$	6	3	$A(\overline{H}) \wedge \overline{A(\overline{\overline{H}})} \wedge \overline{F(\overline{H})} \wedge F(\overline{\overline{H}})$	3
15	$\overline{A(H)} \wedge A(\overline{H}) \wedge \overline{F(H)} \wedge F(\overline{H})$	7	2	$A(\overline{H}) \wedge \overline{A(\overline{\overline{H}})} \wedge F(\overline{H}) \wedge \overline{F(\overline{\overline{H}})}$	2
16	$\overline{A(H)} \wedge A(\overline{H}) \wedge \overline{F(H)} \wedge \overline{F(\overline{H})}$	8	4	$A(\overline{H}) \wedge \overline{A(\overline{\overline{H}})} \wedge \overline{F(\overline{H})} \wedge \overline{F(\overline{\overline{H}})}$	4

The remaining situations build the space for classical fulfillment condition semantics. We can rename the remaining situations, now standing for possible values, leaving out 5 to 12. So we receive the truth value table for negation in classical fulfillment condition semantics as given above.

Conceptualist pragmatic condition:

PC1. $F(p) \Longleftrightarrow F(\sim p)$

p is conceptualistically free, iff $\sim p$ is conceptualistically free.

If this condition holds, situations 2, 3, 6, 7, 10, 11, 14, 15 are excluded as there we don't have $F(H) \Longleftrightarrow F(\sim H)$. The remaining situations build the space for tautological first–degree analytic implication and the systems of logical entailment in the sense of Sinowjew. We can rename these situations (now standing for possible values) leaving out 2, 3, 6, 7, 10, 11, 14, 15. In this way we receive the truth value table for negation in tautological conceptualist systems as given above:

	H			$\sim H$	
1	$A(H) \wedge \overline{A(\overline{\overline{H}})} \wedge F(H) \wedge F(\overline{\overline{H}})$	1	7	$\overline{A(\overline{\overline{H}})} \wedge A(\overline{\overline{\overline{H}}}) \wedge F(\overline{\overline{H}}) \wedge F(\overline{\overline{\overline{H}}})$	13
4	$A(H) \wedge \overline{A(\overline{\overline{H}})} \wedge \overline{F(H)} \wedge \overline{F(\overline{\overline{H}})}$	2	8	$\overline{A(\overline{\overline{H}})} \wedge A(\overline{\overline{\overline{H}}}) \wedge \overline{F(\overline{\overline{H}})} \wedge \overline{F(\overline{\overline{\overline{H}}})}$	16
5	$A(H) \wedge A(\overline{\overline{H}}) \wedge F(H) \wedge F(\overline{\overline{H}})$	3	3	$A(\overline{\overline{H}}) \wedge A(\overline{\overline{\overline{H}}}) \wedge F(\overline{\overline{H}}) \wedge F(\overline{\overline{\overline{H}}})$	5
8	$A(H) \wedge A(\overline{\overline{H}}) \wedge \overline{F(H)} \wedge \overline{F(\overline{\overline{H}})}$	4	4	$A(\overline{\overline{H}}) \wedge A(\overline{\overline{\overline{H}}}) \wedge \overline{F(\overline{\overline{H}})} \wedge \overline{F(\overline{\overline{\overline{H}}})}$	8
9	$\overline{A(H)} \wedge \overline{A(\overline{\overline{H}})} \wedge F(H) \wedge F(\overline{\overline{H}})$	5	5	$\overline{A(\overline{\overline{H}})} \wedge \overline{A(\overline{\overline{\overline{H}}})} \wedge F(\overline{\overline{H}}) \wedge F(\overline{\overline{\overline{H}}})$	9
12	$\overline{A(H)} \wedge \overline{A(\overline{\overline{H}})} \wedge \overline{F(H)} \wedge \overline{F(\overline{\overline{H}})}$	6	6	$\overline{A(\overline{\overline{H}})} \wedge \overline{A(\overline{\overline{\overline{H}}})} \wedge \overline{F(\overline{\overline{H}})} \wedge \overline{F(\overline{\overline{\overline{H}}})}$	12
13	$\overline{A(H)} \wedge A(\overline{\overline{H}}) \wedge F(H) \wedge F(\overline{\overline{H}})$	7	1	$A(\overline{\overline{H}}) \wedge \overline{A(\overline{\overline{\overline{H}}})} \wedge F(\overline{\overline{H}}) \wedge F(\overline{\overline{\overline{H}}})$	1
16	$\overline{A(H)} \wedge A(\overline{\overline{H}}) \wedge \overline{F(H)} \wedge \overline{F(\overline{\overline{H}})}$	8	2	$A(\overline{\overline{H}}) \wedge \overline{A(\overline{\overline{\overline{H}}})} \wedge \overline{F(\overline{\overline{H}})} \wedge \overline{F(\overline{\overline{\overline{H}}})}$	4

We obtain *semantics for the classical systems of Sinowjew* and first–degree analytic implication, if we impose both conditions PC0 and PC1. Then the remaining possible situations are those characterized by the values 1, 4, 13, 16 out of the given 16 values. We can now rename the remaining situations 1, 4, 13, 16, now standing for possible values, and we obtain the truth value table for negation in the *original (classical) conceptualist systems* as given above:

	H			$\sim H$	
1	$A(H)\wedge\overline{A(\overline{H})}\wedge F(H)\wedge F(\overline{H})$	1	3	$\overline{A(\overline{H})}\wedge A(\overline{\overline{H}})\wedge F(\overline{H})\wedge F(\overline{\overline{H}})$	13
4	$A(H)\wedge\overline{A(\overline{H})}\wedge\overline{F(H)}\wedge\overline{F(\overline{H})}$	2	4	$\overline{A(\overline{H})}\wedge A(\overline{\overline{H}})\wedge\overline{F(\overline{H})}\wedge\overline{F(\overline{\overline{H}})}$	16
13	$\overline{A(H)}\wedge A(\overline{H})\wedge F(H)\wedge F(\overline{H})$	3	1	$A(\overline{H})\wedge\overline{A(\overline{\overline{H}})}\wedge F(\overline{H})\wedge F(\overline{\overline{H}})$	1
16	$\overline{A(H)}\wedge A(\overline{H})\wedge\overline{F(H)}\wedge\overline{F(\overline{H})}$	4	2	$A(\overline{H})\wedge\overline{A(\overline{\overline{H}})}\wedge\overline{F(\overline{H})}\wedge\overline{F(\overline{\overline{H}})}$	4

Classical and Tautological pragmatic condition:

PC2. $\forall pF(p)$

One could take this as saying that classical and tautological thinking takes it for granted that in every situation everything is pragmatically free. Or, from the conceptualist point of viwe, it could be said that these logics presuppose that there are no taboos: all is free. And now you could start to blame these logics for their silly pragmatics. On the other hand, and more adequately, one can take this as saying that the classical thinking says nothing specific about pragmatic values.

If we impose the condition PC2 on the space of situations given in the 16–valued semantics, the only possible situations which remain are those, which are indicated by the values 1, 5, 9, 13, as is characteristic for tautological entailments:

	H			$\sim H$	
1	$A(H)\wedge\overline{A(\overline{H})}\wedge F(H)\wedge F(\overline{H})$	1	4	$\overline{A(\overline{H})}\wedge A(\overline{\overline{H}})\wedge F(\overline{H})\wedge F(\overline{\overline{H}})$	13
5	$A(H)\wedge A(\overline{H})\wedge F(H)\wedge F(\overline{H})$	2	2	$A(\overline{H})\wedge A(\overline{\overline{H}})\wedge F(\overline{H})\wedge F(\overline{\overline{H}})$	5
9	$\overline{A(H)}\wedge\overline{A(\overline{H})}\wedge F(H)\wedge F(\overline{H})$	3	3	$\overline{A(\overline{H})}\wedge\overline{A(\overline{\overline{H}})}\wedge F(\overline{H})\wedge F(\overline{\overline{H}})$	9
13	$\overline{A(H)}\wedge A(\overline{H})\wedge F(H)\wedge F(\overline{H})$	4	1	$A(\overline{H})\wedge\overline{A(\overline{\overline{H}})}\wedge F(\overline{H})\wedge F(\overline{\overline{H}})$	1

And if we impose both conditions PC0 (classical assignment condition) and PC2 (classical pragmatic condition), the remaining relevant situations are indicated by the values 1 and 13. In this way, we arrive at the classical propositional logic:

	H			$\sim H$	
1	$A(H)\wedge\overline{A(\overline{H})}\wedge F(H)\wedge F(\overline{H})$	1	2	$\overline{A(\overline{H})}\wedge A(\overline{\overline{H}})\wedge F(\overline{H})\wedge F(\overline{\overline{H}})$	13
13	$\overline{A(H)}\wedge A(\overline{H})\wedge F(H)\wedge F(\overline{H})$	2	1	$A(\overline{H})\wedge\overline{A(\overline{\overline{H}})}\wedge F(\overline{H})\wedge F(\overline{\overline{H}})$	1

In *Entailment*, Vol. II, Anderson/Belnap/Dunn describe the orthodox relevantist: "In logic we find the ordinary truth–functional connectives, in particular conjunction, disjunction, and negation, and what classical logic tells us about is partly true and partly false. The true part is what it tells us about which sentences involving only the truth functions (and quantifiers) are logically true. The false part is what it tells us about which such sentences entail one another ... The relevantist presumes to be talking about the same truth–functional connectives (and quantifiers) as the classical logician, but urges that a different, tighter relationship be taken as entailment."[17]

If we consider only premise relevant entailments, we could end this citation in a slightly extended way: ... The relevantist presumes to be talking about the same truth–functional connectives (and quantifiers) as the classical logician, but urges that different situations (not only classical ones) have to be taken into account if defining the entailment relationship, which is the same for all these situations. What divides classical logic and the different relevant logics treated here are neither different logical connectives for the particular logics nor different entailment relations for these logics but the kind of situations taken into account by these different logics as situations which are possible and worth remembering (and in this sense relevant).

On the one hand, we in fact have different entailments in premise relevant, conclusion relevant and maximal relevant entailments. There difference, on the other hand, collapses in Tautological Entailment to maximal relevance in the sense that we have

neither $p \wedge \sim p \rightarrow q$ (conclusion irrelevance in the tautological sense)
nor $q \rightarrow p \vee \sim p$ (premise irrelevance in the tautological sense).

References

Anderson, A.R., & Belnap Jr., N.D.: *Entailment*, Vol. 1. Princeton 1975.

Anderson, Alan R., Belnap Jr., N.D., & Dunn, J.M.: *Entailment*, Vol. 2. Princeton 1992.

Angell, R.B.: Three Systems of Analytic Containment. *Journal of Symbolic Logic* 42 (1977), 147.

Angell, R.B.: Deducibility, Entailment and Analytic Containment. In: *Directions in Relevant Logic*, ed. by J. Norman & R. Sylvan, Dordrecht 1989, 119–143.

Dunn, J.M.: A modification of Parry's Analytic Implication. *Notre Dame Journal of Formal Logic* 13 (1972), 195–205.

Dunn, J.M.: *The algebra of intensional logics.* PhD thesis, University of Pittsburgh, 1966. Ann Arbor (University Microfilms).

[17](Anderson/Belnap/Dunn 1992, 492f.).

Epstein, R.L.: *The Semantic Foundations of Logic*, Vol. 1: *Propositional Logic*. Dordrecht 1990.

Gettier, E.L.: Is Justified True Belief Knowledge?. *Analysis* 23 (1963), 121–123.

Parry, W.T.: Ein Axiomensystem für eine neue Art von Implikation (analytische Implikation). *Ergebnisse eines mathematischen Kolloquiums* 4 (1933), 5–6.

Parry, W.T.: Analytic Implication. Its History, Justification and Varieties. In: *Directions in Relevant Logic*, ed. by J. Norman & R. Sylvan, Dordrecht 1989, 101–118.

Routley, R., Plumwood V., Meyer R.K., & Brady R.T.: *Relevant Logics and Their Rivals. Part I. The Basic Philosophical and Semantic Theory*. Atascadero 1982.

Schurz, G.: Relevant Deduction. *Erkenntnis* 35 (1991a), 391–437.

Schurz, G.: Relevant Deductive Inference: Criteria and Logics. In: *Advances in Scientific Philosophy*, ed. by G. Schurz & G.J.W. Dorn, Amsterdam 1991b, 57–84.

Sinowjew, A.A.: *Logika wiskasiwanii i teorija wiwoda*. Moscow 1962.

Sinowjew, A.A.: *Osnowi logiceskoi teorii naycnich snanii*. Moscow 1967.

Sinowjew, A.A.: *Kompleksnaja logika*. Mosow 1970a.

Sinowjew, A.A.: *Komplexe Logik*. Berlin 1970b.

Sinowjew, A., & Wessel H.: *Logische Sprachregeln*. Berlin 1975.

Stelzner, W.: Relevant Deontic Logic. *Journal of Philosophical Logic* 21 (1992a), 193–216.

Stelzner, W.: Types of Relevance in A. A. Sinov'ev's Systems of Logical Entailment. In: *Wissenschaftliche Zeitschrift der Humboldt-Unversität zu Berlin, Reihe Geistes- und Sozialwiss.* 41 (1992b), 13–22.

Stelzner, W.: Relevanz, Konsistenz & Entailment. In: *Analyomen 1.*, ed. by G. Meggle & U. Wessels, Berlin 1994, 146–167.

Walton, D.N.: Philosophical Basis of Relatedness Logic. *Philosophical Studies* 36 (1979), 115–136.

Weingartner, P.: Remarks on the Consequence-Class of Theories. In: *The Role of Experience in Science*, ed. by E. Scheibe, Berlin 1988, 161–180.

Weingartner, P., & Schurz, G.: 1986, Paradoxes Solved by Simple Relevance Criteria. *Logique et Analyse* 113 (1986), 3–40.

Wessel, H.: 1979, Ein System der strikten Folgebeziehung. In: *Begriffsschrift. Jenaer Frege-Konferenz*, Jena 1979, 505–518.

Wessel, H.: *Logik*. Berlin 1984.

Wittgenstein, L.: *Wittgenstein's Lectures on the Foundation of Mathematics*, ed. by C. Diamond, Ithaca 1976.

Zinov'jev, A.A.: *Complex Logic*. Dordrecht 1982.

Part II
Negation in Linguistics

Categorial Grammars with Negative Information

Wojciech Buszkowski

1 Introduction

In this paper we discuss some possibilities of introducing negative information in the formalism of categorial grammar. Traditionally, formal grammars, including categorial grammars, admit positive information only. For instance, one postulates the rule S⇒NP,VP but not −S⇒NP,−VP; here $-A$ stands for the negation (complement) of category A.

Negative information is approached in later language formalisms of Noam Chomsky in the form of different filtration principles which are to repair overgeneration of the basic grammar (Chomsky 1982). However, in these approaches negative postulates are applied to the ready corpus produced by the grammar; they do not affect the generation process, and consequently, have no impact on the grammar formalism.

Another domain in which negative information takes a rather essential part is learning theory. It is known that learning procedures admitting negative information are more efficient than those restricted to purely positive data (Gold 1967, Kanazawa 1994). In (Buszkowski/Penn 1990, Marciniec 1994) there is elaborated an algorithm for learning a categorial grammar from a (structured) linguistic data sample, and here admitting negative postulates strongly improves the adequacy of outcome grammars.

Here we consider three issues connected with negation in categorial grammar. In section 2 we introduce a generalized notion of a categorial grammar where the initial type assignment is defined not only for atoms (language units) but for possibly complex expressions. The parsing process for such grammars uses unification techniques from (Buszkowski 1986a, Buszkowski/Penn 1990, Marciniec 1994); also see (Kanazawa 1994) where these techniques underly a mature learning theory for categorial grammars (restricted to positive data). Our generalized categorial grammars admit negative postulates of the form 'expression X should not have type A' together with positive ones 'expression Y should have type B'. We briefly report main properties of this formalism and its linguistic aspects.

Sections 3 and 4 are concerned with categorial grammars based on more advanced logical systems: our working system is the Lambek Calculus, introduced in (Lambek 1958). In section 3 we describe a method of inserting filtration principles (like those in Chomsky's theory) in the metatheory of the Lambek Calculus. In our setting, negative postulates select a corpus of 'possibly grammatical' expressions, and the syntactic analysis is restricted to this corpus, which influences the meaning of functor categories $A\backslash B$ and A/B. In this way one can eliminate certain 'paradoxes' of grammatical description which result from overgeneration. From the logical point of view, that amounts to an interpretation of the Lambek Calculus in more general frames than powerset frames on free semigroups (standard frames). This approach has already been announced in the author's (1986b, 1987); generalized frames of that form have also been used in (Buszkowski 1982b) to prove the finite model property of the (product-free) Lambek Calculus.

In the above negative information is still *external* to the logical systems of categorial grammar; it lives on the level of type-assignments in the former setting, and on the level of models in the latter. Section 4 discusses *internal* negation: one introduces negation as a logical constant to formal systems of type change. We reformulate the approach of section 2 in this framework and point out some interesting facts; even for Basic Categorial Grammars with Negation the consistency problem is undecidable (but it is decidable for deterministic grammars). Further, we consider a system of the Lambek Calculus with De Morgan negation (fulfilling Double Negation and Transposition; see (Dunn 1993b)). We prove some completeness theorems for this system with respect to powerset residuated semigroups with negation defined according to Bialynicki-Birula/Rasiowa (1957) (here we follow some ideas of Dunn (1993a,b)). We also show that this system is not complete with respect to standard models (negation interpreted as Boolean complement).

2 Learning Categorial Grammars

Recall that a *categorial grammar* is a triple $G = (V_G, I_G, C_G)$, such that V_G is a finite lexicon (alphabet), I_G is the *initial type-assignment* which to any $v \in V_G$ assigns finitely many types, and C_G is a type change system (*the system of* G).

Types are formed out of *atomic types* $p, q, r, \ldots$ by means of the following formation rule: if A, B are types, then also A/B and $A\backslash B$ are types. *The Basic System* (**B**) is defined by the axioms:

(A0) $A \rightarrow A$

(A1) $(A/B)B \rightarrow A$

(A2) $A(A\backslash B) \rightarrow B$

and the only inference rule:

$$\text{(CUT)}\ \frac{\Gamma, A, \Delta \to B;\ \Phi \to A}{\Gamma, \Phi, \Delta \to B}.$$

Here A, B range over types and Γ, Δ, Φ over finite sequences of types. So, **B** produces sequents $\Gamma \to A$ with Γ nonempty. The intended meaning of sequent $A_1, \ldots, A_n \to A$ is: if expressions $X_1, \ldots, X_n$ are of type $A_1, \ldots, A_n$, respectively, then the concatenated expression $X_1 \ldots X_n$ is of type A. The expression X of type A/B (resp. $B\backslash A$) is meant to be a right-looking (resp. left-looking) functor which together with every Y of type B forms the complex XY (resp. YX) of type A. Clearly, **B** is sound with respect to this interpretation.

The product-free Lambek Calculus (**L**) arises from **B** after one has added the new rules:

$$(/\mathrm{R})\ \frac{\Gamma, B \to A}{\Gamma \to A/B}$$

$$(\backslash \mathrm{R})\ \frac{A, \Gamma \to B}{\Gamma \to A\backslash B}$$

with Γ nonempty (dropping this constraint yields a stronger system $\mathbf{L}^*$). A Gentzen-style form of **L** replaces (A1) and (A2) with the rules:

$$(/\mathrm{L})\ \frac{\Gamma, A, \Delta \to C;\ \Phi \to B}{\Gamma, A/B, \Phi, \Delta \to C}$$

$$(\backslash \mathrm{L})\ \frac{\Gamma, B, \Delta \to C;\ \Phi \to A}{\Gamma, \Phi, A\backslash B, \Delta \to C}.$$

A classical result of Lambek (1958) shows that (CUT) can be eliminated in the latter system. Dropping (/R) and (\L) one gets an analogous cut-free form of **B**.

Categorial grammars based on **B** are called *Basic Categorial Grammars* and those based on **L** are called *Lambek Categorial Grammars.* For a thorough discussion of their logical and linguistic aspects the reader is referred to collection volumes (Buszkowski/Marciszewski/van Benthem 1988, Oehrle/Bach/Wheeler 1988) and books (Buszkowski 1989, Moortgat 1988, van Benthem 1991).

A key notion is *the terminal type-assignment* of the grammar G: we write $v_1 \ldots v_n \to_G A$, where $v_i \in V_G$, if there are $A_i \in I_G(v_i)$, $i = 1, \ldots n$, such that sequent $A_1, \ldots, A_n \to A$ is derivable in C_G. By $T_G(A)$ we denote the set of all strings X such that $X \to_G A$; this set is called the *string category* of type A in the sense of G.

For the Basic System, each derivation of $\Gamma \to A$ uniquely determines a functor-argument structure on the string Γ. Let us demonstrate it by an example. The derivation:

1. $NP, NP\backslash S \rightarrow S$ (A2)
2. $NP/N, N \rightarrow NP$ (A1)
3. $NP/N, N, NP\backslash S \rightarrow S$ (CUT)

determines (step by step) the structures:

1. $(NP, NP\backslash S)_2$
2. $(NP/N, N)_1$
3. $((NP/N, N)_1, NP\backslash S)_2$.

Accordingly, if the initial type-assignment assigns types NP/N, N and $NP\backslash S$ to words *a, girl* and *sings*, respectively, then the resulting grammar assigns type S to the sentence *a girl sings* and parses this sentence as the structure:

$$((\mathrm{a}, \mathrm{girl})_1, \mathrm{sings})_2.$$

Clearly, the subscript in $(X, Y)_i$ indicates which of the constituents X, Y takes the part of functor (the other is referred to as the argument). Then, essentially, types are assigned to functor-argument structures rather than merely strings. The above grammar assigns no type to the structure:

$$(\mathrm{a}, (\mathrm{girl}, \mathrm{sings})_1)_1,$$

although the latter lives on the same string.

We define $S_G(A)$ as the set of all functor-argument structures X such that X is assigned type A by the grammar G. It follows from the above that these sets are better candidates for *syntactic categories* determined by G than the string-categories $T_G(A)$. For stronger categorial grammars (based on **L** and related systems), one meets the phenomenon of *structural completeness*: $S_G(A)$ contains all possible structures living on strings from $T_G(A)$ (Buszkowski 1988). Consequently, the string interpretation of categories is more adequate for Lambek-style grammars.

The idea of describing a language by assigning types to atomic expressions (words) belongs to the leitmotives of categorial grammar. Nevertheless, it seems to be appropriate for the case when one describes a language whose syntax is perfectly clear and, actually, it has already been defined in a formal way (by another grammar). For the case of natural language, our starting point is not a ready type-assignment defined on words but a language experience which enables us to qualify certain (atomic and complex) expressions to different categories. For instance, we know the following:

1. *a girl sings*
2. *a beautiful girl sings*

3. *a boy jumps*

4. *a tall boy jumps*

are correct sentences; what types should be assigned to words occurring in these sentences is not so clear. As a matter of fact, there are many possible solutions to this question. A reasonable strategy is to look for the most economical solutions (in some sense).

The strategy elaborated in (Buszkowski/Penn 1990) is to assume a functor-argument analysis of sample sentences. So, the initial type-assignment may look as follows:

1. $((\text{a,girl})_1,\text{sings})_2 \to S$
2. $((\text{a,(beautiful,girl)}_1)_1,\text{sings})_2 \to S$
3. $((\text{a,boy})_1,\text{jumps})_2 \to S$
4. $((\text{a,(tall boy)}_1)_1,\text{jumps})_2 \to S$.

Then, we assign different variables to different occurrences of argument substructures of the above structures. Further, functor substructures are assigned types according to the rules:

(R/) if $(XY)_1 \to A$ and $Y \to B$, then $X \to A/B$

(R\\) if $(XY)_2 \to B$ and $X \to A$, then $Y \to A\backslash B$.

The variables assigned to argument substructures are:

1. $(\text{a,girl})_1 \to x$, girl $\to y$
2. $(\text{a,(beautiful,girl)}_1)_1 \to z$, $(\text{beautiful,girl})_1 \to u$, girl $\to v$,

and similarly for 3 and 4 with x', y', z', u', v'. The types (with variables) assigned to functor substructures are:

1. sings $\to x\backslash S$, a $\to x/y$
2. sings $\to z\backslash S$, a $\to z/u$, beautiful $\to u/v$
3. jumps $\to x'\backslash S$, a $\to x'/y'$
4. jumps $\to z'\backslash S$, a $\to z'/u'$, tall $\to u'/v'$.

We collect all types assigned to each word:

- a $\to x/y$, z/u, x'/y', z'/u'
- girl $\to y$, v

- beautiful $\rightarrow u/v$
- sings $\rightarrow x\backslash S$, $z\backslash S$
- boy $\rightarrow y'$, v'
- tall $\rightarrow u'/v'$
- jumps $\rightarrow x'\backslash S$, $z'\backslash S$.

Now, we unify all types assigned to each word; precisely, we do not want to unify all types appearing above but, for each word, we try to unify as many types, as possible. For the word *a*, we substitute x for z, x', z' and y for u, y', u'. For *girl*, we substitute y for v. For *beautiful*, we get type y/y. For *sings*, we get type $x\backslash S$. For *boy*, we substitute y for v' (note that y has been substituted for y'). For *tall*, we get type y/y, and for *jumps*, we get type $x\backslash S$. As a result, we obtain the following deterministic type-assignment:

- a $\rightarrow x/y$
- girl $\rightarrow y$
- beautiful $\rightarrow y/y$
- sings $\rightarrow x\backslash S$
- boy $\rightarrow y$
- tall $\rightarrow y/y$
- jumps $\rightarrow x\backslash S$.

We can interprete $x := NP$ (the type of Noun Phrase), $y := N$ (the type of Common Noun). Then, words *beautiful, tall* are correctly assigned type N/N of Adjective, and the word *a* is assigned type NP/N of Determiner. The resulting grammar produces an infinite language in which, for instance, each expression of the form:

a tall tall ... tall girl jumps

is assigned type S.

The above example explains enough the general procedure, studied in detail in (Buszkowski/Penn 1990), which discovers a categorial grammar compatible with a given finite corpus of functor-argument structures supplied with types. It may happen that there is no substitution which unifies all types assigned to each word; in that case the outcome is a finite family of nondeterministic grammars (one word can be assigned several types). The main result of (Buszkowski/Penn 1990) shows that the outcome provides precisely the most general forms of minimal grammars compatible with the given corpus. (The grammar G is said to be *minimal* in the class of grammars compatible with the corpus, if it is compatible with this corpus, but dropping any type from I_G yields a grammar which is not compatible with this corpus). In other words, every minimal grammar compatible with the given corpus results from a grammar produced by our procedure by substituting certain types for variables.

It follows from the results of Kanazawa (1994) that, for any Basic Categorial Grammar G, there is a finite set $L \subset L(G)$, such that our procedure applied to L yields a grammar equivalent to G. However, the corpus L can be quite large. In practice, the linguist postulates a grammatical description on the basis of a relatively small corpus of 'typical' data. A procedure based on positive data only can lead to wrong outcomes, mainly due to overunification. Let us consider the following corpus:

1. $(\text{Mary sings})_1 \rightarrow S$
2. $(\text{Susan sings})_1 \rightarrow S$
3. $(\text{Mary (sings and dances})_2)_1 \rightarrow S$
4. $((\text{Mary sings})_1 \text{ and } (\text{Susan dances})_1)_2 \rightarrow S.$

In that case, the unification algorithm assigns type S to verbs *sings* and *dances* and type S/S to proper nouns *Mary* and *Susan* by unifying types for two occurrences of *and*, which is a linguistic nonsense. To block this wrong solution, one can add the negative constraint:

5. sings $\not\rightarrow S$.

The unification procedure is modified as follows: one looks for the most general unifier in the family of substitutions fulfilling the negative constraints (that means, each step of the algorithm regards these substitutions only). For our example, we obtain the assignment:

1. Mary, Susan $\rightarrow S/x$
2. sings, dances $\rightarrow x$
3. and $\rightarrow x\backslash x/x,\ S\backslash S/S,$

which is linguistically correct (interpret x as the type of Verb Phrase).

The above is a merely telegraphic announcement of the leitmotive of our learning procedure (for details and more advanced examples, see (Buszkowski/Penn 1990, Marciniec 1994). Again, the outcomes are precisely the most general forms of minimal grammars compatible with the given (positive and negative) corpus. Introducing negative data improves the adequacy of outcomes without any essential modification of the very method.

The above justifies the following generalized notion of a categorial grammar. By a *generalized categorial grammar* we mean a pair $G = (P_G, N_G, C_G)$, such that P_G and N_G are finite collections of postulates $X \to A$ and $Y \not\to B$, respectively, where X, Y are functor-argument structures over a finite lexicon and A, B are types, and C_G is a type change system. There are two steps of processing such a grammar. First, using unification techniques, one determines a most general grammar compatible with the postulates. Second, one recognizes types and structures of complex expressions with the aid of the latter grammar. The new notion of a categorial grammar seems to be more realistic with respect to standards of natural language description.

This more general approach to categorial grammars rises new research problems, and many of them are still to be solved. One obvious example is the consistency of data. If $C_G = \mathbf{B}$, the consistency problem is decidable: there is a grammar compatible with the given corpus if and only if the algorithm yields at least one outcome. The problem is open for stronger systems of type change, e.g. the Lambek Calculus.

Linguists may object our assumption the input data are structured expressions instead of merely strings of words. Of course, if we started from strings, the procedure would require to try all possible structures on the strings and to process them as above. Interestingly, for systems comparable with the Lambek Calculus all possible choices of structures are equivalent, in a sense. Due to structural completeness (Buszkowski 1988), if a string is assigned type A on the basis of $\mathbf{L}$, then, for each structure X on this string, there is a derivation of the type A for this string which determines precisely the structure X. Consequently, the outcomes of our procedure corresponding to different structures postulated on the input strings yield equivalent grammars.

3 Restricted language models

In the preceding section we discussed a way to block overunification by introducing negative postulates to the linguistic data corpus. Here we consider another typical trouble with categorial grammars which results from overgeneration. Already in section 2 we met a grammar which accepts all sentences of the form:

a tall tall ... tall girl jumps.

Sentences of this form must be accepted due to the fact that adjectives are assigned type N/N, and all types A/A admit the reduction:

$$A/A,\ A/A, \ldots, A/A,\ A \to A$$

on the basis of **B**. In **L** one derives:

$$(\mathrm{I}) A/A,\ A/A, \ldots, A/A \rightarrow A/A,$$

which leads to similar effects. Lambek (1961) points out so-called paradoxes of syntactic description for the case of pronouns. If one assigns type $(S/PN)\backslash S$ to *him* on the basis of sample sentences like:

John likes him,

and type PN/PN to *poor* on the basis of examples like *poor Jane*, then the pseudosentence:

John likes poor him

is assigned type S in **L**. The remedy proposed in (Lambek 1961) is to drop associativity from **L**; the resulting Nonassociative Lambek Calculus does not accept the latter sentence. On the other hand, our former example is retained, since it requires nothing more than **B**, which is a subsystem of the Nonassociative Lambek Calculus.

In my opinion, most overgeneration phenomena call for a different treatment than modifying the basic grammatical rules or even the very logic, as e.g. in Lambek's proposal. Assigning type N/N to adjectives or type S/S to sentence modifiers like *probably, tomorrow* is justified by semantical reasons, since these expressions denote semantic objects of these types. The semantic functions corresponding to these expressions admit iteration, and consequently, rules like (I) are semantically sound ((I) expresses some obvious properties of function composition). Actually, if we applied the procedure from section 2 to a corpus containing positive data:

tall girl $\rightarrow N$, beautiful tall girl $\rightarrow N$

and negative data:

tall tall girl $\not\rightarrow N$, beautiful beautiful girl $\not\rightarrow N$,

then the outcome grammar would assign a different type to each adjective, say type x/N to *tall* and type y/N to *beautiful*, which is not satisfactory (in fact, each adjective would be assigned a finite family of different types).

The approach we propose here roughly follows the ideas of Noam Chomsky to supply the basic grammar with filtration principles which eliminate ill-formed expressions from the language (Chomsky 1982). For the case of the Lambek Calculus, filtration principles can be introduced on the level of language models in a quite natural way. We present this notion for the full Lambek Calculus: besides $\backslash$ and $/$ there appears *the product operation* $\times$, and the corresponding introduction rules are:

$$(\times L)\ \frac{\Gamma, A, B, \Delta \rightarrow C}{\Gamma, A \times B, \Delta \rightarrow C}$$

$$(\times R)\ \frac{\Gamma \rightarrow A;\ \Delta \rightarrow B}{\Gamma, \Delta \rightarrow A \times B}.$$

The meaning of $A \times B$ is the set of expressions XY such that X is of type A and Y is of type B.

A language model is a pair $M = (V, (L_A)_{A \in T})$ such that V is a finite lexicon, and $(L_A)_{A \in T}$ is an indexed family of subsets of V^+ (i.e. sets of nonempty finite strings over V), where T stands for the set of types, satisfying the following conditions:

(M$\times$) $L_{A \times B} = \{XY : X \in L_A, Y \in L_B\}$

(M/) $L_{A/B} = \{X \in V^+ : (\forall Y \in L_B)\, XY \in L_A\}$

(M\) $L_{A \backslash B} = \{X \in V^+ : (\forall Y \in L_A)\, YX \in L_B\}$.

The sequent $A_1, \ldots, A_n \rightarrow A$ is said to be *true* in the model M, if for all $X_1 \in L_{A_1}, \ldots, X_n \in L_{A_n}$ there holds $X_1 \ldots X_n \in L_A$. It has been proven by Pentus (1993) that theorems of **L** (with product) are precisely the sequents true in all language models (for the product-free Lambek Calculus, an analogous completeness result has been proven in (Buszkowski 1982a), and for the full Lambek Calculus and a more general class of semigroup models, in (Buszkowski 1986b)).

Language models correspond to the standard interpretation of the meaning of types in categorial grammar. Below we define a more realistic notion of a restricted language model which relativizes clauses (M$\times$), (M/) and (M\) to a fixed set of 'non-ill-formed expressions'. We believe that natural language obeys some syntactic constraints which are esentially non-categorial, hence they should not be reflected in the type assignment. A constraint of that kind is Non-Iteration: for some strings X, no well-formed expression contains the subinterval XX. Another example is the distribution of articles *a* and *an* which is grounded in phonetics. Negative Polarity Items are expressions which may appear in negative contexts only, as e.g. *at all, a bit, anything* (in the meaning of *something*) (Sánchez Valencia/van der Wouden/Zwarts 1993). Also in that case, one can block the wrong distribution of such expressions by means of appropriate negative constraints. Given a fixed list of constraints, one defines the set $K \subset V^+$, of all expressions obeying these constraints. Members of K are potentially non-ill-formed expressions according to the constraints.

A restricted language model is a triple $M = (V, K, (L_A)_{A \in T}))$ such that V is a finite lexicon, $K \subset V^+$, and $(L_A)_{A \in T}$ is an indexed family of subsets of K, satisfying the following conditions:

(RM$\times$) $L_{A \times B} = \{XY \in K : X \in L_A, Y \in L_B\}$

(RM/) $L_{A/B} = \{X \in K : (\forall Y \in L_B)\, XY \in (-K) \cup L_A\}$

(RM\) $L_{A\backslash B} = \{X \in K : (\forall Y \in L_A) YX \in (-K) \cup L_B\}$.

Clearly, the condition defining $L_{A/B}$ can be read: for all $Y \in L_B$, if $XY \in K$ then $XY \in L_A$, and similarly for $L_{A\backslash B}$. The sequent $A_1 \ldots A_n \rightarrow A$ is said to be *true* in the model M if, for all $X_1 \in L_{A_1}, \ldots, X_n \in L_{A_n}$, if $X_1 \ldots X_n \in K$ then $X_1 \ldots X_n \in L_A$ (of course, $X_1 \ldots X_n$ denotes the juxtaposition of strings $X_1, \ldots, X_n$).

Not all restricted language models validate the rules of the Lambek Calculus. What is needed to validate the associativity law:

$$(A \times B) \times C \leftrightarrow A \times (B \times C),$$

is precisely the following condition to be fulfilled by K:

(AS) for all $X, Y, Z \in K$, if $XYZ \in K$ then $XY \in K$ iff $YZ \in K$.

It is easy to show that the Lambek Calculus is sound with respect to restricted language models fulfilling (AS). Since language models are a special class of such restricted language models (with $K = V^+$), it follows from the afore-mentioned completeness theorem that the Lambek Calculus is also complete with respect to restricted language models satisfying (AS). Consequently, restricted language models can be regarded as a legitimate formalization of a family of syntactic categories compatible with the logic of the Lambek Calculus.

(AS) is satisfied, if the set K fulfills one of the stronger conditions:

(SI) if $XYZ \in K$ and $Y \neq \Lambda$, then $Y \in K$

(CON) if $X, Y \in K$, then $XY \in K$;

(SI) means that K is closed under nonempty subintervals, and (CON) means that K is closed under concatenation. If K is defined as the set of all strings which do not contain any subinterval $Y \in P$, where P is a fixed set of ill-formed strings, then K satisfies (SI) (clearly, Non-Iteration can be imposed in this way).

Accordingly, a possible way of processing negative information in categorial grammar is to select a set of ill-formed strings which cannot appear in any well-formed string and to define K as above. Then, one assigns types to words on the basis of a finite corpus of positive data (the procedure can follow the method of section 2). For the first example above, although adjectives *tall* and *beautiful* are assigned type N/N, if strings *tall tall girl* and *beautiful beautiful girl* are included in the set of ill-formed strings, then in spite of the fact that (I) remains a valid logical rule we do not assign type S to sentences like:

a tall tall ... tall girl jumps,

since these sentences are beyond K.

It should be emphasized that in the above discussion we have formalized a language as a (restricted) language model rather than a categorial grammar. Problems of approximation of a model by means of a grammar were considered (for language models) in (Buszkowski 1982a, 1985). The theory of an adequate description of restricted language models by means of categorial grammars has not been elaborated as yet. There are reasons to hope restricted language models admit a more complete description than it is the case for language models (an incompleteness theorem of (Buszkowski 1982a) shows that no language model is completely generated by any Lambek Grammar).

Finite language models are restricted language models in which the set K is finite. The product-free Lambek Calculus is complete with respect to finite language models, and consequently, it possesses the finite model property (Buszkowski 1982b). This result seems to be the most interesting logical application of restricted language models (see also (Pentus 1993) for some technical application of similar structures in a completeness proof).

We can imagine a linguistic usage of finite language models, as well. Let V_n denote the set of all strings over V of length at most n. We set $K_n = K \cap V_n$. If K fulfills (SI), then all sets K_n do so, for $n \geq 1$. Starting from a small $n \geq 1$ such that V_n contains reasonably many positive and negative data, one can obtain a complete categorial description of the restricted language model based on K_n. For some $n' > n$, the corresponding model may disagree with the former one; then, we take $K_{n'}$ as our basis. For a sufficiently big number m, we should eventually produce a finite language model whose type assignment would be totally adequate with the infinite (restricted) language model corresponding to the language under consideration.

4 Internal negation

In this section we discuss type change systems with internal negation. The linguistic motivation for regarding such systems seems quite clear. Negative postulates can be written $X \rightarrow -A$ instead of $X \not\rightarrow A$. It also seems reasonable to admit more involved postulates $X \rightarrow (-A)/B$, $X \rightarrow A\backslash((-B)/C)$, and so like. For instance, not only:

$$\text{sings} \rightarrow -S,$$

but also:

$$\text{sings} \rightarrow (-S)/NP$$

is an admissible linguistic assumption.

Types with negation are defined in an obvious way; the recursive condition is: if A, B are types, then $A \times B$, $A\backslash B$, A/B and $-A$ are types.

The notion of a categorial grammar is generalized to admit types with negation. For Basic Categorial Grammars, the type change system is **B**; negation may appear in types, but the rules of **B** do not depend on it. A grammar G is *consistent*, if both $X \rightarrow_G A$ and $X \rightarrow_G -A$ hold for no string X and type A.

The consistency problem for Basic Categorial Grammars is undecidable. It easily follows from the fact that the emptiness problem for the intersection of two context-free languages is undecidable. Given two context-free grammars G_1, G_2, one constructs Basic Categorial Grammars G_1', G_2' such that $L(G_1) = L(G_1')$ and $L(G_2) = L(G_2')$ (the Gaifman theorem; see (Bar-Hillel/Gaifman/Shamir 1960)). Recall that, for a categorial grammar G, $L(G) = T_G(S)$, where S is a distinguished atomic type (*the principal type*). We make the atomic types involved in G_1' different from those involved in G_2'. Let S, S' be the principal types of G_1', G_2', respectively. In the initial type-assignment of G_2' we substitute $-S$ for S' at each place. The final step is joining G_1' and G_2' in one grammar (that means, to each word we assign all types assigned to this word by G_1' or G_2'). Clearly, $L(G_1) \cap L(G_2) = \emptyset$ if and only if the latter grammar is consistent.

On the contrary, all Deterministic Basic Categorial Grammars are consistent. (A categorial grammar is deterministic, if it assigns at most one type to each word.) That is an obvious consequence of the consistency of **B**. Every sequent derivable in **B** contains an even number of negation symbols, hence both $\Gamma \rightarrow A$ and $\Gamma \rightarrow -A$ cannot be derivable. The same holds for the Lambek Calculus with Negation to be introduced below. As a result, all Deterministic Lambek Categorial Grammars (with negation) are consistent. This observation is important from the linguistic point of view. A non-deterministic grammar can always be transformed in a deterministic one by introducing syntactic copies of words. For grammars with negation, multiplying syntactic atoms rather than types assigned to one atom is a proper strategy, since it guarantees the consistency.

Learning procedures, discussed in section 2, can be extended towards types with negation. In the form given above, they only admit postulates $X \rightarrow A$ and $Y \rightarrow -B$, where A, B are negation-free. There seems to be no essential trouble with admitting succedent types of the form $(-A)/B$, $((-A)/B)/C$ and so on (still negation appears only once). For the general case, of arbitrary succedent types, these procedures should be more involved, and we defer them for further research.

We pass to the final topic of this section: introducing negation in the Lambek Calculus. The Lambek Calculus is widely recognized as a basis for categorial grammar sensitive to dynamic phenomena in syntax and semantics (van Benthem 1988, 1991), hence it is interesting to see how can it handle negative types. One obvious way is to interpret $-A$ as $A\backslash 0$, where 0 is a distinguished primitive type, which follows the treatment of negation in constructive logics (Wansing 1993). The failure of this constructive negation is the lack of any clear linguistic sense. The meaning of negation suggested by the above discussion is Boolean negation: $-A$ is assigned to those strings which are not assigned type A. Accordingly, negation should satisfy the following principles:

$$\text{(DN)}\ A \to -(-A),\ -(-A) \to A$$

$$\text{(TR)}\ \frac{A \to B}{-B \to -A},$$

that means, the axioms of Double Negation and the rule of Transposition. If $\to$ stands for a lattice ordering, then each operation $-$ fulfilling (DN) and (TR) must also satisfy De Morgan laws, hence it is called a De Morgan negation (Dunn 1993b).

The system **LN** (*the Lambek Calculus with Negation*) can be axiomatized by affixing (DN) and (TR) to the 'algebraic' axiomatization of the Lambek Calculus, given in (Lambek 1958):

$$\text{(L0)}\ A \to A,$$

$$\text{(L1)}\ (A \times B) \times C) \to A \times (B \times C),\ A \times (B \times C) \to (A \times B) \times C,$$

$$\text{(L2)}\ \frac{A \times B \to C}{A \to C/B},\ \text{(L3)}\ \frac{A \times B \to C}{B \to A\backslash C},$$

$$\text{(L4)}\ \frac{A \to C/B}{A \times B \to C},\ \text{(L5)}\ \frac{B \to A\backslash C}{A \times B \to C},$$

$$\text{(L6)}\ \frac{A \to B;\ B \to C}{A \to C}.$$

The system **LN** is a conservative extension of **L**. That follows from the completeness of **L** with respect to powerset residuated semigroups over arbitrary semigroups (Buszkowski 1986b). Given a semigroup $(M, \circ)$, we define operations $\times, \backslash, /$ on subsets of M as follows:

$$A \times B = \{a \circ b : a \in A,\ b \in B\},$$

$$A\backslash B = \{c \in M : (\forall a \in A)\, a \circ c \in B\},$$

$$A/B = \{c \in M : (\forall b \in B)\, c \circ b \in A\}.$$

One easily shows that the structure $(P(M), \subset, \times, \backslash, /)$ is a residuated semigroup. Recall that a residuated semigroup is a structure $(A, \leq, \times, \backslash, /)$ such that $(A, \leq, \times)$ is a partially ordered semigroup (i.e. $\leq$ is a partial order, and $\times$ is monotone in both arguments), and $\backslash, /$ satisfy the equivalences:

$$(\forall x, y, z \in A)(y \leq x\backslash z \Leftrightarrow x \times y \leq z \Leftrightarrow x \leq z/y).$$

The structure $(P(M),\ldots)$ is called the powerset residuated semigroup over the semigroup $(M,\circ)$. Clearly, each powerset residuated semigroup can be expanded to a model for **LN** by defining $-A$ as the Boolean complement of A, which yields the conservativity of **LN** over **L**. Actually, since **L** is even strongly complete with respect to powerset residuated semigroup (that means, each axiomatic extension of **L** is complete with respect to models satisfying the axioms), then each axiomatic extension of **LN** such that new axioms are negation-free is conservative over the analogous extension of **L**.

Let us notice that **LN** is different from the system of Non-Commutative Linear Logic introduced in (Yetter 1990) which is also conservative over **L** (actually, Yetter considers a system conservative over $\mathbf{L}^*$, but it is easy to modify it for **L**). In Yetter's system formulas $A\backslash B$ and $-(-B\times A)$ are equivalent, which is not true for **LN** (these formulae can have different denotations in powerset residuated semigroups with Boolean complement).

The system **LN** can equivalently be presented with dual operations $\times_l$, $\times_r$, $\backslash_d$, $/_d$ which can be defined from $\times, \backslash, /$ as follows:

$$A \times_l B = -(-A \times B),\ A \times_r B = -(A \times -B),$$
$$A\backslash_d B = -(A\backslash -B),\ A/_d B = -(-A/B),$$

by the following axioms and rules:

(L1') $(A \times_l B) \times_r C \to A \times_l (B \times_r C)$, $A \times_l (B \times_r C) \to (A \times_l B) \times_r C$,

(L2') $\dfrac{A \to B \times_l C}{B\backslash_d A \to C}$

(L3') $\dfrac{A \to B \times_r C}{A/_d C \to B}$

(L4') $\dfrac{B\backslash_d A \to C}{A \to B \times_l C}$

(L5') $\dfrac{A/_d C \to B}{A \to B \times_r C}$

together with (L0), (L6), (DN) and (TR). Operations $\times, \backslash, /$ can be defined in an obvious way. Notice that $\backslash_d$ and $/_d$ are natural duals of $\backslash$ and $/$ in powerset residuated semigroups:

$$A\backslash_d B = \{c \in M : (\exists a \in A)\, a \circ c \in B\},$$
$$A/_d B = \{c \in M : (\exists b \in B)\, c \circ b \in A\}.$$

In mathematical linguistics, these dual residuals have been considered more often than $\backslash, /$, and they seem also be of interest for categorial grammar in relation to *partial functors*: expressions which form an expression of category B with some (not necessarily all) arguments of category A.

We pass to the question of adequate semantics for **LN**. First, we prove that the product-free **LN** is strongly complete with respect to powerset residuated semigroups with quasi-Boolean complement. Quasi-Boolean complement was introduced in (Bialynicki-Birula/Rasiowa 1957) for a representation theorem for quasi-Boolean algebras, i.e. distributive lattices with De Morgan negation. Here we consider a semigroup $(M, \circ)$ and a bijection $f : M \mapsto M$. In the powerset residuated semigroup $(P(M), \ldots)$ we define a quasi-Boolean complement by setting:

$$-A = M \setminus f[A], \text{ for all } A \subset M,$$

where $f[A]$ denotes the image of A under f and $\setminus$ stands for set difference. Clearly, $-$ satisfies (DN) and (TR), hence **LN** is sound with respect to such frames.

To get completeness, we observe that **LN** is strongly complete with respect to residuated semigroups with De Morgan negation. For the canonical model consisting of equivalence classes of formulas with respect to the provable equivalence belongs to this class. Then, it is enough to prove the following representation theorem: for each residuated semigroup with De Morgan negation $(M, \leq, \times, \backslash, /, -)$, the reduced structure $(M, \leq, \backslash, /, -)$ is isomorphically embeddable into some powerset residuated semigroup with quasi-Boolean complement. So, fix a residuated semigroup with De Morgan negation $(M, \ldots)$. Let N be the set of all cones in the p.o. set $(M, \leq)$. For $X, Y \in N$, we set:

$$X \circ Y = \{b \in M : (\exists a \in X)\, a\backslash b \in Y\} = \{b \in M : (\exists a \in Y)\, b/a \in X\};$$

the second equality follows from the valid conditions:

$$a \leq (b/a)\backslash b, \; a \leq b/(a\backslash b).$$

Using associativity of slashes $(a\backslash b)/c = a\backslash(b/c)$, one gets $(X \circ Y) \circ Z = X \circ (Y \circ Z)$. Further, $X \circ Y$ is a cone, if X, Y are so. Consequently, $(N, \circ)$ is a semigroup. We consider the powerset residuated semigroup $(P(N), \ldots)$, and a quasi-Boolean complement is defined by using the bijection $f : N \mapsto N$, defined as follows:

$$f(X) = M \setminus -X, \text{ where } -X = \{-a : a \in X\}.$$

If X is a cone, then $-X$ is a dual cone, hence $f(X)$ is a cone; obviously, $f(f(X)) = X$, hence f is bijective. We define a mapping $h : M \mapsto P(N)$:

$$h(a) = \{X \in N : a \in X\}.$$

Clearly, h is 1-1. The homomorphism conditions:

$$h(a\backslash b) = h(a)\backslash h(b), \; h(a/b) = h(a)/h(b), \; h(-a) = -h(a),$$

$$a \leq b \Leftrightarrow h(a) \subset h(b),$$

can easily be verified.

The above construction does not yield an embedding of the whole structure $(M, \ldots)$ into the powerset structure $(P(N), \ldots)$, since we only get:

$$h(a) \times h(b) \subset h(a \times b).$$

To obtain a representation theorem for full residuated semigroups with De Morgan negation we must modify the construction of $P(N)$ according to the lines of Dunn (1993a) and Došen (1985). Let $(M, \circ, \leq)$ be a partially ordered semigroup. By $C(M)$ we denote the set of all cones in $(M, \leq)$. For $X, Y \in C(M)$, we define:

$$X \times Y = \{c \in M : (\exists a \in X)(\exists b \in Y)\, a \circ b \leq c\},$$

and $X \backslash Y$, X/Y are defined as for $(P(N), \ldots)$. Clearly, $(C(M), \subset, \times, \backslash, /)$ is a residuated semigroup; it will be called the residuated semigroup of cones over the p.o. semigroup $(M, \circ, \leq)$. Now, if $f : M \mapsto M$ is bijective and antitone, that means, $a \leq b$ entails $f(b) \leq f(a)$, then the quasi-Boolean complement in $P(M)$, defined above, is an operation in $C(M)$. For assume $X \in C(M)$. Then, $-X = M \setminus f[X]$, and we show $-X \in C(M)$. Let $a \in -X$ and $a \leq b$. Then, $f(a) \notin X$ and $f(b) \leq f(a)$, hence $f(b) \notin X$, and consequently, $b \in -X$. The so-defined structure $(C(M), \ldots, -)$ will be called the residuated semigroup of cones with quasi-Boolean complement (over the p.o. semigroup $(M, \circ, \leq)$).

We prove that each residuated semigroup with De Morgan negation is isomorphically embeddable into some residuated semigroup of cones with quasi-Boolean complement. Fix a residuated semigroup with De Morgan negation $(M, \leq, \times, \backslash, /, -)$. Again, N denotes the set of all cones in $(M, \leq)$. For $X, Y \in N$, we define $X \circ Y$ as above. Now, $(N, \circ, \subset)$ is a p.o. semigroup. We consider the residuated semigroup of cones $(C(N), \subset, \times, \backslash, /)$ over the p.o. semigroup $(N, \circ, \subset)$. The bijection $f : N \mapsto N$, defined above, is antitone, hence it gives rise to a quasi-Boolean complement $-$ in $C(N)$. It follows that $(C(N), \subset, \times, \backslash, /, -)$ is a residuated semigroup of cones with quasi-Boolean complement. The embedding h defined above satisfies $h : M \mapsto C(N)$ and is a homomorphism with respect to $\leq, \times, \backslash, /$ and $-$.

We note that the above results are closely related to Dunn's representation theorems (restricted to neagtion-free structures) using so-called concrete residuated semigroups (Dunn 1993a), but the former sound a bit stronger than the latter. Concrete residuated semigroups are residuated semigroups of sets in which operations are defined by means of a ternary relation over the basic universe, whereas powerset residuated semigroups use powerset operations over the basic structure. Precisely, powerset residuated semigroups are concrete residuated semigroups in which the ternary relation $R(a, b, c)$ is defined as $a \circ b = c$. We believe, the former are a bit more 'concrete' than the latter and more close to linguistically intended models.

Consequently, **LN** is strongly complete with respect to residuated semigroups of cones with quasi-Boolean complement. We leave it as an open problem if **LN** is complete with respect to powerset residuated semigroups with quasi-Boolean complement. The methods of (Buszkowski 1986b) which yield the completeness of **L** with respect to powerset residuated semigroups (actually, the isomorphic embeddability of each residuated semigroup into some powerset residuated semigroup) use certain special natural deduction systems, and we do not see any clear way to adapt them for negation.
Unfortunately, **LN** is not complete with respect to linguistically intended models, i.e. powerset residuated semigroups with Boolean complement over semigroups. For $C = B/B$, the following formula:

$$-A/(C/C) \rightarrow -(A/(C/C))$$

is generally valid in this narrow semantics but not provable in **LN**. Under any valuation, C/C is nonempty, hence any element belonging to the set denoted by the left-hand type must also belong to the set denoted by the right-hand type. Accordingly, the above formula is valid in all intended models. On the other hand, there is a powerset residuated semigroup with quasi-Boolean complement in which this formula is not valid. Let $V = \{a, b\}$, and consider the free semigroup of all nonempty finite strings on V. The bijective mapping f assigns the mirror image X^* to any string X. Let B denote the set of all nonempty b-strings, and let A denote the set of all strings whose first symbol is a. One easily checks $-A/(C/C)$ consists of all nonempty strings, and $-(A/(C/C))$ consists of all strings whose last symbol is b. Consequently, the left-hand set is not contained in the right-hand one. Therefore, the formula in question is not provable in **LN**.

There are many fundamental problems concerning **LN** and related systems which still remain open, as e.g. the axiomatizability of the theory of intended models, the possibility of cut-elimination in an appropriate Gentzen style system, and the decidability of **LN** and its extensions. It would also be interesting to find interpretations of negation other than quasi-Boolean complement, more compatible with the linguistic sense of negative information. We must leave these questions to further research.

References

Bar-Hillel, Y., Gaifman, C., & Shamir, E.: On categorial and phrase structure grammars. *Bulletin Research Council Israel* F 9 (1960), 155–166.

van Benthem, J.: The Lambek Calculus. In: *Categorial Grammars and Natural Language Structures*, ed. by R.T. Oehrle, E. Bach & D. Wheeler, Dordrecht 1988, 35–68.

van Benthem, J.: *Language in Action. Categories, Lambdas and Dynamic Logic.* Amsterdam 1991.

Bialynicki-Birula, A., & Rasiowa, H.: On the Representation of Quasi-Boolean Algebras. *Bulletin de l'Académie Polonaise des Sciences* 5 (1957), 259–261.

Buszkowski, W.: Compatibility of a Categorial Grammar with an Associated Category System. *Zeitschrift für mathematische Logik und Grundlagen der Mathematik* 28 (1982a), 229–238.

Buszkowski, W.: Some Decision Problems in the Theory of Syntactic Categories. *Zeitschrift für mathematische Logik und Grundlagen der Mathematik* 28 (1982b), 539–548.

Buszkowski, W.: Algebraic Models of Categorial Grammars. In: *Foundations of Logic and Linguistics: Problems and Their Solutions*, ed. by G. Dorn & P. Weingartner, New York 1985.

Buszkowski, W.: Typed Functorial Languages. *Bulletin of the Polish Academy of Sciences, Mathematics* 34 (1986a), 495–505.

Buszkowski, W.: Completeness Results for Lambek Syntactic Calculus. *Zeitschrift für mathematische Logik und Grundlagen der Mathematik* 32 (1986b), 13–28.

Buszkowski, W.: Discovery Procedures for Categorial Grammars. In: *Categories, Polymorphism and Unification*, ed. by E. Klein & J. van Benthem, University of Amsterdam, 1987.

Buszkowski, W.: Generative Power of Categorial Grammars. In: *Categorial Grammars and Natural Language Structures*, ed. by R.T. Oehrle, E. Bach & D. Wheeler, Dordrecht 1988, 69–94.

Buszkowski, W., Marciszewski, W., & van Benthem , J. (eds.): *Categorial Grammar*, Amsterdam 1988.

Buszkowski, W.; *Logical Foundations of Ajdukiewicz-Lambek Categorial Grammars*, (in Polish). Warszawa 1989.

Buszkowski, W., & Penn, G.: Categorial Grammars determined from Linguistic Data by Unification. *Studia Logica* 49 (1990), 431–454.

Chomsky, N.: *Concepts and Consequences of the Theory of Government and Binding*, Cambridge/MA. 1982.

Došen, K.: A completeness theorem for the Lambek calculus of syntactic categories, Zeitschrift für mathematische Logik und Grundlagen der Mathematik 31 (1985), 235–241.

Dunn, J.M.: Partial-Gaggles Applied to Logics with Restricted Structural Rules. In *Substructural Logics*, ed. by P. Schroeder-Heister & K. Došen, Oxford 1993a, 63–108.

Dunn, J.M.: Perp and Star: Two Treatments of Negation. In: *Philosophical Perspectives* (*Philosophy of Language and Logic*) 7 (1993b), ed. by J. Tomberlin, 331–357.

Gold, E.M.: Language identification in the limit. *Information and Control* 10 (1967), 447–474.

Kanazawa, M.: *Learnable Classes of Categorial Grammars.* PhD thesis, Stanford University, 1994.

Lambek, J.: The mathematics of sentence structure. *American Mathematical Monthly* 65 (1958), 154–170.

Lambek, J.: On the calculus of syntactic types. In: *Structure of Language and Its Mathematical Aspects*, ed. by R. Jakobson, American Mathematical Society, Providence 1961, 166–178.

Marciniec, J.: Learning categorial grammars by unification with negative constraints. *Journal of Applied Non-Classical Logics* 4 (1994), 181–200.

Moortgat, M.: *Categorial Investigations. Logical and Linguistic Aspects of the Lambek Calculus.* Dordrecht 1988.

Oehrle, R.T., Bach, E., & Wheeler D. (eds.), *Categorial Grammars and Natural Language Structures*, Dordrecht 1988.

Pentus, M.: Lambek Calculus is L-complete, *ILLC Prepublication Series*, University of Amsterdam, 1993.

Sánchez Valencia, V., van der Wouden, T., & Zwarts, F.: Polarity, veridicality, and temporal connectives. In: *Proceedings of the Ninth Amsterdam Colloquium*, ed. by P. Dekker & M. Stokhof, Amsterdam 1994, 587–606.

Wansing, H.: *The Logic of Information Structures*, LNAI 681. Berlin 1993.

Yetter, D.N.: Quantales and (Noncommutative) Linear Logic. *The Journal of Symbolic Logic* 55 (1990), 41–64.

NEGATIVE CONCORD AND 'MODE OF JUDGEMENT'

William A. Ladusaw*

My goal in this paper is to explore the interface between clause structure and semantic interpretation in natural languages as it pertains to the expression of negation, concentrating on the analysis of languages whose clauses show negative concord, the morphological expression at multiple points in a clause of its semantic negative polarity. By doing so, I hope to defend the thesis that the best analysis of clausal descriptive negation in natural language is one which treats affirmation and negation symmetrically as 'modes of judgement' rather than asymmetrically, with negation treated as a propositional operator.

I will begin with a prolegomenon laying out some of my assumptions about the relation between syntactic structure and interpretation and then turn to an overview of the negative concord phenomena. I then try to show that under certain assumptions negative concord languages seem to pose a puzzle about compositional interpretation and lay out two strategies which resolve this puzzle. I conclude by sketching a framework in which both concord and non-concord languages can be viewed as instantiating the same fundamental type of semantic analysis of negation, with differences between them attributed to parameters in the syntax-semantics interface.

1 Autonomous syntactic constraints

Let us approach the problem from the general perspective of a linguist doing so-called 'Montague grammar', convinced that to a great extent, the units and relations of grammatical description provide a suitable basis for formal semantic interpretation, and that the general form of a semantic interpretation for a natural language is an assignment of semantic interpretations of various types to the lexical items of the language and a set of composition principles which project those interpretations into interpretations for syntactically complex expressions.

Nonetheless, this Montagovian purity must be alloyed by the linguist's realization that the syntactic structures of a language reveal patterns and constraints which are simply independent of semantic interpretation or pragmatic function. This I take to be the tenet of the 'autonomy of syntax' fundamental to the theory of generative grammar. Because this structural

*The research reported here was supported by NSF Grant BNS-9021398 and grants from the Academic Senate of the University of California, Santa Cruz.

autonomy is revealed in (at least) two fundamental ways which affect our view of the interpretation of negative clauses, I want to say more about what I mean by this.

The structure of German and Dutch provide a clear example of the kind of autonomous structural generalization I am referring to: the requirement in so-called root clauses that the tense-bearing verb be in second position in the clause. This is a fundamental aspect of clause structure in these languages which in and of itself has no semantic motivation to enforce it, at least in so far as the formulation of the semantic content of a sentence, the proposition expressed by the clause. This is not to say that there are not semantic consequences of which clausal constituent actually appears before the tensed verb, particularly for discourse coherence constraints. But the proposition expressed by the assertion of (1) is presumably the same one that that Hans reportedly believes in the assertion of (2).

(1) Marie liebt Franz.

(2) Hans glaubt, dass Marie Franz liebt.

The composition principles which govern the saturation of predicates and the formulation of propositions simply look through the difference in position of the verb in these cases. Two effects of structural autonomy are relevant to our discussion of negative concord. The first is the fact that a language has a recognizable way of marking a clause as negative and the grammatical properties of the markers of negation are fixed in terms of identity and position regardless of what semantic or pragmatic function the negation serves. Variation exists from language to language in whether those markers are treated by the syntax of the language as heads which govern the nature of their complements or as morphologically inert phrases ("adverbs", if you will) which are adjunct to other phrases.

In English, the morphology of clausal negation is mixed, realized either as an inflectional suffix on a restricted class of auxiliary verbs or as a phrasal adjunct to the complement of the finite head of the clause. The former structure is illustrated by the examples in (3); the latter by those in (4).

(3) a. Mary hasn't finished her supper.
b. John won't eat his broccoli.
c. Susan isn't interested in semantics.
d.*I amn't interested in phrenology.
e. Children don't like brussels sprouts.

(4) a. Mary has not finished her supper.
b. John will not eat his broccoli.

c. Susan is not interested in semantics.
d. I am not interested in phrenology.
e. Children do not like brussels sprouts.

These two English paradigms illustrate the involvement of arbitrary grammatical constraints in two ways. The inflectional paradigms of the auxiliary verbs in English are morphologically irregular; the negative forms of the auxiliaries *will* and *do* are not phonologically predictable and a century or so of prescriptivist effort has created the arbitrary gap in the inflectional paradigm of the first person singular copula for the standard dialect.

The requirement that the phrasal *not* be adjoined to the complement of the finite head of the clause interacts with the split between auxiliary and lexical verbs in English to force the occurrence of so-called 'empty *do*' in the (e) examples. An auxiliary verb whose motivation is structural and not semantic in any way.

I take us through these grammatical details in order to make the following point: these constraints are enforced on English clause structure regardless of the pragmatic function of negation in particular utterances of these sentences. Any of the well-formed sentences in (3) and (4) can be used to assert a negative proposition or to deny the prior assertion or implication of the corresponding affirmative proposition.

This uniformity of form across pragmatic function extends even to their use as metalinguistic denials, in the sense of (Horn 1989), building on (Ducrot 1972). As Horn points out, the range of things which can be taken as the reason for uttering a grammatically negative sentence extend to conversational implicata (as in (5a)), choice of lexical form (as in (5b)), or even pronunciation.

(5) a. Chris didn't manage to solve *some* of the problems — he managed to solve *all* of them.
b. I didn't manage to trap two mongeese — I managed to trap two mongooses.

I will follow Horn in recognizing the pragmatic ambiguity of the negative marking in a clause and will distinguish between this metalinguistic interpretation of negative sentences and the descriptive interpretation where the morphology indicates the polarity of the clause. The fact that metalinguistic uses of negation are, in Horn's terms, "in the clause, but not of it" (p. 397) is seen by the way in which it interacts with so-called polarity items. Metalinguistic uses of negation treat the sentence that they are a part of as affirmative, allowing affirmative polarity items as illustrated by the examples in (6), where (a) and (b) must be used as a denial of the previous assertion that the Red Sox have already clinched the pennant or that Sue would rather be living in Chicago:

(6) a. The Red Sox have**n't** *already* clinched the pennant.
b. Sue *would***n't** *rather* be living in Chicago

As a consequence, metalinguistic negation fails to license negative polarity items, as illustrated by the lack of a metalinguistic denial reading of the sentences in (7):

(7) a. The Red Sox have**n't** clinched the pennant *yet.*
b. The train does**n't** *ever* pick up *any* passengers in Benicia.

The negative concord constructions which are my concern in the rest of this paper are constructions of descriptive negation and the arguments and analyses pursued here are intended to be limited in application to the account of clausal descriptive negation, where there are enough issues to keep us busy.

The second grammatical constraint of relevance in our discussion of negative concord is the requirement in many languages that every clause have a visible subject argument phrase. This requirement is enforced even at the cost of creation of meaningless expletive pronouns as subjects, as in weather verbs and existential constructions and extraposed structures, as in (8):

(8) a. It is raining in Berlin.
b. There are patients in the waiting room.
c. It is obvious that some clause subjects are meaningless.

It is this grammatical requirement which has fed the centuries long discussion about the relation between the semantic notion of subject and predicate and the syntactic notion. The syntactic requirement that clauses have subjects regardless of their semantic needs creates constructions like that in (9), which are termed 'raising' constructions in generative analyses because their interpretation must allow for the subject to be interpreted as if it appeared as the subject of the embedded infinitival clause, opaque with respect to the matrix verb and modulo differences in mood, equivalent to the propositions expressed by the sentences in (1).

(9) a. Two students appear to be absent from class.
b. A unicorn seems to be approaching.

(10) a. It appears that two students are absent from class.

b. It seems that a unicorn is approaching.

The near synonymy between the sentences of (9) and (10) supports an analysis of the sentences in (9) which assumes that the apparent subject of the verb *seem* bears no semantic role as argument to that predicate but rather saturates the embedded predicate only. The occurrence of the phrase *two students* as subject in (9a) is seen as a consequence of semantically irrelevant constraints on case marking and clausal structure, so that the principles of composition must be allowed to ignore its surface position and treat it as interpreted as an internal subject.

The relevance of this constraint for our purposes is that contrasts between internal and external interpretations of subjects can be seen even in monoclausal negative sentences. In such cases, the internal interpretation of the subject is one which treats it only as saturating an argument in the predicate. A prime instance of this is with negation, where this distinction is generally discussed as a scope difference between the subject and negation. Consider the sentences in (11):

(11) a. Some papers were not read on Monday.
b. Three students didn't attend the lecture.
c. Many people weren't interested in the topic.
d. More Americans aren't studying foreign languages.

Based upon our understandings of sentences like these, we might conclude that the subject position is consistently interpreted externally, having wider scope than negation. Certainly the (a) and (d) sentences lack the readings that would be expressed by the sentences in (12), where the syntactic form of the clause reinforces an internal interpretation of the subject.[1]

(12) a. There weren't any papers read on Monday.
b. There weren't three students who attended the lecture.
c. There weren't many people interested in the topic.
d. There aren't more Americans studying foreign languages.

However there are also monoclausal sentences in which the subject seems

[1]The availability of an internal interpretation for the subject phrase in these cases is determined by the status of some of these phrases as affirmative polarity items and the properties of the predicate. The possibility of such a construal even in these cases is reinforced by the fact that in certain embedded contexts ('double negative' contexts, in the sense of (Baker 1970)) the scope relation in these cases is reversed. Cf. the judgements on embedded these examples in a matrix like *It's a shame that* ... or *I'm not saying that*

readily or even obligatorily internally construed:

(13) a. A train didn't arrive for an hour.
b. Three boxes weren't sufficient to hold all the books.
c. Every student isn't interested in semantics.

Though the difference between the interpretations of (12) and (13) is generally talked about in terms of the scope of negation and sometimes these cases are taken to distinguish between and internal and an external negation, recent work in syntactic theory has in my opinion supported the point of view taken here: that these differences are differences in subject construal and not differences in how negation is to be interpreted. Where subject phrases are internally construed, interpreted only as saturating a predicate's argument position, the content of the subject phrase will be in the scope of clausal negation. Where the subject phrases are externally construed, where they both saturate a predicate and satisfy some other semantic or discourse-based pragmatic function, their content will appear to have scope over negation.

The purpose of this initial section is to prepare the reader for the possibility that the interface between visible syntactic structure and semantic interpretation contains a good deal of freedom, and that though phrasal structure is an excellent guide to certain aspects of composition, natural languages obscure certain semantic generalizations in responding to demands for structural well-formedness which are not sensible semantically. Though this is not a property only of negation, it is certainly a core consideration when we turn to considering the issues posed for the interpretation of negative concord.

2 Negative Concord

We turn now to an examination of the phenomenon of negative concord and the problem it poses for compositional interpretation. Negative concord is illustrated in the sentences below from (nonstandard) English, Italian, and Catalan respectively.

(14) a. **Nobody** said **nothing (to nobody)**
Nobody said anything (to anybody).
b. John did**n't** tell **nobody** (**nothing**)
John didn't tell anybody (anything).

(15) Mario **non** ha parlato con **nessuno** (di **niente**)
Mario neg has spoken with nobody (about nothing)
Mario hasn't spoken with anybody (about anything).

(16) **No** m'ha telefonat **ningú**
neg me-has telephoned nobody
Nobody has telephoned me.

(17) Maria did**n't** say **nothing** (to **nobody**).
Maria didn't say anything (to anybody).

In each of these clauses, there are at least two items (indicated in bold-face) which could be considered expressors of negation, in that sentences can be found which clearly express negation but in which the only source of the negation is a single occurrence of one of the bold-faced items. However they are all interpreted as containing a single negation. In negative concord clauses, there seem to be multiple occurrences of expressions of negation, though their interpretation requires us to treat these occurrences as in some sense 'idempotent', together expressing only a since negation.

One attitude toward negative concord is alluded to by van der Wouden and Zwarts (1993) when they say (p. 1) that "it is not uncommon in natural languages that negation seems to behave in an illogical manner." What assumptions do we make that would lead us to say of such a common natural language construction that it is 'illogical'? That is the question to be investigated in some detail now. The sense of 'illogicality' in their statement is not unsystematicity, as their analysis makes clear; rather it is the fact that unlike formal logical languages, predictions of logical equivalence cannot be read off the syntax of these sentences. Rather, negative concord is one of those constructions that the syntax semantics must be able to 'see through' in order to give a sensible account of the meaning.

But first some structural notes. The apparent expressors of negation in the sentences above fall into two groups: those which are associated with the clause head (*didn't* in English, *non* in Italian, *no* in Catalan, and those which are argument expressions: *nobody, nothing, niente, nessuno, ningú*. These two classes of negative expressions are the basis for stating several possible parameters of variation between languages. To use the terminology of van der Wouden and Zwarts (attributed by them to den Besten), (14a) shows only negative spreading because all of the indefinite argument phrases in the clause are marked as negative, but there are no clause-head markers of negation in it. By contrast in (14b) we see both negative spreading across the direct object and object of *to*, but also negative doubling, because the head of the clause is marked as negation as well. The same is true of (c) from Italian, but (d) in their terms shows only negative doubling, as the *no* seems to be doubling the negative subject.

The distinction between doubling and spreading, between marking by clause-head negative elements or in argument positions, is important in understanding the details of structural variation between languages, but I will continue to use the cover term negative concord here to concentrate on the central semantic puzzle posed by the interpretation of these clauses.

That puzzle arises from very general assumptions about assignment of lexical interpretations to basic expressions and composition principles. We can see the 'illogicality' of negative concord arise very quickly in considering how to interpret the sentences in (18) in the variety of negative concord English on which (c) is synonymous with (b).

(18) a. Mary didn't talk to John.
b. Nobody talked to John.
c. Nobody didn't talk to John.

In interpreting (18a) we find that *didn't* must be assigned a meaning which is sufficient to express negation, i.e. to make the clause express the proposition that is the contradiction of the proposition expressed by the claim that Mary talked to John. This can be accomplished by assuming that the subject of (18a) is interpreted internally as saturating the predicate *talked to John* to form the latter proposition and treating *didn't* as providing both a tense and a negation propositional operator to yield a final proposition expressed. (That is, we assign the negative element the Fregean negation 'opposite of'; see below.)

By a similar line of reasoning, we want (18b) to express the contradiction of the claim that somebody talked to John. This can be done in one of two ways. We may preserve the unity of the subject phrase *nobody* in the semantic composition by treating it as a quantifier (i.e. the generalized quantifier denoted by the intensional logic expression $\lambda P[\neg\exists x[person'(x) \wedge Px]]$). Or we can decompose the phrase into two element units in the composition: an indefinite existential quantifier to serve as the saturator of the predicate and the aforementioned propositional negator. In the latter case, we must also enforce on the composition in some way the fact that the indefinite must be interpreted within the scope of the negation. However in either way, we derive the correct prediction of the truth-conditions of the sentence.

The compositionality problem posed by negative concord is that by the time we have made these assumptions we already predict that (18c) should entail that everyone talked to John,[2] and this is not what the sentence would mean

[2]or that somebody talked to John, depending upon the relative scope given the two negations; this is orthogonal to my concern.

in the relevant dialect.[3]

The essential problem is highlighted by (14a), where only 'spreading' has occurred. If we assume that *nobody* is capable of expressing negation (i.e. of adding to the composition an 'opposite of' operation), then we do not understand why both occurrences of that term do not express negation in (a). The puzzle has arisen because we standardly make the two assumption in (19) about semantic interpretation of these elements.

(19) a. Morphologically negative expressions are assigned lexical denotations which 'express negation' by contributing a negation operator to the composition of the content of the sentences in which they occur.

b. The assignment of lexical denotations is to expressions qua linguistic types, not tokens.

The first is made in identifying *nobody* with the given generalized quantifier or compound of a propositional operator and an indefinite existential and *didn't* with the propositional operator. It seems clear that to provide a compositional account of the interpretation of these clauses, some way must be found around this result.

If negative concord is going to cease appearing 'illogical', which I believe it must, then one of the two assumptions must go. Some way must be found of saying that negative expressions sometimes do not express negation.

The most common reaction to this type of puzzle in generative linguistic analysis is to assume that the structure which guides composition is removed from the visible segmentation and 'surface' clause structure by a transformational derivation. The classic version of such an account for this data is the approach of Klima (1964), in which a single abstract negative formative is postulated in the interpreted structure and the apparent multiplicity of expressors of negation is the result of morphological agreement among terms in the clause. (Hence the name *negative concord.*) More modern accounts, such as that of Haegeman and Zanuttini (cf. Haegeman 1995) take a similar approach by gathering up instances of morphological negatives into single constituents in 'logical form'. I do not intend to argue against this strategy here; indeed the analysis that I will base my discussion on (Ladusaw 1992) can be viewed as a particularly conservative version of such an analysis as

[3]The same issue will arise if we consider negative indefinites in the verb phrases of all of the negative concord dialects illustrated above. The restriction to subject position negative indefinites is not crucial to the concern but highlights the problem in simple sentences.

well. But even if this strategy is taken, the assumptions in (19) are useful in classifying the resulting strategies.

Of the two assumptions in (19), the second one seems to me to be a fairly deep assumption about the relation between lexicon and phrasal syntax, and between the arbitrariness of lexical meaning and the compositionality of phrasal interpretation. If meaning assignment to basic expressions of a language can be relativized to structural facts about the position of the token of the expression in the structure being interpreted, then many unnatural possibilities become available. (For example, one could create verbs which express one relation when they are in embedded clauses but a different relation when they are in a root clause.)

Hence I will assume that it is the first of these two assumptions which must be abandoned. Van der Wouden and Zwarts (1993) explore the abandonment of the second approach while maintaining as much as possible the first assumption. On their account, negative spread is to be accounted for by assuming that negative argument expressions are assigned either negative or positive interpretations, based upon the semantic properties of its embedding context. Under their system, the occurrences of one negation-expressing argument expression, say in the preverbal subject position, will create the semantic context which will require all of the other occurrences of these expressions to be interpreted by complementary, nonnegation expressor interpretations. By doing this, they can continue to assume that one of the obviously negative expressions in the clause does have a lexical interpretation which is recognizably negative (e.g. antiadditive, monotone decreasing, etc.).

In (Ladusaw 1992) I have sought to defend the other approach, keeping lexical interpretation assignment context free but abandoning the assumption that in negative concord clauses[4] the negation is directly expressed in the lexical interpretation of one of the negative phrases. This approach leads to an analysis in which semantic interpretation must be applied to the language in a more abstract way than in the van der Wouden and Zwarts account.

But one of the goals of my prolegomenon was to argue that there are independent reasons for recognizing a certain abstractness in the 'logical forms' which are the basis of interpretation of natural language clauses. I will sketch that account here and follow out the argument based upon it. In doing so I do not argue directly against the van der Wouden and Zwarts account because I believe the difference between the two approaches are better seen as whether the variation between languages should be parameterized in

[4]and implicitly, in all clauses including those in nonnegative concord languages.

terms of lexical meaning assignment or parochial structural constraints on the syntactic licensing of negative terms.

In brief, my approach to the interpretation of negative concord is to analogize — but not reduce — the negative argument terms to the negative polarity indefinites. Negative concord dialects of English are helpful in showing that semantically there seems to be no difference between the apparent negation expressing terms like *nobody* and the negative polarity terms like *anybody* which can never seem to be expressors of negation. (This difference is made clear in that negative terms — *n-words*, to adopt the terminology of (Laka 1990) — may be single-word negative answers to constituent questions but negative polarity indefinites cannot be.)

Negative polarity indefinite terms like *anything* on my analysis are treated as non-quantificational indefinite terms, in the sense of (Heim 1982) where their content is treated as restricting values for an individual parameter but not introducing quantificational force. The quantificational force of an indefinite term is always inherited from some other operator, either by direct binding or by existential closure. In my discussion I will occasionally unify these two ways of closing an indefinite parameter by saying that a parameter is *roofed* by its binder or the item which triggers existential closure of the domain containing the indefinite.

Negative polarity indefinites (NPIs) are distinguished in being items of limited distribution, items which are subject to licensing conditions on their well-formed occurrence. For NPIs, there are two sorts of licensing conditions: a semantic condition on its roof and a structural requirement on its occurrence structurally inferior to an appropriately negative expression.

(20) A Negative Polarity indefinite must be both (i) roofed (i.e. existentially closed) by an appropriately negative operator in interpretation and (ii) c-commanded by an appropriate license in clause structure.

The analogy between these NPI indefinites and the so-called n-words involved in negative concord leads to the assumptions that n-words are interpreted as restricted parameters which are subject to similar semantic licensing conditions in being existentially closed in the scope of a negation, but they are not subject to the structural condition. Hence they may occur anywhere that they will be properly licensed semantically. This approach assumes that the constraints on the occurrence of these expressions in clauses arise from factors irreducible, or rather not yet reduced, to the analysis of their truth-conditional contributions. They share at least the property of being allowed only in the scope of a negation.[5]

[5]Though they may well differ in the strength of negation required to license them.

N-words and NPI indefinites are distinguished by the syntax, however. The latter are subject to a structural licensing condition as well. The difference between the presence or absence of this syntactic licensing requirement is what distinguishes the sentences below.

(21) **Anybody* did**n't** say *anything.*

(22) *Nobody* did**n't** say *nothing.*

The ungrammaticality of (21) is due to the fact that the subject NPI is not properly c-commanded by the licensing negation in the head of the clause. In (22), the subject n-word is properly licensed when interpreted as an indefinite in the scope of the negation introduced by the inflectional morphology of the auxiliary. Since it is not also subject to the structural requirement, the sentence is well-formed.

For this theoretical framework to provide an analysis of negative concord clauses in most languages, however, a fundamental question about clauses like (23) must be addressed: How are clauses with no visible clause-head marker of negation interpeted as negative?

(23) Nobody said nothing to nobody.

If argument terms do not introduce negation in the composition of the proposition expressed directly, as I am assuming, what does introduce it when there are only n-word terms and no clause-head marker?

The answer I propose is that in place of the requirement that n-words be licensed syntactically, they are syntactic licensors of an abstract negative marker associated with the head of the clause. So the derivation of an interpretation for such clauses proceeds as follows. As it stands, the clause is illformed because there is no negation to license the n-word terms. If the additional assumption is made that the head of the clause is interpreted as negative, then the clause would have a well-formed meaning. Allowing n-words the ability to license the postulation of a negative clause head completes the picture, albeit at the cost of a very fine distinction. That distinction is between whether the n-word terms directly introduce negation or only indirectly introduce negation through this licensing property.

One reason to consider this distinction plausible is that it would explain the observation of Zanuttini (1991) that the preponderance of negative concord languages show the following property: the clause is well-formed only if some negative expression governs the head of the clause. That is, in all of the negative concord under discussion here, the clauses of the form (24) are illformed.

(24) *John said nothing to nobody. (on a negative concord reading)

This is to be expected on the strict interpretation of the structural licensing requirement. In (24), neither of the two n-words govern the head of the clause and so the postulation of an abstract negative head is not allowed.[6]

The issue of how to interpret negative concord then, on the view expressed here, is one which bears heavily on assumptions about the relation between syntactic structure and the compositional formulation of propositional content. Every account must provide a way of distinguishing the n-words from non-licensing negative polarity items and also provide an account of why *grosso modo*, n-words seem to be negation expressors only when they are relatively high in the clausal structure.

I do not claim to have made the case for the treatment of n-words as indefinites with licensing properties here. From this point the argument would proceed to examine the utility of indefinites and their binding and closure properties to provide a view of negative concord and its relation to quantification theory. But in this paper I want to concentrate on assumptions about clausal negation.

3 Negation as a Mode of Judgement

In the foregoing discussion, I have used the term 'clausal negation' fairly freely, inviting the reader to import a standard account of what is meant by that, roughly a one-place propositional operator. In the remainder of this paper I want to take the comments about negative concord and use it to justify a different analysis of clausal negation.

Central to the problem posed by negative concord is the need to pick out a structural domain within which there is at most one occurrence of an expression of negation. The apparent idempotence of morphologically negative expressions within a clause in these languages is accounted for in every account by insuring that within a particular syntactic domain, only one negation gets expressed. Van der Wouden and Zwarts for example enforce this by their assumption that the relevant items are always assigned nonnegative interpretations in the scope of a negative element. Absorption accounts like Haegeman and Zanuttini's (cf. Haegeman 1995) gather the idempotent elements together into one logical constitutent before interpretation. Hence there is no way for there to be more than one expressor of negation in the domain within which this context sensitive assignment of meaning is assumed to be operational.

[6]By the same token, negative concord languages like Piedmontese (and perhaps standard English on one view) are problematic for the account.

On the account of (Ladusaw 1992), the uniqueness of negation in a clausal domain is enforced by assuming that negation is not associated directly with the formatives of the clause but rather with an abstract feature of clause structure. If it is the abstract feature which is the expressor of clausal negation, then there can be only one negation per clausal domain. The idempotence follows from stipulations which invite two questions under the assumption that clausal negation is simply the instantiation in the algebra of propositions of a boolean complement operation.

A. Why should n-word terms not express negative quantifiers?

B. Why should negation be associated with the formation of a propositional content for the clause in a way which favors only one negation?

In standard approaches to interpreting formal languages, the denotational algebras have a boolean structure. When applied to natural languages (as in (Keenan/Faltz 1985)), this means that denotations are available for 'negations' of items in every denotation type. The boolean complement operation is a one-place operation within the algebra. As a result it is semantically 'endocentric' in a sense, extending a term from traditional grammatical descriptions. (Endocentric constructions are those which have the same distribution as their heads.) The result of applying the complement operation to an element from a given algebra is another item within the same algebra to which the operation can apply. To assume that negative morphology in natural languages evidences a semantic operation like boolean complement is to assume that there is no semantic reason why negation cannot iterate with the kind of semantic oscillation captured by the logical laws of double negation. It further assumes that negative morphology in any category — term phrase, adjective phrase, verb phrase — which is interpreted in a type that has a boolean structure has a readily available negative interpretation.

The questions A and B raised above seem to be directly related to these two consequences of interpreting negation in natural language as a direct expression of boolean complement. An endocentric, cross-type characterization of natural language negation makes the existence of negative concord appear 'illogical'. I take this as an indication that a competing 'exocentric' account of natural language negation should be explored and adopted.

An exocentric negation operator would be one which could not take its output as its input. That is, construable as a function whose domain is disjoint from its range. This might be either because it is a function of two arguments or else simply because the type of its domain and range are different.

If natural language negation is exocentric, then it would follow that the underlying interpretive ontology would provide a natural semantic place for negation to be expressed. "Below" that point in the interpretation, morphology associated with negation simply wouldn't be interpretable, providing a semantic pressure against clausal constituents with negative interpretations. "Above" that point, other interpretive matter or operations would have to be included in the composition for the negation to iterate, as it could not apply immediately to its output. There is a good fit between the profile an exocentric negation operator and the desiderata for a theory of interpretation of negative concord.

For concreteness, consider the following illustrations of endocentric and exocentric negation operators. Let us assume that the output of a clausal negation function is a proposition.

The standard (contradictory) logical negation would be a function from propositions into propositions. This is the operator which is justified in Frege's discussion of negation as one which for any thought, provides the "opposite thought". This is an endocentric negation. In adopting this operator as the model for natural language negation, we take what Horn (1989) calls an 'asymmetrist' position with respect to the relation between negation and affirmation. Logics with a negation operator need no affirmation operator because by default propositions are considered "affirmative". Of course, as semantic objects (modeled say as functions from world indices to truth values), there is no way of classifying propositions as either affirmative or negative. Affirmative and negative are classifications of formulas or sentences in a language.[7]

By contrast, exocentric negation operators seem to commit one to the "symmetrist" position on the relation between negation and affirmation. An operator which combined a predicate and an individual to form a proposition denying the predicate of the individual would be an exocentric negative operator because its result (a proposition) is not a possible argument for it. However a system which contained such an operator would, like a term logic, require a corresponding affirmative operator, which combined an individual and a predicate to form the proposition which affirmed the predicate of the individual.

Thus it seems to me that the discussion in (Horn 1989) between the term logic approach and the propositional logic approach to negation is very relevant to our consideration of negative concord. On the term logic approach, negation and affirmation are modes of judgement or predication. If we look at negative concord through the assumption that natural language seman-

[7]And hence $\neg\neg p$ must count as negative even though it is logically equivalent to p, an affirmative.

tics should treat negation (and hence affirmation) as a mode of predication, then we can see that questions A and B have natural answers.

Beyond the traditional categorical mode of predication which would relate a predicate to an (individual) subject, it is also necessary (for reasons explored in (Ladusaw 1994)) to postulate a unary mode of predication which combines with descriptions (equivalent for our purposes to predicates on individuals or eventualities) to form propositions. Like categorical predication, this "Thetic" mode of predication can be either affirmative (in which it is equivalent to unrestricted existential quantification) or negative (unrestricted negative existential quantification). This mode of predication is exocentric as well, because its argument, a description, is not identical to its value, a proposition. This is the mode of predication which appears in event sentences in which the subject argument phrase is interpreted within the scope of the the affirmation or negation of the description.

To return to the discussion that began the paper then, we can view the sentences in (25) with new eyes.

(25) a. Mary didn't talk to John.
b. Nobody talked to John.
c. Nobody didn't talk to John.

For (25a), the interpretation can be viewed as containing either categorical or thetic predication. In the categorical predication, the propety of having talk to John is denied (rather than affirmed) of Mary. In the thetic rendering of this proposition, the existence of any instance of the eventuality type of Mary talking to John is denied. In either case, the fact that the mode of predication is negative is based upon the morphological indication on the head of the clause that the mode of predication is negative.

In (25b), however, the head of the clause gives no indication that the mode of predication is negated. Further, the subject n-word is interpreted as an indefinite argument, saturating the relational predicate *talk to*. In this case, the most obvious interpretation of the sentence is as a thetic denial of the existence of any instantiations of the event type of someone talking to John. That the interpretation is negation rather than affirmation is only indirectly expressed, through the fact that the subject term is able to license the introduction of the (thetic) negative mode of judgement to form a proposition.

Finally in (25c), the crucial negative concord case, the reasoning proceeds in precisely the same way, except that now there are multiple indications that the mode of predication is the negative one. Both the morphological marking on the head of the clause and a superior licensing n-word are available to justify the interpretation of the clause with a negative mode of predication. However the semantic assumptions — that the n-word is not inherently a negative quantifier and that the proposition is formulated crucially with the use of a single mode of predication — explain why negative concord should be a natural, logical result. We do not generate two tokens of a negative expressor, but rather the language cooperates in doubly marking the use of a negative mode of predication.

References

Baker, C.L.: Double negatives. *Linguistic Inquiry* 1 (1970), 169–86.

Ducrot, O.: *Dire et ne pas dire.* Paris 1972.

Haegeman, L.: *The syntax of negation.* Cambridge 1995.

Heim, I.: *The semantics of definite and indefinite Noun Phrases.* PhD thesis, University of Massachusetts/Amherst, 1982.

Horn, L.R.: *A Natural History of Negation.* Chicago 1989.

Keenan, E.L. & Faltz, L.M.: *Boolean semantics for natural language.* Dordrecht 1985.

Klima, E.S.: Negation in English. In: *The structure of language*, ed. by J.A. Fodor & J.J. Katz, Englewood Cliffs 1964, 246–323.

Ladusaw, W.A.: Expressing negation. In: *SALT II: Proceedings of the conference on Semantics and Linguistic Theory 2*, ed. by Ch. Barker & D. Dowty, Columbus/Ohio (The Ohio State University) 1992, 237–259.

Ladusaw, W.A.: Thetic and categorical, stage and individual, weak and strong. In: *SALT IV: Proceedings of the conference on Semantics and Linguistic Theory 4.* Ithaca/NY (Cornell University, Dept. of Linguistics) 1994, 220–229.

Laka, I.: *Negation in syntax: On the nature of functional categories and projections.* PhD thesis, MIT 1990.

van der Wouden, T., & Zwarts, F.: A semantic analysis of negative concord. In: *SALT III: Proceedings of the third conference on semantics and linguistic theory*, ed. by U. Lahiri & A.Z. Wyner, Ithaca/NY (Cornell University, Dep. of Modern Languages and Linguistics) 1993, 202–219.

Zanuttini, R.: *Syntactic Properties of Sentential Negation: A Comparative Study of Romance Languages.* PhD thesis, University of Pennsylvania, 1991.

LITOTES AND DOWNWARD MONOTONICITY

Ton van der Wouden

1 Introduction

This paper deals with the linguistic analysis of the rhetorical figure of litotes. In litotes, a positive statement is expressed by means of the not uncommon strategy of negating a negative one. Taking the analysis of Horn (1989, 1991) as a starting point, I will argue that litotes is not restricted to negation per se, as it occurs in other negative-like contexts as well. It will be shown that Horn's analysis of the phenomenon can be augmented and extended in a natural way to also cover these other contexts.

The paper is organized as follows. In section 2, I supply a working definition of litotes, arguing that litotes is not to be confused with related phenomena such as understatement. Next, I summarize the semantico-pragmatic analysis of Horn's in section 3. Section 4 contains the main contribution of this paper, viz., that litotes is not triggered by negation alone, but rather by all downward entailing expressions, at least in principle. Apparent or real counterexamples to this claim are dealt with in section 5. The last section contains some concluding remarks.[1]

2 What is litotes?

Litotes is the rhetorical figure in which an affirmative is expressed by the negative of the contrary (Jespersen 1917, 62 f.). The phenomenon is exemplified below in (1):

(1) a. He's a *not un*happy man

b. He lifted his hat with respect, and *not without* gallantry

c. She does*n't* look *too bad*

---

[1]The paper elaborates on sections of my 1994 dissertation. Earlier versions were presented at the TABU-dag (Groningen, June 25, 1993) and the Workshop on Negation at *Analyomen 2* (Leipzig, September 7–10, 1994). Thanks are due to the audiences at these conferences and to Peter Blok, Jack Hoeksema, Larry Horn, Hotze Rullmann, Víctor Sánchez Valencia and Frans Zwarts for discussion and comments, and to Sharon Parry for correcting my English; none of them is to be held responsible for any error or opinion in this paper. The work reported here is part of a larger project entitled *Reflections of Logical Patterns in Language Structure and Language Use*, which is supported by the Netherlands organization for scientific research (NWO) within the framework of the *PIONIER*-program (PGS 30-329).

In each of the examples, the meaning of the expression containing a double negation is not completely equivalent to the one without negations: sentence (1a) does not precisely mean 'he is a happy man' but is somewhat vague as regards the subject's position on the happiness scale: it may be anywhere between reasonably happy and absolutely ecstatic. A comparable vagueness is found in the other examples: it is left unspecified in (1b), whether the subject is rather gallant or extremely gallant. Finally (1c) can be used to describe both a stunning beauty and an average female person. This type of vagueness seems to be an essential feature of litotes (cf. below).

Note that this usage of the term 'litotes' is not the only one to be found in the literature (Hoffmann 1987; Horn 1991): the term is also used for logical double negations as exemplified below:

(2) a. It is *not impossible* that we will visit you tomorrow
'It is possible that we will visit you tomorrow'

b. You are wrong: she is *not un*married!
'She is married'

Logically, the first example is fully equivalent to 'it is possible', as there exist no degrees of possibilities in standard modal logic — although people may tend to try and interpret the double negation as a litotes construction after all (cf. below). The unmarked interpretation for the second example is as an explicit denial of an (explicit or implicit) statement that the subject might be unmarried (cf. van der Wouden 1994, 2.4).

Moreover, the term 'litotes' is also used as a synonym for 'understatement' or 'meiosis', in which the speaker uses a weaker term than (s)he might without violating the truth, and the listener is aware of this (Berg 1978; Hübler 1983):

(3) a. The performance is *satisfactory*
'superb'

b. He is *rather well off*
'filthy rich'

In this paper, however, I will restrict my attention to cases where a negative term occurs in a negative context (to be defined below) and the resulting meaning, although affirmative, is not (necessarily) completely equivalent to the expression without negation.[2]

[2]As Larry Horn pointed out to me, the phenomenon of (so-called) Neg-Raising — as in *I don't want to eat it* — might be subsumed under the general heading of litotes, at least to some extent. This suggestion will be pursued at another occasion.

3 Horn's analysis of litotes

Horn's analysis has two parts, a semantic one and a pragmatic one. These parts will be dealt with in the following sections.

3.1 The semantic part

The semantic part of Horn's explanation of litotes crucially involves the distinction between contradictory and contrary opposition, which dates back as far as Aristotle. Schematically, the difference may be depicted as follows:

	contradictory opposition		contrary opposition		
	not G	not F	not G		
				not F	
(4)	F	G	F		G
	white	non-white	white	neither F nor G	black
	odd	even	poor		rich
	male	female	happy		unhappy

This schema should be read as follows: the contradictory opposites *odd* and *even* cannot be both true of an individual, nor can they be false at the same time of such an individual. The contrary opposites *white* and *black*, however, can again not both be true of an individual, but they can both be false of such an individual — for instance, if it is red or blue. In other words (Horn 1989, 1991):

(5) a. contradictory opposites are mutually exhaustive as well as mutually inconsistent

b. contrary opposites do not mutually exhaust there domain

And in yet other words we can say the following — again following Horn very closely:

(6) a. contradictory opposition is governed by the Law of Contradiction (LC) and the Law of Excluded Middle (LEM)

b. contrary opposition is governed by LC but not by LEM.

(7) a. LC: for any x in the relevant domain, $\neg(Fx \,\&\, Gx)$

b. LEM: for any x in the relevant domain, $(Fx \vee Gx)$

(8) a. contradictory (contrary (P)) $\neq$ P

b. contradictory (contradictory (P)) $=$ P

That is to say, in all cases *not P* refers to everything that is not P. In the case of a contradictory opposition such as *odd–even*, everything that is *not even* is *odd*, and everything that is *not odd* is of course *even*. In the case of a contrary opposition such as *poor–rich*, however, someone who is *not poor* is not necessarily *rich*, and someone who is *not rich* doesn't have to be *poor*: (s)he may very well be somewhere between these two extremes. Thus, the intersection of *not even* and *not odd* necessarily denotes the empty set, whereas the intersection of *not rich* and *not poor* may have a considerable number of members. And this difference explains the contrast between the following two sentences. The first one is a contradiction, or necessary false, and therefore the continuation *it's somewhere in between* is odd (to say the least); the second, however, may be true or false; the continuation *they're somewhere in between* only makes the reference to the grey area between the two extremes explicit.

(9) a. This number is neither odd nor even (*it's somewhere in between)

b. These people are neither rich nor poor (they're somewhere in between)

Only in the case of (nonclassical) contrary negation can we get litotes readings, which explains why *not impossible* and *not unmarried* cannot get a litotes reading in sentences like (2).[3]

3.2 The pragmatic part

The pragmatic part of Horn's explanation of litotes is explained by means of a variation on Grice's maxims (Grice, 1989):

(10) **Division of Pragmatic Labor** The use of a longer, marked expression in lieu of a shorter expression involving less effort on the part of the speaker tends to signal that the speaker was not in a position to employ the simpler version felicitously. (Horn 1991)

Let me explain the interplay semantics pragmatics in litotes first, before adding complications. Assume that a speaker utters sentence (11):

(11) It is not unwise to take precautions

Upon hearing this utterance, the cooperative listener will reason as follows. Literally, the speaker says that it is 'not not-wise' to take precautions. Logically, this is equivalent to saying that it is wise to take precautions. But

[3]Compare, however, the next section.

the speaker doesn't say this, so, according to (10), and assuming he is reasonable and cooperative as well, he'll have his reasons for saying what he wants to say in this roundabout way. So probably he doesn't want to claim that it is wise to take precautions. Evidently, he doesn't want to claim that it is unwise to take precautions either, given the fact that that statement is explicitly denied. Presumably, then, the speaker wants to refer to the grey zone (Sapir 1944), somewhere between *wise* and *unwise*, comparable to expressions such as 'it is rather wise to take precautions', 'it is pretty wise to take precautions', 'it is reasonably wise to take precautions', 'it is neither wise nor unwise to take precautions', etc.

The following picture may help to see what is going on (the dotted areas are not available for interpretation):

(12)

<table>
<tr><td>a</td><td>unwise</td><td>..........</td><td>wise</td></tr>
<tr><td>b</td><td colspan="2">not wise</td><td>....</td></tr>
<tr><td>c</td><td>......</td><td colspan="2">not unwise</td></tr>
<tr><td>d</td><td>......</td><td>not unwise</td><td>....</td></tr>
</table>

The a-row depicts a scale of wisdom, going from unwise on the left to wise on the other extreme. There is an area in between the two extremes where neither *wise* nor *unwise* apply (of course, there is no sharp boundary between the extremes and the middle part). The b-row shows the logical denotation of *not wise*: it covers all of the scale that is not covered by *wise*. The c-row is a picture of the logical denotation of *not unwise*: that expression covers all of the scale that is not covered by *unwise*. The d-row, finally, depicts the pragmatic possibilities of *not unwise*: the principle in (10) restricts the usage of the expression to the middle area, the grey zone in between the two extremes.

Do not think that this approach predicts that the meaning of *not wise* is either the same as that of *not unwise*, or of *wise*. To see this, note that *not unwise* refers only to the middle area in the picture above and *unwise* only refers to the left column, whereas *not wise* covers the two columns to the left, i.e., the complete area not covered by *wise*.

From this analysis, it follows that negated nongradable or absolute predicates cannot function as litotes, as no middle area is available between the predicate and its opposite. Consider the following cases:

(13) a. ?A *not unmarried* woman entered the room

b. That girl is *not unmarried*

c. It is *not impossible* that I will attend the meeting

The oddity of example (13a) is caused by the fact that a person is either married or unmarried, and that there is no grey zone in between the two for the litotetic construction to allude to. Hence the sentence forms an unfelicitous utterance. The following picture may help:

(14)

married	unmarried
not unmarried	not married

The truth conditions of (13b), already discussed above, are the same as that of the simple expression *that girl is married*. The only difference between the doubly negated variant and the simple one is that (13b) explicitly denies expectations or presupposition of the opposite. For these purposes the sentence is fine, but not for conveying a meaning along the lines of 'that girl is somewhere in the grey area between married and not married' or 'that girl is lightly married', i.e., a litotetic meaning, since such a grey area does not exist. In terms of usage possibilities, this means that this statement can be used felicitously as a case of denial (van der Wouden 1994), i.e., to react to a certain statement:

(15) You are wrong: that girl is *not unmarried*!

In order to be able to make sense of (13c) the listener must (and usually will) construct a nonlogical scale of possibilities on which 'not impossible' is somewhat less possible than 'possible'. That is to say, *not impossible* doesn't have its logical meaning ($\Diamond$) here but rather refers to a certain degree of (im)probability.

3.3 On the meaning of litotetic expressions

There is some uncertainty about the exact meaning of litotes. The rhetorical tradition (e.g. (Erasmus 1512), (Lausberg 1973)), states that the litotes construction is strongly positive, whereas the linguistic tradition (Jespersen 1917; Bolinger 1972; Horn 1991) tells us that the doubly negated expression is somewhat weaker than the straightforwardly positive one. The analysis of litotes that is adopted here points in the direction of the correctness of the latter option. The truth-functional meaning of a lexical item such as *unwise* is vague: it covers the area between *rather wise* and *extremely wise*. The independent mechanism of understatement (Berg 1978; Hübler 1983) that was already mentioned must be held responsible for the fact that litotes constructions may be used occasionally to express strong positive statements. Thus, via understatement, weakly positive expressions such as *nice* and *not bad* can be used to express a very positive attitude of the speaker: 'extremely beautiful' or something along these lines.

To get a clear view on the way litotes and understatement interact, consider the following (Dutch) example:

(16) Het is niet niks
It is not nothing

Truthfuctionally, this sentence is either equivalent to 'it is something' or 'it is everything'. Via Horn's principle (10) these are not the readings we get, as there are simpler ways to express them. The readings we do get are 'it is at least something' and 'it is quite a lot'. Analogously to the case of *not impossible* in (13c), *niks* 'nothing' is not read as member of any of the binary oppositions *everything – nothing* or *something – nothing*, but rather as the lower end point of a scale such as the following:

(17)

nothing	something	a lot
	not nothing	

In other words, the interpretation of *niet niks* 'not nothing' is vague as regards the exact location to the right of 'nothing'. We typically get the 'something' reading in a context such as the following: My friend desperatately needs 1000 dollar by tomorrow. I offer to help him with a $ 100 bill. He then can say *Het is niet niks, alle beetjes helpen, dankjewel* ('it is not nothing, all small bits help, thank you'). The other reading, 'a lot' we get, for example, in the following situation. I am being interviewed just after having finished my first marathon. Gasping for air, I say, with tongue in cheek, i.e., with a lot of understatement: *Het is niet niks, zo'n marathon* 'it is not nothing, such a marathon', before I collapse.

4 More than negation

It is often assumed that litotes is restricted to negative particles such as *not*. This position is also taken in a recent monograph on Latin litotes (Hoffmann 1987). The relevant part of Hoffmann's view on litotes is given below (Hoffmann 1987, 216):

(18) **Hypothesis** (Hoffmann): In litotes, the process of negation is effected by NEG particles only

According to this hypothesis, the second of the following construction types is excluded from the discussion: only the first one is supposed to be an instance of litotes:

(19) a. I do*n't deny* she is right — litotes
'Of course she is right'

b. *Nobody denies* she is right — no litotes
'Of course she is right according to everyone'

The restriction to NEG is unsatisfactory for several reasons. The first problem is already noted by Hoffmann herself (Hoffmann 1987, 229 n. 41):

> For [...] expressions in other languages adaptations will be necessary. In e.g. English, Dutch and German, litotes expressions are used that contain zero-quantifiers, for instance: *He is no fool*, *Dat is geen gek idee* ('That's no foolish idea'), *Er ist kein Tor* ('He is no fool'). In Latin, expressions of the form **nullus stultus est* are ungrammatical if used as [litotes] expressions.

Although Latin expressions of the form **nullus stultus est* are ungrammatical if used as litotes expressions, constructions of the type *nemo negat* ('nobody denies') are perfectly well-formed (Erasmus 1512). They are, however, still excluded from the discussion. The following table depicts this ad hoc and unsatisfactory division between what counts as litotes and what not:

(20)

(Hoffmann 1987)	
litotes	no litotes
He is *not* a *bad* guy	He is *no fool*
Non nego	*Nemo negat*
I do*n't deny*	*Nobody denies*

The second problem is that negative adverbs such as *never, nowhere* and *not at all* show the same effect, just like the negative conjunct *neither*; these are also left out from Hoffmann's discussion:[4]

(21) a. I *never denied* his claim.

b. Of course, some acts may be *neither unjust nor immoral*

c. A fact *nowhere* conceded in this series

To see that we are indeed dealing with litotes, note that the possible readings of (21a) cover the whole range from not explicitly contradicting the claim to strongly affirming it, i.e., the vagueness we met in earlier cases of litotes is present here as well. The same holds for example (21b): the sentence denies that the acts are unjust or immoral, but whether they are very just and completely moral or rather somewhere between the extremes is left open.

[4]Many examples in this paper, including the ones below, were not made-up for the present discussion but found in actual texts. Most are from Jack Hoeksema's text corpora.

A comparable story holds for (21c): according to this sentence, the fact is never admitted in this sentence, but neither is it denied expressly.

A third argument against restricting discussion of litotes to negation *not* and *n't* is constituted by the fact that weak negatives such as *seldom*, *hardly* and Dutch *weinig* 'little, few' may also trigger the effect of litotes, with the by now well-known vagueness in meaning:

(22) a. I recognize that this is *scarcely tightly* formulated
'This is rather/very loosely formulated'

b. The weather was *seldom uncomfortable*
'The weather was pretty/very comfortable most of the time'

c. This is *scarcely little less* than infanticide
'This is almost the same as/worse than infanticide'

d. Met deze soep is *weinig mis* (Dutch)
With this soup is little wrong
'This is (extremely/rather) nice soup'

A fourth argument against this approach needs some background. An analysis in terms of an abstract negative operator NEG used to be quite popular, among other things for the explanation of other natural language phenomena that seem to involve negation. For example, this is essentially the type of analysis Klima (1964) and Baker (1970) propose for the distribution of so-called negative polarity items (NPIs). Such NPIs, for example the indefinite *any*, verbal idioms such as *have a hope in hell* and 'minimizers' (Bolinger 1972) like *a red penny*, only occur in the scope of negation (and some other things, cf. below and (McCawley 1988, 562)). Consider the following examples:

(23) a. *John drank *any* beer tonight

b. John didn't drink *any* beer tonight

(24) a. *Victor has *a hope in hell* of solving this puzzle

b. Victor doesn't have *a hope in hell* of solving this puzzle

(25) a. *Sue wanted to *give a red penny* to help the refugees

b. Sue didn't want to *give a red penny* to help the refugees

Sentence negation *not* or *n't* is not the only possible licenser of NPIs, as the sentences below demonstrate:

(26) a. *Nowhere* did John drink any beer

b. *Neither* author was a scientist, *nor* had they made any pretense of investigating
c. Victor *never* had a hope in hell of solving this puzzle
d. Sue *scarcely* wanted to give a red penny to help the refugees

One might suggest that elements such as *never*, *nowhere* and *scarcely* contain a negative particle after all, as they are more or less equivalent in meaning to *always not*, *everywhere not* and *not quite*, respectively. This, however, may easily lead to circular reasoning. Moreover, this analysis becomes more problematic in the case of certain other contexts that license NPIs, such as comparatives, conditionals, and relatives depending on universal quantifiers:

(27) a. John drank *more* beer *than* any of his friends
b. *If* I'd had a hope in hell of solving this puzzle, I would have continued
c. *Everyone who* gives a red penny to help the refugees deserves a medal

Rather than trying to force all these cases into the Procrustean bed of an analysis involving underlying negations, Ladusaw (1979) was the first one to point out that all contexts that license NPIs possess the logical property of being downward entailing.[5]

(28) **Definition** An expression δ is downward-entailing if and only if $\forall X \forall Y (X \subseteq Y) \rightarrow (\delta'(Y) \subseteq \delta'(X))$.[6]

Ladusaw hypothesized that this property is crucial in the distribution of negative polarity items. In other words (Ladusaw 1979, 113):

(29) δ is a trigger for NPIs if and only if δ is downward-entailing.

This generalization – let me call it Ladusaw's hypothesis – is attractive for its elegance. Although not completely unproblematic (Linebarger 1980; von Bergen/von Bergen 1993; Progovac 1994), it has defined a very fruitful research paradigm (Zwarts 1981; Kas 1993; Landman 1993; Sánchez Valencia *et al.* 1994; van der Wouden 1994; Israel 1994).

Downward entailing contexts allow one to reason from sets to subsets. The validity of the following reasoning shows that the prototypical trigger of NPIs, sentence negation, is downward entailing:

[5] Ladusaw built on earlier work by Fauconnier (e.g. his (1975)). In the literature the crucial property is also known as 'downward monotonic', 'monotone decreasing' (Hoeksema 1983) and 'antitone' (Dunn 1993).

[6] δ' denotes the interpretation of δ.

(30) John doesn't eat vegetables
$[\![\text{spinach}]\!] \subseteq [\![\text{vegetables}]\!]$

John doesn't eat spinach

Application of this test shows that the adverb *scarcely* denotes a DE function as well (in the other contexts, an analogous test applies):

(31) John scarcely eats vegetables
$[\![\text{spinach}]\!] \subseteq [\![\text{vegetables}]\!]$

John scarcely eats spinach

Now the time has come to give my fourth argument against an analysis of litotes in terms of a negative particle: given that the same contexts may license negative polarity items and give rise to litotes readings, and given that downward monotonicity has been used to explain other phenomena that seem to be governed by negation as well (van der Wouden/Zwarts 1993), let us assume that the same logical principles govern both phenomena. Ladusaw's hypothesis in terms of downward monotonicity is quite successful in explaining a lot of the intricacies connected with natural language negation (van der Wouden 1994). Why not investigate whether the same approach works for litotes as well? That is to say, let me pursue the possibility that litotes may in principle occur in all types of negative contexts. In other words, I will assume that the hypothesis in (18) be replaced by the following:

(32) **Hypothesis** Downward monotonicity triggers litotes

If this is anywhere near correct, I have to explain how Horn's analysis of litotes may be revised and expanded in order to cope with the monotone decreasing contexts that trigger litotes. Before doing so, however, I will further investigate the parallels between contexts that license polarity items and those that may function as negatives for litotes.

4.1 Other negative contexts

The hypothesis in (32) predicts that litotes may occur in all downward entailing contexts. This is not completely born out by the facts: although litotes certainly occurs in many monotone decreasing contexts, it doesn't show up in all. Two options are possible here: either one takes the point of view that litotes occurs with negation only — but then one needs a story about the cases of litotes in other monotone decreasing contexts — or that litotes may, in principle, occur in all monotone decreasing contexts — and then one needs to explain why it doesn't always do so. I will defend the latter option.

In order to stress the parallels between polarity phenomena and litotes, I will now discuss various contexts where negative polarity items may show up, whereas many positive polarity items do not occur there. Litotes occurs in these contexts as well. It is a well-known fact that not all NPIs have the same distribution (Klima 1964; Zwarts 1981; van der Wouden 1994): there is not only variation in the type of negative context the elements are sensitive to, there are additional restrictions that must be held responsible for the fact that certain NPIs do not collocate with all downward entailing expressions of the appropriate type.

It will turn out that a comparable situation exists in the case of litotes. Not all negative predicates give rise to a litotetic meaning when they are combined with just any MD operator. An account will be given for several of such cases.[7]

4.1.1 Conditionals

Conditionals are well-known places for negative polarity items. Not all predicates, however, give rise to litotes in this context. Let me illustrate this with the adjective *onverdienstelijk* 'unmeritorious', that occurs in litotes constructions only (cf. the large *Woordenboek der Nederlandsche Taal* or *WNT* (1882–)). The following examples show that *onverdienstelijk* is a negative polarity item, but that it yields ungrammatical results in the antecedent clause of a conditional:

(33) a. Hij is een niet *onverdienstelijk* schilder
He is a not unmeritorious painter
'He is a painter not without merit'

b. *Hij is een *onverdienstelijk* schilder
He is a unmeritorious painter

c. *Als hij een *onverdienstelijk* schilder is, ontsla hem dan
If he is a painter without merit, fire him!

The meaning of sentence (33a) shows the vagueness we met in earlier cases of litotes: the sentence can be used to express both a moderate and a high estimation of the subject's pictorial abilities.

[7]The examples discussed are mainly Dutch because the judgements are sometimes rather subtle and my judgements concerning Dutch are more trustworthy — although not all informants agree on all of them — than those concerning English. The results, however, seem to carry over to English and other languages.

A lexical item comparable to *onverdienstelijk*, *onbetuigd*, is also restricted to litotes constructions. As could be expected, sentence (34a) covers the whole range between minimal and maximal activity.

(34) a. Hij liet zich *niet onbetuigd*
He left himself not un-testified
'He acquitted himself well'

b. *Hij liet zich *onbetuigd*

A difference, however, between *onverdienstelijk* and *onbetuigd* is that the latter word is acceptable in conditionals, with the litotetic meaning vagueness.

(35) Als je je onbetuigd gelaten had waren er ongelukken gebeurd
If you hadn't acquitted yourself well accidents would have happened

It is not totally surprising that only some *on*-formations give rise to litotes readings in conditionals, as not all negative polarity items are well-formed in this context either: *hoeven* 'need' and *meer* 'anymore' do not occur here. On the other hand, the fact that the NPI *ook maar* 'at all' is fine in this context suggests that the context is monotone decreasing (Sánchez Valencia *et al.* 1994).

(36) a. Als er *ook maar iets* gebeurt moet je me bellen
If there at all something happens must you me call
'If anything happens at all, call me'

b. *Als je *hoeft* te werken, kun je daar gaan zitten
If you need to work, can you there go sit

c. *Als je koffie *meer* wilt, daar is de thermoskan
If you coffee anymore want, there is the thermos

To complicate matters even more, note that *hoeven* does occur in conditional sentences, provided a negation is present as well (van der Wouden 1994). One can also get litotes in the antecedent of a conditional if an extra negation is present:

(37) a. Als je *niet hoeft* te werken morgen kun je nog wat drinken
If you not need to work tomorrow can you still something drink
'If you don't have to work tomorrow you can take another drink'

b. Waarom ontsla je hem als hij *geen onverdienstelijk* werk aflevert?
Why fire you him if he not undeserving work produces?
'If he is doing a good job, why fire him?'

The data involving *onbetuigd* leave no other conclusion than that the conditional construction may, in principle, trigger litotetic effects. Often, however, it doesn't: in the following example the readings 'somewhere between good and bad' and 'neither wise nor unwise' are very hard to get:

(38) a. If this is a *bad* idea, don't do it

b. If it is *unwise* to take precautions, why bother?

In section 5, I will return to the question why not all occurrences of a negative element in a downward entailing context lead to litotes readings.

4.1.2 Before

Litotes isn't found too often in *before*-clauses, although they are provably downward entailing (Landman 1991) and even anti-additive (Sánchez Valencia *et al.* 1994). The Dutch NPI *ooit* 'ever' is fine in *before*-clauses, whereas *meer* 'anymore' cannot occur there:

(39) a. Voordat je *ooit* naar Frankrijk gaat moet je dit boek lezen
Before you ever to France go must you this book read
'You must read this book before you ever go to France'

b. *Voordat de gasten koffie *meer* willen moeten we maar bijzetten
Before the guests coffee anymore want must we but more-make

The next example shows that this context may produce litotetic effects in principle too: although it doesn't work with *onverdienstelijk*, it does with *slecht* 'bad' (van der Wouden 1995):

(40) a. *Voordat je *onverdienstelijk* werk levert word je ontslagen
Before you undeserving work produce are you fired

b. Voordat Frans *slecht* werk aflevert moet er heel wat gebeuren
Before Frans produces inferior work, a lot has to happen
'Frans will never produce work that is below standards'

The *before*-context is averidical in the sense that it follows from sentence (40b) that Frans will not produce inferior work (Sánchez Valencia *et al.* 1994). However, we again find the vagueness in meaning which is typical for litotes constructions: the exact quality of Frans's work is not articulated in this sentence nor can it be calculated from it.

4.1.3 Without

Without-clauses trigger negative polarity items and litotes alike:

(41) a. Karel verliet het gebouw *zonder ook maar* iets te zeggen
Karel left the building without at all something to say
'Karel left the building without saying anything'

b. Karel verliet het gebouw *zonder* zijn pasje te *hoeven* laten zien
Karel left the building without his badge to need let see
'Karel left the building without having to show his badge'

c. We zullen u helpen *zonder* een middel *onbeproefd* te laten
We will you help without a means untried to leave
'We will help you with all means'

Many negative polarity items are fine in clauses headed by *zonder* 'without'[8] Certain items that flourish in other litotetic constructions are hard to get in *zonder*-clauses; others, however, seem to create a litotetic effect in this context:

(42) a. ?*Zonder* een *onverdienstelijk* schilder te zijn krijg je geen prijs
Without an unmeritorious painter to be get you no prize
'Without being an unmeritorious painter you'll get no prize'

b. *Zonder* te willen beweren dat Jan een *slecht* mens is, zou ik toch geen tweedehands auto van hem kopen
Without to want say that Jan a bad man is, would I no second-hand car of him buy
'I would say that Jan is a reasonable guy, but I don't want to buy a second hand car from him'

The vagueness of the last sentence (we know that Jan is 'not bad' , but we do not know, and we cannot know, whether he is 'a reasonable guy' or 'an angel') shows that we are once again dealing with litotes here.
In section 5, I will address the question why litotes sometimes doesn't occur in *without* clauses.

4.1.4 Comparatives

Negative polarity items occur in (certain) comparatives as well (Hoeksema 1983; Rullmann 1994) (cf. sentence (27a)). Litotes seems to be possible here as well:

(43) Het is beter dan niks
It is better than nothing

[8]Certain negative polarity items require that their context possess additional logical properties that *zonder* 'without' lacks, which explains why a class of negative polarity items does not occur in this context cf. (Zwarts 1993) and (van der Wouden 1994).

This last sentence is quite a striking case of litotes without anything negative. Depending on the context in which it is uttered, it may both convey an extremely positive meaning and refer to somewhere in the middle area. In a situation where my friend needs \$ 1000, my offer of helping him with a \$ 100 dollar bill may elicit this utterance, and then the message is 'well, it's not exactly what I wanted, but it's at least something'. But if my friend wins \$ 1000000 in a lottery, the same sentence can be used (with the help of understatement or conventionalized irony) to express the meaning 'it is an erormous amount of money'. In other words, the sentence exhibits exactly the same vagueness which is typical for litotes constructions and which I discussed in section 3.3.[9]

4.2 Extending Horn's hypothesis

If it is indeed the case that litotetic effects may be caused by weak negations, i.e., if the hypothesis in (32) is correct, the explanation of the semantics and pragmatics of litotes needs some revision.

First an approach that is incorrect. Consider a combination such as *hardly doubt*. Given a scale on which *doubt* is an end point, the modification with *hardly* might mean that the term *doubt* is hardly applicable to the situation. That is to say, this analysis boils down to metalinguistic negation (Horn 1989), in which the appropriateness of (parts of) an utterance is denied.

To see that this cannot be the right way of looking at things, take a look at an uncontroversial case of metalinguistic negation, in which the speaker corrects the pronunciation of a proper name (44a), and compare this to an undisputable case of litotes (44b):

(44) a. It is not *BernstEEn*, it is *BernstAIn*
b. It is not unwise to take precautions

Firstly, the reference to a grey area in between two extremes that is typical for litotes readings is completely lacking in the case of example (44a): the sentence just cannot be used to express the meaning 'it is somewhere in the grey area between *BernstEEn* and *BernstAIn*'.

Note secondly that the intonation in the two examples is different: in (44a) the proper name necessarily bears stress, whereas this negated constituent is unstressed in (44b).[10] If it is stressed in the latter case, one feels the need to offer an alternative, which yields a metalinguistic reading. That reading,

[9]Compare also a sentence such as *His performance was less than ideal*, which can be used to criticize the subject's behavior both mildly ('it was rather good') and severely ('it was terrible').

[10]Cf. (Seuren 1976) on the role of intonation in metalinguistic readings.

however, is radically different from the original one, as is illustrated by the schemes in (45b) and (45d).

(45) a. It is not unwise to take precautions

b.

unwise	....	wise
......	not unwise	....

c. It is NOT UNWISE to take precautions, it is downright stupid

d.

stupid	unwise		wise
NOT UNWISE	......	..	....

That means that the litotetic interpretation is different from the one where the presence of metalinguistic negation is beyond doubt. But if the interpretations are different, it cannot be metalinguistic negation in both cases, which implies that metalinguistic negation is not the crucial factor in litotes.

A more fruitful approach may be to attack the problem of why weak negations may trigger litotetic readings from the semantic side. Consider the scale in (45b) and assume we want to find a place for the term *stupid* on it. The result would be something like the following:

(46)

unwise		more or less wise	wise
stupid	unwise but not stupid	more or less wise	wise

All downward entailing environments, that is, negations weak and strong, allow for reasoning from sets to subsets, i.e., from predicates to more specific predicates. In other words, logic allows the following reasoning:

(47) It is hardly unwise to take precautions
$[\![\text{unwise}]\!] \Rightarrow [\![\text{stupid}]\!]$

It is hardly stupid to take precautions

From this, it follows that the interpretation of litotetic constructions involving weak negations such as *hardly* must be unambiguous, in principle. Notwithstanding the fact that the double negation is not on an end point of a scale, logic excludes reasoning away from the extreme, so only the other direction is left over. That is to say: *hardly unwise* is completely comparable to *not unwise* in as far as the two expressions denote the same degree of wisdom.

This state of affairs is depicted in the following illustration:

(48)

	a	stupid	unwise	more or less wise	wise
(48)	b	hardly unwise	......	hardly unwise	
	c	....	......	hardly unwise	

The a-row just repeats the scale of wisdom already familiar from (46). The b-row shows the interpretations of *hardly unwise* that are logically possible: all intervals of the scale that are not *unwise*. The c-row depicts the inference that is allowed according to (47): because of the fact that *hardly* creates downward entailing, *hardly unwise* entails *hardly stupid*. That leaves only the right hand part of the scale as a possible interpretation. That is to say, the meaning of *hardly unwise* exhibits the well-known vagueness in that it ranges from *more or less wise* to *wise*.

If this is the right approach to litotes under weak negation, an obvious question to ask is: if *not unwise* and *hardly unwise* refer to the same part of the scale of wisdom, that is, if pragmatics causes them to mean the same, why would one ever prefer the more complex form *hardly unwise* over the simple *not unwise*?

The answer to this question will be pragmatic. Consider a linguistic context where litotes is used very often. A "normal" litotes construction such as *not unwise* may become fossilized in such a way that it is hardly appropriate anymore. For example, frequent use of such a combination may have weakened its subtilities; if a pragmatic principle — such as tact (Leech 1980) — asks for a very subtle wording, the expression *hardly unwise* may be useful, although it is more costly and more elaborate than *not unwise*, but also more friendly and less worn out. But these situations are rare, and so are litotes constructions with weak negations such as *hardly*.

Another obvious question to ask is whether this reinterpretation of Horn's theory is also applicable to downward monotonous contexts that are less obviously negative. The answer to this question is positive, and can be derived from the fact that the monotone decreasing contexts are a subset of the nonveridical contexts.[11] In other words, the occurrence of a statement p in a nonveridical context implies *not* p. Now given that, in the cases we are talking about, p is a negative element, this results in a double negation, which is a rather roundabout, or marked, way of saying things. This markedness is noticed by the listener, and (s)he will therefore apply Horn's principle (10), which will lead him/her to a litotes interpretation.

[11]See (Zwarts 1994) in this volume.

5 Some problematic cases

It is clear from the last section that litotes might occur in all monotone decreasing contexts. Often, however, it doesn't. Why not?

To get a clear view of the situation, compare the distribution of litotes over various downward entailing environments with that of several negative polarity items. Consider the following chart, in which data from the last section (with subtleties put aside) are combined with findings concerning NPIs from (van der Wouden 1994, Ch. 1):

(49)

item → ↓ context	*hoeven* 'need'	*ook maar* 'at all'	*meer* 'anymore'	*een beetje* 'a bit'	litotes
DE operators	ok	*	ok	*	ok
conditionals	*	ok	*	ok	ok
relatives to *alles* ('everything')	ok	ok	*	ok	*
comparatives	ok	ok	*	ok	ok
before	ok	ok	*	ok	ok

Clearly, there is no complete parallelism between the distributional pattern of any negative polarity item and that of litotes. That means that either various types of litotes have to be distinguished, or that an explanation for the restricted distribution of litotes should be found elsewhere.

Note firstly that collocational effects play a role.[12] Many litotetic expressions are subject to fossilization or grammaticalization, sometimes even to the degree that the negated element becomes a negative polarity item (Hoeksema 1994). Cases such as *not bad*, *it is beyond doubt* and *weinig mis* 'little wrong' have developed into fixed cliches or idioms.

Apart from that, an explanation of the remaining cases (or even most cases) may be found along the following lines. Horn's analysis of the semantics and pragmatics of litotes, which I follow here, crucially involves scales: an expression such as *not unwise* means what it means because of the fact that a scale can be constructed on which *wise* and *unwise* are end points, and the double negation denotes a grey area in the sense of Sapir (1944) between these end points.

[12]Langendoen/Bever (1973) and Aitchison/Bailey (1979) discuss a number of syntactic restrictions on the usage of *not un-* formations. Bolinger (1980) re-interprets these restrictions in pragmatic terms. Perhaps some of them carry over to other types of litotes as well, which would offer a partial explanation of this instance of collocational behavior (cf. (van der Wouden 1994, Ch. 3)).

The negated predicate, however, is not the only element that is associated with a scale. Negation or, in general, any downward entailing context, is associated with a scale itself. In the case of simple negation (*not, n't*) this is a trivial scale of truth values on which 0 ('not') and 1 ('true') are end points. This scale is compatible with almost any scale a predicate may be associated with, which explains why litotes with normal negation is the unmarked case (Hoffmann 1987). The weak negations *hardly, scarcely, barely* are associated with the same scale of truth values, so these elements may occur in litotetic constructions with many predicates as well. There exist, however, pragmatic restrictions on litotes with these elements: I discussed these at the end of section 4.3. These restrictions explain why litotes with weak negations is comparatively rare.

Most other downward entailing operators, however, have more semantic content. They are associated with scales that may be incompatible with the scale of the negated predicate. It appears to be reasonable to assume that this scale clash or category mistake is responsible for the fact that one often cannot make a litotes construction by putting some negative predicate in just any downward entailing context.[13] If this is right, this is an explanation for the observation of Jespersen (1924, 332) that the two negations in the double negation construction known as litotes have to refer to the same idea. This can be restated as follows:

(50) **Hypothesis** In litotes constructions, the two negations have to be associated with compatible scales.

6 Concluding remarks

Just like other types of multiple negations in natural language, litotes is found in almost any downward entailing context. I have shown that a slight reinterpretation of Horn's analysis of litotes accounts for this in an elegant way. The fact that litotes is traditionally seen as an instance of double negation has various reasons: firstly, negation is the most frequent, prototypical type of monotone decreasing context; secondly, pragmatic reasons may forbid usage of a weak negation in many cases where a normal negative is already effective, thirdly, collocational effects play a role in the sense that many litotes constructions involve rather fixed combinations of lexical elements, and finally, the semantics of downward entailing context and negated expression may be incompatible. These factors notwithstanding, the possibility of litotes in these context cannot be denied. The main conclusion,

[13]Cf. also van der Wouden (1995) on incompatible scales in the case of polarity items.

therefore, must be that it is hardly unwise to extend the study of litotes beyond the cases that involve double negations.

References

Aitchison, J., & Bailey, G.: Unhappiness about not unhappy people. *Journal of Linguistics* 15 (1979), 245–266.

Baker, C.L.: Double negatives. *Linguistic Inquiry* 1 (1970), 169–86.

Berg, W.: *Uneigentliches Sprechen. Zur Pragmatik und Semantik von Metapher, Metonymie, Ironie, Litotes und rhetorischer Frage.* Tübingen 1978.

von Bergen, A., & von Bergen, K.: *Negative Polarität im Englischen.* Tübingen 1993.

Bolinger, D.: *Degree words.* The Hague 1972.

Bolinger, D.: A not impartial review of a not impeachable theory: some new adventures of ungrammaticality. In: *Language use and the uses of language*, ed. by R.W. Shuy & A. Shnukal, Washington D.C. 1980.

Dunn, J.M.: Perp and Star: Two Treatments of Negation. In: *Philosophical Perspectives* (*Philosophy of Language and Logic*) 7 (1993), ed. by J. Tomberlin, 331–357.

Erasmus, D.: *De copia verborum ac rerum.* Paris 1512. Cited after: *Opera omnia Desiderii Erasmi Roterodami, recognita et adnotatione critica instructa notisque illustrata.* Amsterdam 1969–; I, 6 (1988).

Fauconnier, G.: Polarity and the scale principle. In: *CLS 11* (Chicago Linguistic Society), Chicago 1975, 188–199.

Grice, P.: *Studies in the way of words.* Cambridge/Mass. 1989.

Hoeksema, J.: Negative polarity and the comparative. *Natural Language and Linguistic Theory* 1 (1983), 403–34.

Hoeksema, Jack.: On the grammaticalization of negative polarity items. Paper delivered at the meeting of the Berkeley Linguistic Society, 1994.

Hoffmann, M.E.: *Negatio contrarii. A study of Latin litotes.* Assen 1987 (PhD thesis, University of Amsterdam).

Horn, L.R.: *A Natural History of Negation.* Chicago 1989.

Horn, L.R.: *Duplex negatio affirmat ...*: the economy of double negation. In: *CLS 27-II: Papers from the parasession on negation* (Chicago Linguistic Society), Chicago 1991.

Hübler, A.: *Understatements and hedges in English.* Amsterdam 1983.

Israel, M.: Polarity sensitivity as lexical semantics. Unpublished ms., University of California at San Diego 1994.

Jespersen, O.: Negation in English and other languages. *Historisk-filologiske Meddeleser* 1 (1917), 1–151. Reprinted in: *Selected writings of Otto Jespersen*, London 1962.

Jespersen, O.: *The philosophy of grammar.* London 1924.

Kadmon, N., & Landman, F.: Any. *Linguistics and Philosophy* 16 (1993), 353–422.

Kas, M.: *Essays on Boolean functions and negative polarity.* PhD thesis, University of Groningen, 1993.

Klima, E.S.: Negation in English. In: *The structure of language*, ed. by J.A. Fodor & J.J. Katz, Englewood Cliffs 1964, 246–323.

Ladusaw, W.A.: *Polarity sensitivity as inherent scope relations.* PhD thesis, University of Texas at Austin, 1979.

Landman, F.: *Structures for semantics* (SLAP 45). Dordrecht 1991.

Langendoen, D., & Bever, Th.G.: Can a not unhappy person be called a not sad one? In: *A festschrift for Morris Halle*, ed. by St.R. Anderson & P. Kiparsky, New York 1973, 392–409.

Lausberg, H.: *Handbuch der literarischen Rhetorik: Eine Grundlegung der Literaturwissenschaft.* München 1973.

Leech, G.N.: Language and tact. In: *Explorations in semantics and pragmatics*, Chapter 4. Amsterdam 1980.

Linebarger, M.: *The grammar of negative polarity.* PhD thesis, MIT, 1980.

McCawley, J.D.: *The syntactic phenomena of English.* Chicago 1988.

Progovac, L.: *Negative and positive polarity. A binding approach.* Cambridge 1994.

Rullmann, H.: The ambiguity of comparatives with "less". Paper presented at ESCOL '94. University of South Carolina, 1994. To appear in: J. Fuller, H. Han & D. Parkinson (eds.), Proceedings of ESCOL 1994.

Sánchez Valencia, V., van der Wouden, T., & Zwarts, F.: Polarity, veridicality, and temporal connectives. In: *Proceedings of the Ninth Amsterdam Colloquium*, ed. by P. Dekker & M. Stokhof, Amsterdam 1994, 587–606.

Sapir, E.: Grading: A study in semantics. *Philosophy of Science* 11 (1944), 93–116.

Seuren, P.A.M.: Echo: een studie in negatie. In: *Lijnen van taaltheoretisch onderzoek. Een bundel oorspronkelijke stukken aangeboden aan prof. dr. Henk Schultink*, ed. by G. Koefoed & A. Evers, Groningen 1976, 160–184.

WNT. *Woordenboek der Nederlandsche taal*; red. M. de Vries & L.A. te Winkel [*et al.*]. 's-Gravenhage 1882–.

van der Wouden, T.: *Negative Contexts.* PhD thesis, University of Groningen, 1994.

van der Wouden, T.: Moeilijk is (not) difficult. Paper delivered at the TINdag, Utrecht, January 21, 1995. To appear in: *Linguistics in the Netherlands 1995*, ed. by M. den Dikken & K. Hengeveld, Amsterdam, 1995, pp. 203–14.

van der Wouden, T., & Zwarts, F.: A semantic analysis of negative concord. In: *SALT III: Proceedings of the third conference on semantics and linguistic theory*, ed. by U. Lahiri & A.Z. Wyner, Ithaca/NY (Cornell University, Dep. of Modern Languages and Linguistics) 1993, 202–219.

Zwarts, F.: Negatief polaire uitdrukkingen 1. *Glot* 4 (1981), 35–132.

Zwarts, F.: Three types of polarity. Ms. Groningen, to appear in: *Plural Quantification*, ed. by F. Hamm & E. Hinrichs, Dordrecht, 1995.

Zwarts, F.: A Hierarchy of Negative Expressions. Ms. Groningen 1994. This volume.

A Hierarchy of Negative Expressions

Frans Zwarts*

1 Introduction

The study of negative polarity items in natural language has led to the discovery of various classes of negative expressions. In what follows, a detailed analysis of the logical behavior of these elements will be presented. It turns out that there exists, in fact, a hierarchy of negative expressions, some of which are rather weak, whereas others display one or more of the semantic features typically associated with classical negation. We will see that an analogy can be drawn with nonclassical logic, where various treatments of negation have been proposed which show a close resemblance to natural language phenomena. Independent evidence for the linguistic relevance of these notions comes from the area of negative concord.

2 Negative Polarity Items

The term polarity item, as applied to language, allows us to describe the behavior of certain words and phrases with respect to negation. One class of such expressions, usually referred to as the class of negative polarity items, requires the presence of a negative element in the sentence. As an illustration, consider the examples in (1)

(1) a. None of the children noticed anything

b. * Each of the children noticed anything

The ungrammaticality of (1b) proves that the presence of the noun phrase *each of the children* is not sufficient to justify the occurrence of the polarity item *anything*. Apparently, it is only negative expressions such as *none of the children* that are capable of licensing such elements. We must not suppose that this is a peculiar feature of English. Similar patterns can be found in Dutch, German and Spanish, as shown by the examples in (2), (3) and (4), which feature the negative polarity items *hoeven* 'need', *auch nur irgendetwas* 'anything (at all)' and *mover un dedo* 'lift a finger'.

*I would like to thank Víctor Sánchez-Valencia and Ton van der Wouden for valuable comments and support. And also Pauline Erickson, who helped me correct the proofs.

Dutch

(2) a. Niemand zal zulk een beproeving hoeven te doorstaan
No one will such an ordeal need to go through
'No one need go through such an ordeal'

b. * Iedereen zal zulk een beproeving hoeven te doorstaan
Everyone will such an ordeal need to go through
* 'Everyone need go through such an ordeal'

German

(3) a. Keiner von diesen Leuten hat auch nur irgendetwas bemerkt
None of these people has anything (at all) noticed
'None of these people noticed anything (at all)'

b. * Jeder von diesen Leuten wird auch nur irgendetwas erleben
Each of these people will anything (at all) expericence
* 'Each of these people will experience anything (at all)'

Spanish

(4) a. Juan se fué antes que María moviera un dedo para ayudarle
Juan left before María lifted a finger to help him
'Juan left before María lifted a finger to help him'

b. Juan se fué después que María moviera un dedo para ayudarle
Juan left after María lifted a finger to help him
*'Juan left después María lifted a finger to help him'

Although the contrasts between these sentences may well seem perplexing at first, Ladusaw (1979) has shown that they can be explained in terms of the monotonicity properties associated with various words and phrases. By way of illustration, consider the conditional *At least one villager sang loudly* ⇒ *At least one villager sang.* Provided that the structure of the universe is such that the class of individuals associated with the verb phrase *sang loudly* (VP_1) is a subset of the class of individuals associated with the verb phrase *sang* (VP_2), we may legitimately pass from the proposition *At least one villager sang loudly* to *At least one villager sang*. What this means is that noun phrases of the form *at least n N* are monotone increasing: if *NP*

VP_1 and $VP_1 \subseteq VP_2$, then *NP* VP_2. The same test shows that expressions of the forms *some N*, *every N* and *both N* are also upward monotonic. For if the predicate *ate fish* applies only to what the predicate *ate* also applies to, then the following conditionals are all valid: *Some porters ate fish* ⇒ *Some porters ate*, *Every child ate fish* ⇒ *Every child ate*, *Both lawyers ate fish* ⇒ *Both lawyers ate*. That the temporal connective *after* is also monotone increasing can be seen from the validity of *The mayor left after the judge sang loudly* ⇒ *The mayor left after the judge sang*.

It turns out that monotone increasing noun phrases have a decreasing counterpart. To demonstrate this, we begin by considering the conditional *At most one villager sang* ⇒ *At most one villager sang loudly*. Whenever the state of affairs in the universe is such that the class of individuals associated with the verb phrase *sang* (VP_1) is a superset of the class of individuals associated with the verb phrase *sang loudly* (VP_2), we may legitimately pass from the proposition *At most one villager sang* to *At most one villager sang loudly*. This is important because it entails that noun phrases of the form *at most n N* are monotone decreasing: if *NP* VP_1 and $VP_2 \subseteq VP_1$, then *NP* VP_2. In a similar manner, one easily shows that expressions of the forms *not every N*, *no N* and *neither N* are also downward monotonic. For if the predicate *ate* applies to whatever the predicate *ate fish* applies to, then the following conditionals are all valid: *Not every woman ate* ⇒ *Not every woman ate fish*, *No attorney ate* ⇒ *No attorney ate fish*, *Neither connoisseur ate* ⇒ *Neither connoisseur ate fish*. The temporal connective *before* belongs to the class of monotone decreasing elements as well, since we may legitimately pass from *The mayor left before the judge sang* to *The mayor left before the judge sang loudly*.

In the light of the distinction between upward and downward monotonic expressions, the contrasts in (1) - (4) admit only one explanation: the class of elements which are capable of licensing the occurrence of negative polarity items is coextensive with the class of monotone decreasing expressions. This conclusion is corroborated by the contrasting examples in (5).

(5) a. At most five of the children noticed anything

b. * At least five of the children noticed anything

Of the two phrases *at most five of the children* and *at least five of the children*, it is only the first that can act as a licensing expression for the negative polarity item *anything* - a state of affairs which must be attributed to the circumstance that *at most five of the children* belongs to the class of monotone decreasing noun phrases, and *at least five of the children*, to the class of monotone increasing noun phrases.

3 Weak and strong polarity items

Negative polarity items can be either of the weak, or of the strong, type.[2] In order to get a clear view of the content of this distinction, one does well to take the following Dutch and German examples into consideration.

Dutch

(6) a. Hoogstens één kind zal zichzelf hoeven te verantwoorden
At most one child will himself need to justify
'At most one child need justify himself'

b. Niemand zal zulk een beproeving hoeven te doorstaan
No one will such an ordeal need to go through
'No one need go through such an ordeal'

(7) a. Hoogstens zes agenten hebben ook maar iets bemerkt
At most six cops have anything (at all) noticed
'At most six cops noticed anything (at all)'

b. Niemand heeft van de regenbui ook maaar iets bemerkt
No one has of the rain anything (at all) noticed
'No one noticed anything (at all) of the rain'

German

(8) a. Höchstens eine Frau wird sich zu verantworten brauchen
At most one woman will herself to justify need
'At most one woman need justify herself'

b. Keiner wird solch eine Prüfung durchzustehen brauchen
No one will such an ordeal to go through need
'No one need go through such an ordeal'

(9) a. Höchstens zehn Kinder haben auch nur irgendetwas bemerkt
At most ten children have anything (at all) noticed
'At most ten children noticed anything at all'

---

[2] A more elaborate discussion of weak and strong polarity can be found in (Van der Wouden 1994) and (Zwarts 1995).

b. Keiner von diesen Leuten hat auch nur irgendetwas bemerkt
None of these people has anything (at all) noticed
'None of these people noticed anything (at all)'

The contrast between (6) and (8), on the one hand, and (7) and (9), on the other, proves that expressions such as *ook maar iets* and *auch nur irgendetwas* ('anything (at all)') place stronger restrictions on their environments than the negative polarity items *hoeven* and *brauchen* ('need'). As the ungrammatical sentences in (7) and (9) show, neither Dutch *ook maar iets* nor German *auch nur irgendetwas* is satisfied with the presence of a monotone decreasing expression of the form *hoogstens* (*höchstens*) *n N* 'at most n N'. Instead, both seem to require an *n*-word like *niemand* (*keiner*) 'no one' or *keiner von diesen Leuten* 'none of these N'.[3] As a matter of fact, the distinction between weak and strong forms of negative polarity appears to correspond with that between monotone decreasing and anti-additive noun phrases. Whereas Dutch *hoeven* and German *brauchen* are content with a monotone decreasing expression like *at most n N* as licensing element, *ook maar iets* and *auch nur irgendetwas* require the presence of an anti-additive expression like *no one* or *none of these N*.

4 A closer look at monotonic noun phrases

In order to get a clear view of the content of these notions, we do well to give a more precise account of the semantic interpretation assigned to noun phrases. Let us assume that each verb phrase will receive a subset of some universe U as its semantic value. Such a way of portraying the matter entails that the universe of possible denotations of verb phrases may henceforth be equated with $P(U)$ - that is, the first power set of U. Since the collection $P(U)$ with the usual set-theoretical operations of union, intersection, and complementation can be regarded as a Boolean algebra, we shall from now on speak of the *algebra of verb phrases*[4]. We assume as well that each noun phrase receives a collection of subsets of U as its semantic value. This implies that the universe of possible denotations of noun phrases may be equated

[3]The notion of an *n*-word stems from Laka Mugarza (1990), who uses the term to describe universal negatives like *nadie* 'no one', *nada* 'nothing' and *nunca* 'never' in Spanish. Though Laka herself regards these expressions as existential polarity items, Zanuttini (1991) argues that they should be treated as universal negatives. Van der Wouden and Zwarts (1994) maintain that Romance *n*-words are at times polarity items and at times universal negatives, a point of view which was advanced earlier in (Zanuttini 1989). For present purposes, an *n*-word is simply a universal negative which has the semantic structure of what will hereinafter be referred to as an anti-additive expression.

[4]Keenan and Faltz (1985) offer a detailed study of the role of Boolean algebras in the semantical analysis of natural languages.

with $P(P(U))$ - that is to say, the second power set of U. Since this set also displays the characteristic features of a Boolean algebra, it is said to be the *algebra of noun phrases*. The algebraic nature of the categories *NP* and *VP* enables us to give a precise formulation of a notion which is frequently used in the linguistic literature - to wit, that of a quantifier. Henceforth, what is meant by *a quantifier on a Boolean algebra B* is simply a subset of B. Such a definition immediately explains why it is natural to regard noun phrases as quantifiers. For if these expressions receive a collection of subsets of U as their semantic value, they can semantically be equated with a subset of $P(U)$ - that is to say, with a subset of the algebra of verb phrases. However, this entails that noun phrases could just as well be regarded as quantifiers on the algebra of verb phrases. It is now possible to say how the interpretation of an expression like *at most one villager* differs from that of other noun phrases. For it is easily seen that each set which contains at most one villager, is a member of the quantifier associated with the expression *at most one villager*. This turns out to be important because it entails that the quantifier in question has the property of being closed under subsets: if $X \in Q$ and $Y \subseteq X \subset U$, then $Y \in Q$. Such quantifiers are commonly referred to as *monotone decreasing quantifiers*[5]. For the sake of clarity we record this in the form of a definition.

Definition 1 Let B be a Boolean algebra. A quantifier Q on B is said to be *monotone decreasing* iff for each two elements X and Y of the algebra B: if $X \in Q$ and $Y \subseteq X$, then $Y \in Q$.

Noun phrases which invariably receive a monotone decreasing quantifier on the VP algebra as their semantic value will henceforth be called monotone decreasing noun phrases. From the fact that expressions of the form *at most n N* are downward monotonic it follows immediately that the conditional *At most one villager sang* ⇒ *At most one villager sang loudly* is valid when the predicate *sang* is true of whatever the predicate *sang loudly* is true of. An expression like *at least one villager* differs from *at most one villager* in that the associated quantifier has the property of being closed under supersets: if $X \in Q$ and $X \subseteq Y \subseteq U$, then $Y \in Q$. Such quantifiers are usually called *monotone increasing quantifiers*.

Definition 2 Let B be a Boolean algebra. A quantifier Q on B is said to be *monotone increasing* iff for each two elements X and Y of the algebra B: if $X \in Q$ and $X \subseteq Y$, then $Y \in Q$.

Noun phrases which invariably receive a monotone increasing quantifier as their semantic value will accordingly be referred to as monotone increasing noun phrases. Given that expressions of the form *at least n N* are upward

[5]See (Barwise 1979) and (Barwise and Cooper 1981).

monotonic, one easily proves that the conditional *At least one villager sang loudly* ⇒ *At least one villager sang* is valid when the predicate *sang loudly* applies only to what the predicate *sang* also applies to. This does not exhaust the matter, for a closer look reveals that there are several alternative ways to determine whether a particular noun phrase is upward or downward monotonic. In fact, monotonic noun phrases can be given a number of logically equivalent characterizations. The next theorem provides the relevant details.

Facts
A noun phrase is monotone increasing iff the following two schemata are logically valid:

(10) a. $NP\ (VP_1\ and\ VP_2) \Rightarrow (NP\ VP_1\ and\ NP\ VP_2)$

b. $(NP\ VP_1\ or\ NP\ VP_2) \Rightarrow NP\ (VP_1\ or\ VP_2)$

A noun phrase is monotone decreasing iff the following two schemata are logically valid:

(11) a. $NP\ (VP_1\ or\ VP_2) \Rightarrow (NP\ VP_1\ and\ NP\ VP_2)$

b. $(NP\ VP_1\ or\ NP\ VP_2) \Rightarrow NP\ (VP_1\ and\ VP_2)$.

On the basis of these tests one can arrive at fairly accurate judgments concerning the presence of monotonicity properties. It is readily established, for instance, that expressions of the forms *many N*, *most N*, and *several N* are all upward monotonic. The class of monotone decreasing noun phrases, on the other hand, can be shown to include expressions of the forms *few N*, *no N*, and *not all N*. For the sake of clarity, the outcomes of the tests have been collected in table 1. The forty-four classes of noun phrases mentioned there must all be regarded as being either upward or downward monotonic.

By way of illustration we consider some examples. It is clear, for instance, that the conditionals in (12) and (13) are both valid - a state of affairs which in view of the schemata in (11) admits no other explanation than that expressions of the forms *few N* and *only NP* belong to the class of monotone decreasing noun phrases.

(12) Few hangmen complained or resisted ⇒ Few hangmen complained and few hangmen resisted

(13) Only judges resign or only judges get strangled ⇒ Only judges resign and get strangled

Table 1: Forty-four monotonic noun phrases

Monotone increasing	Monotone decreasing
every N	not every N
all N	not all N
each N	not each N
some N	no N
sm N	only a few N
nearly all N	almost no N
both N	neither N
at least n N	at most n N
many N	few N
several N	only n N
more than n N	no more than n N
the n N	none of the n N
the more than n N	none of the more than n N
the N [pl]	none of the N
the N [sg]	not a single N
most N	not one N
everything	not everything
something	nothing
everyone	not everyone
someone	no one
not only NP	only NP
proper names	negated proper names

However, if we consider the conditionals in (14) and (15), then it is immediately clear that neither can be accepted as valid.

(14) More than nine nuns prayed or knelt ⇒ More than nine nuns prayed and more than nine nuns knelt

(15) More men than women escaped or died ⇒ More men than women escaped and more men than women died

For even though there may be more than nine nuns who prayed, there need not be any who knelt, in which case the antecedent of the conditional in (14) is true, but its consequent false. In an analogous manner, it is easily shown that the conditional in (15) is invalid. For if two man escaped and one woman died, then the antecedent is true, but the consequent false. We must therefore conclude that noun phrases of the forms *more than n N* and *more* N_1 *than* N_2 are not monotone decreasing.

It is easy to see that each of the two schemata in (10) gives us a test for upward monotonicity. As an illustration we note that the conditionals in (16) and (17) are both valid - a circumstance which admits no other explanation than that proper names and definite descriptions are both monotone increasing.

(16) Jonathan called and begged ⇒ Jonathan called and Jonathan begged

(17) The girl sighed or the girl coughed ⇒ The girl sighed or coughed

On the other hand, if we take the conditionals in (18) and (19) into consideration, then it is immediately clear that neither can be regarded as valid.

(18) No fireman said good-bye and left ⇒ No fireman said good-bye and no fireman left

(19) Only women got whipped or only women starved ⇒ Only women got whipped or starved

For if there are precisely two firemen, one of whom said good-bye without leaving and one of whom left without saying goodbye, then the antecedent of the conditional in (18) is true, but its consequent false. In an analogous manner, one easily proves that the conditional in (19) is invalid as well. For if those who got whipped can all be regarded as being women, whereas those who starved not only include women, but also men, then the antecedent is true and the consequent false. This state of affairs leads to the conclusion that noun phrases of the forms *no N* and *only NP* are not monotone increasing.

5 Monotonic expressions as functions

Thus far we have assumed that noun phrases must be regarded as quantifiers on the algebra of verb phrases. It is conceivable, however, that the semantic value assigned to expressions of this type is not a quantifier, but rather the characteristic function of a quantifier. Such a conceptual change immediately raises the question to what extent the usual notion of a function lends itself to the classifications discussed above. Surprisingly, this matter appears not to be as complex as one might at first expect. Let us begin by considering the special case of upward monotonicity. The fact that quantifiers of this type are closed under supersets entails that the associated characteristic function is such that assignment of the value *1* to an element X is invariably accompanied with assignment of the value *1* to each element Y which contains X. Put differently, if Q is a monotone increasing quantifier on the algebra of verb phrases, then its characteristic function K_Q: $P(U) \Rightarrow \mathit{2}$ has the property that the conditions $X \subseteq Y$ and $K_Q(X) = \mathit{1}$ imply $K_Q(Y) = \mathit{1}$. We must not suppose that this exhausts the matter, for in view of the composition of the Boolean algebra *2* it should be obvious that what the property in question amounts to is that whenever X is contained in Y, $K_Q(X)$ is contained in $K_Q(Y)$[7]. This state of affairs leads to the conclusion that, in general, a function f from a Boolean algebra B to a Boolean algebra B^* can be regarded as monotone increasing, just in case it preserves the inclusion relation. That is to say:

Definition 3 Let B and B^* be two Boolean algebras. A function f from B to B^* is said to be *monotone increasing* iff for each two elements X and Y of the algebra B: if $X \subseteq Y$, then $f(X) \subseteq f(Y)$[8].

It is evident that such functions can also be given an alternative characterization. As a matter of fact, corresponding to (10), we have the following result.

Fact
Let B and B^* be two Boolean algebras. The following statements about a function f from B to B^* are equivalent:

(20) a. f is monotone increasing

b. $f(X \cap Y) \subseteq f(X) \cap f(Y)$

[7]The relation of Boolean inclusion in the algebra *2* is defined in such a way that the zero element (*0*) contains only itself, whereas the unit element (*1*) contains both itself and the zero element.

[8]Such functions are sometimes said to be *isotone*. Their monotone decreasing counterparts, defined in (21), are accordingly referred to as *antitone* functions. See (Birkhoff 1967) and (Stoll 1974).

c. $f(X) \cup f(Y) \subseteq f(X \cup Y)$

We must not suppose that this exhausts the variety of available functions, for besides monotone increasing functions we can obviously also distinguish monotone decreasing functions. In that case it is the properties of monotone decreasing quantifiers that serve as our point of departure. Indeed, the fact that these collections are closed under inclusion entails that the associated characteristic function is such that assignment of the value *1* to an element Y is invariably accompanied with assignment of the value *1* to each element X which is contained in Y. In other words, if Q is a monotone decreasing quantifier on the algebra of verb phrases, then its characteristic function K_Q: $P(U) \Rightarrow \mathit{2}$ possesses the property that the conditions $X \subseteq Y$ and $K_Q(Y) = \mathit{1}$ imply $K_Q(X) = \mathit{1}$. In view of the nature of the Boolean algebra *2*, it should be clear that what the property in question amounts to is that whenever X is contained in Y, $K_Q(Y)$ is contained in $K_Q(X)$. More precisely:

Definition 4 Let B and B^* be two Boolean algebras. A function f from B to B^* is said to be *monotone decreasing* iff for each two elements X and Y of the algebra B: if $X \subseteq Y$, then $f(Y \subseteq f(X)$.

Needless to say, such functions can also be characterized in a different way. Analogous to (11) we have the following:

Fact
Let B and B^* be two Boolean algebras. The following statements about a function f from B to B^* are equivalent:

(21) a. f is monotone decreasing

b. $f(X \cup Y) \subseteq f(X) \cap f(Y)$

c. $f(X) \cup f(Y) \subseteq f(X \cap Y)$

Although the distinction between monotone increasing and monotone decreasing functions may be regarded as an unnecessary complication, one should keep in mind that the concept of a function is much more general than the rather limited notion of a quantifier. This is clear from the fact that natural languages as a rule have a variety of expressions which cannot be associated with a quantifier, but which nevertheless display the characteristic features of monotonicity. The next two conditionals may serve as an illustration.

(22) a. No willow will die slowly ⇒ No blossoming willow will die slowly

b. All jonquils will blossom ⇒ All fertilized jonquils will blossom

Provided that the noun *blossoming willow* applies only to willows and the noun *fertilized jonquils* only to jonquils, the conditionals in (22) must both be accepted as valid. In other words, if the state of affairs in the universe is such that the class of blossoming willows is a subset of the class of willows and the class of unfertilized jonquils a subset of the class of jonquils, then we may legitimately pass from the propositions *No willow will die slowly* and *All jonquils will blossom* to the propositions *No blossoming willow will die slowly* and *All unfertilized jonquils will blossom*. These findings show that the logical behavior of the determiners *all* and *no* resembles that of monotone decreasing noun phrases, in spite of the fact that the semantic value of both expressions cannot be treated as a quantifier. In order to account for this resemblance, we must invoke the notion of a monotonic function. Indeed, it is not uncommon to find determiners being portrayed as functions which carry nouns into noun phrases. In accordance with this, such expressions are often identified semantically with a function from the algebra of nouns to the algebra of noun phrases - a state of affairs which admits no other interpretation than that a determiner associates with each element of the power set $P(U)$ a uniquely determined element of the power set $P(P(U))$. Once it is recognized that determiners are to be treated as functions, we have an explanation for the similarities between the logical behavior of monotone decreasing noun phrases and that of expressions like *all* and *no*. For if the semantic value of determiners is functional in nature, then the validity of the conditionals in (22) must be attributed to the circumstance that the functions associated with *all* and *no* are downward monotonic. As an illustration, we consider the case of the determiner *all*. It is easy to see that the monotone decreasing character of this element entails that the extension of the noun phrase *all jonquils* is contained in the extension of *all fertilized jonquils*. Stated differently, each set which is a member of the quantifier associated with *all jonquils* is also a member of the quantifier associated with *all fertilized jonquils*. From this it follows immediately that someone who considers the proposition *All jonquils will blossom* to be true must also accept the truth of *All fertilized jonquils will blossom*. For that reason, we may say that the validity of the conditional in (22b) is a consequence of the monotone decreasing nature of the determiner *all*. Needless to say, such reasoning is not limited to this single case, for in a completely analogous fashion the validity of the implication in (22a) can be attributed to the downward monotonic nature of the determiner *no*.

These and similar findings with regard to the laws which govern the logical behavior of determiners have been collected in table 2. The eighteen classes of expressions which are mentioned there must all be regarded as monotonic. It should be obvious that the logical properties of the determiner may be quite unlike those of the corresponding noun phrase. As a case in point, we consider the expressions *every* and *all*. When the semantical properties of these determiners are compared with those of noun phrases of the forms *every N* and *all N*, it is at once clear that one has downward monotonicity in the first case and upward monotonicity in the second. Things are different, however, when, instead of *every* and *all*, we consider their negations *not every* and *not all*, for now the determiners are monotone increasing and the corresponding noun phrases monotone decreasing.

6 De Morgan's laws

When we compare the logical behavior of monotone decreasing noun phrases with that of the sentential prefix *it is not the case that*, it becomes clear that downward monotonicity represents a very weak form of negation. By way of illustration, we consider the biconditionals in (23) and (24).

Table 2: Eighteen classes of monotonic determiners

Monotone increasing	**Monotone decreasing**
not every	every
not all	all
not each	each
sm	no
at least n	at most n
some	not a single
several	only n
more than n	no more than n
some but not all	not one

(23) It is not the case that Jack ate or Jill ran ⇔It is not the case that Jack ate and it is not the case that Jill ran

(24) It is not the case that Jack ate and Jill ran ⇔ It is not the case that Jack ate or it is not the case that Jill ran

One sees immediately that the equivalences in (23) and (24) must both be accepted as valid - a state of affairs which admits no other explanation than that the operation in question is governed by the laws of De Morgan. This observation is important because it has frequently been argued that the logical patterns in (23) and (24) characterize the use of negation. Although such a conclusion is correct with respect to the sentential prefix *it is not the case that* and similar elements, it must be regarded as misleading when it comes to describing the negative force associated with downward monotonic expressions. Not only does natural language contain a variety of such elements, their logical behavior is also not the same as that of the prefix *it is not the case that*. In order to convince ourselves of this fact, we need only appeal to the formulas in (21). As an illustration, consider the conditionals in (25)-(28).

(25) $\models$ Few trees will blossom or will die $\Rightarrow$ Few trees will blossom and few trees will die

(26) $\not\models$ Few trees will blossom and few trees will die $\Rightarrow$ Few trees will blossom or will die

(27) $\not\models$ Few trees will blossom and will die $\Rightarrow$ Few trees will blossom or few trees will die

(28) $\models$ Few trees will blossom or few trees will die $\Rightarrow$ Few trees will blossom and will die

From these examples it is clear that the monotone decreasing noun phrase *few trees*, though undoubtedly a negative expression, differs substantially from the prefix *it is not the case that*. Of the four conditionals presented above, only two are valid: the one in (25) and the one in (28). In other words, the logical behavior of downward monotonic noun phrases of the form *few N* is governed by one half of the first law of De Morgan and one half of the second law of De Morgan.[9] What this suggests is that such expressions must be regarded as being weakly negative. We can now explain what the difference is between a monotone decreasing expression and one which is anti-additive. Anti-additive phrases like *no one*, *nothing*, *neither N* and *none of the N* embody a stronger form of negation than downward monotonic ones in that they are governed by the first law of De Morgan as a whole. This logical difference is reflected in the validity of the following four conditionals.

(29) $\models$ No man escaped or got killed $\Rightarrow$ No man escaped and no man got killed

(30) $\models$ No man escaped and no man got killed $\Rightarrow$ No man escaped or got killed

(31) $\not\models$ No man escaped and got killed $\Rightarrow$ No man escaped or no man got killed

[9]The laws in question are, of course, the familiar set-theoretical identities $-(X \cup Y) = -X \cap -Y$ and $-(X \cap Y) = -X \cup -Y$. Quine (1952) speaks in this connection of the first and the second law of De Morgan, respectively.

(32) $\models$ No man escaped or no man got killed $\Rightarrow$ No man escaped and got killed

From these examples we may conclude that the anti-additive noun phrase *no man*, regarded as a negative expression, differs considerably from the monotone decreasing expression *few trees*. Of the four conditionals presented above, no less than three must be counted as valid: the one in (29), the one in (30), and the one in (32). The conclusion must therefore be that anti-additive expressions embody a stronger form of negation than downward monotonic ones, though not as strong as the type of negation expressed by the sentential prefix *it is not the case that*.

7 A closer look at anti-additive expressions

A more accurate inspection of the conditionals in (29) - (32) shows that the validity of (29) and (32) is a consequence of the monotone decreasing nature of the expression *no man*. This explanation does not hold, however, for the proposition in (30). The fact that we have again a valid implication, must instead be attributed to the circumstance that the monotone decreasing quantifier which is associated with *no man* is in addition closed under (finite) unions. Such quantifiers will henceforth be called *quasi-ideals*[10]. More precisely:

Definition 5 Let B be a Boolean algebra. A quantifier Q on B is said to be a *quasi-ideal* iff for each two elements X and Y of the algebra B:

1. if $X \cup Y \in Q$, then $X \in Q$ and $Y \in Q$
2. if $X \in Q$ and $Y \in Q$, then $X \cup Y \in Q$.

Noun phrases which invariably receive a quasi-ideal on the algebra of verb phrases as their semantic value will be referred to as anti-additive noun phrases. From the fact that the conditionals in (29) and (30) are both valid, it follows immediately that expressions of the form *no* N are associated with a quasi-ideal as their semantic value and must therefore be classified as anti-additive. With the aid of the foregoing test, one easily proves that noun phrases of the forms *nothing*, *neither* N and *none of the* N belong to this category as well. To show that the notion of a quasi-ideal has a counterpart, it is sufficient to consider the conditionals in (33) and (34).

[10]The notion of an ideal is, of course, familiar from the theory of Boolean algebras. It involves a distinguished type of nonempty subset meeting the conditions in Definition 5. See (Sikorski 1969) and (Stoll 1974). The notion of a quasi-ideal is slightly more general in that it can also involve the empty set.

(33) All monks rob and kill ⇒ All monks rob and all monks kill

(34) All monks rob and all monks kill ⇒ All monks rob and kill

The validity of (33) is a consequence of the monotone increasing nature of the expression *all monks*. That the reverse proposition in (34) also represents a valid entailment must, however, be attributed to the fact that the monotone increasing quantifier which acts as the denotation of *all monks*, is closed under (finite) intersections as well. Such quantifiers will henceforth be referred to as *quasi-filters* - a term which may not be customary within the theory of Boolean algebras, but which has gained some currency in modal logic[11]. For the sake of clarity we record this in the form of a definition.

Definition 6 Let B be a Boolean algebra. A quantifier Q on B is said to be a *quasi-filter* iff for each two elements X and Y of the algebra B:

1. if $X \cap Y \in Q$, then $X \in Q$ and $Y \in Q$
2. if $X \in Q$ and $Y \in Q$, then $X \cap Y \in Q$

Noun phrases which are invariably interpreted by means of a quasi-filter on the algebra of verb phrases will be called multiplicative noun phrases. The validity of the conditionals in (33) and (34) makes it clear that expressions of the form *all N* are associated with a quasi-filter as their semantic value and must consequently be classified as multiplicative. With the aid of the schemata in Definition 6, one can easily show that proper names and expressions of the form *the n N* belong to this category as well.

Table 3: Twenty-two classes of multiplicative and anti-additive noun phrases

Multiplicative	**Anti-additive**
every N	no N
all N	neither N
each N	none of the N
both N	none of the more than n N
the n N	none of the N
the more than n N	not a single N
the N [pl]	not one N
the N [sg]	nothing
everything	no one
everyone	only NP
proper names	negated proper names

[11]See (Chellas 1980).

By means of the available tests, we can usually arrive at trustworthy judgments when we must decide whether a given monotonic noun phrase is multiplicative or anti additive. For ease of survey, the outcomes of these tests have been collected in table 3. The twenty-two classes of noun phrases which are mentioned there must all be regarded as being either multiplicative or anti-additive.

Anti-additive and multiplicative noun phrases can both be given a functional characterization. The relevant definitions are given below.

Definition 7 Let B and B^* be two Boolean algebras. A function f from B to B^* is said to be *anti-additive* iff for each two elements X and Y of the algebra B: $f(X \cup Y) = f(X) \cap f(Y)$.

Definition 8 Let B and B^* be two Boolean algebras. A function f from B to B^* is said to be *multiplicative* iff for each two elements X and Y of the algebra B: $f(X \cap Y) = f(X) \cap f(Y)$.

There are several determiners whose logical behavior resembles that of an anti-additive function. By way of illustration, we consider the biconditionals in (35) and (36).

(35) Every goat or donkey will be killed ⇔ Every goat and every donkey will be killed

Table 4: Five anti-additive determiners.

every	not a
all	not one
no	

(36) Not a priest or baker will be fired ⇔ Not a priest and not a baker will be fired

Anyone who regards one of the members of the biconditionals in (42) and (43) as true, must also accept the truth of the other member - a state of affairs which admits no other interpretation than that the determiners *every* and *not a* are associated with anti-additive functions. For if one regards the determiner as a functor and the common noun as its argument, then the valid schema in (37) results.

(37) $(DET\ (N_1\ or\ N_2))\ VP \Leftrightarrow (DET\ (N_1)\ and\ DET\ (N_2))\ VP$

It is immediately clear that the above formula exhibits the characteristic features of an anti-additive function and can be used as a test to determine whether a given expression belongs to this category. Our findings have been collected in table 4. The five determiners mentioned there must all be understood as being anti-additive. Curiously enough, it is very difficult to find determiners which exhibit the characteristic properties of a multiplicative function. It seems that natural language excludes such expressions on principle. Given the semantic constraints on the expressive nature of determiners proposed in (Van Benthem 1986), one can in fact prove that the observed gap is a logical one.[12]

The logical difference between anti-additive expressions and downward monotonic ones appears to be reflected in the behavior of the negative polarity items in (6)-(9). Whereas Dutch *hoeven* and German *brauchen* ('need') are content with a monotone decreasing expression like *hoogstens* (*höchstens*) *n N* ('at most n N') as licensing element, *ook maar iets* and *auch nur irgendetwas* ('anything (at all)') require the presence of an anti-additive phrase like *niemand* ('no one') or *keiner von diesen Leuten* ('none of these people'). Accordingly, we may formulate the following hypothesis concerning the distribution of negative polarity items.

(38) a. Only sentences in which a monotone decreasing expression occurs can contain a negative polarity item of the weak type.

b. Only sentences in which an anti-additive expression occurs can contain a negative polarity item of the strong type.

By (38a), the presence of a monotone decreasing expression is a necessary condition for the appearance of weak polarity items. On the other hand, (38b) stipulates that strong polarity items require the presence of an anti-additive expression as licensing element. To forestall any misunderstanding, we note that every anti-additive expression is monotone decreasing as well. It follows that weak polarity items can occur in sentences containing an anti-additive expression.

8 Antimultiplicative expressions

Independently of the class of anti-additive expressions, there exists a class of elements which also represent a stronger form of negation than the downward monotonic ones. The members of this class are called antimultiplicative and are typically associated with the functional pattern in Definition (9).

Definition 9 Let B and B^* be two Boolean algebras. A function f from B to B^* is said to be *antimultiplicative* iff for each two elements X and Y of the algebra B: $f(X \cap Y) = f(X) \cup f(Y)$.

[12] See (Zwarts 1993) for a formal proof to this effect.

It is easy to see that antimultiplicative expressions differ from their anti-additive counterparts in that they validate not the first, but the second law of De Morgan as a whole. Well-known representatives of this group are phrases of the forms ***not all** N*, ***not every** N* and ***not always***. Antimultiplicative determiners, on the other hand, are as difficult to find as multiplicative ones. In fact, Zwarts (1993) has shown that this is not an accidental gap either.

To determine whether a given noun phrase is antimultiplicative or not, it is sufficient to consider a biconditional of the form *NP* (VP_1 *and* VP_2) ⇔ (*NP* VP_1 *or NP* VP_2). The next sentence may serve as an example.

(39) Not all knights rob and kill ⇔ Not all knights rob or not all knights kill

It is clear that (39) must be accepted as valid in virtue of the antimultiplicative quantifier associated with the expression ***not all knights***. This suffices to establish that noun phrases of the form ***not all** N* belong to the class of antimultiplicative elements.

Besides anti-additive expressions there are additive ones as well. What characterizes these elements is the functional pattern in Definition (10).

Definition 10 Let B and B^* be two Boolean algebras. A function f from B to B^* is said to be ***additive*** iff for each two elements X and Y of the algebra B:
$f(X \cup Y) = f(X) \cup f(Y)$.

Among the various expressions whose logical behavior resembles that of an additive function are noun phrases of the form *some N [sg]*, proper names, and the determiners *some [sg]*, *a* and ***not all***. By way of illustration, we consider the conditionals in (40) and (41).

(40) Some fisherman waved or called ⇒ Some fisherman waved or some fisherman called

(41) Some fisherman waved or some fisherman called ⇒ Some fisherman waved or called

Clearly, it follows from the monotone increasing nature of the singular noun phrase *some fisherman* that the conditional in (41) must be accepted as valid. That the reverse is also valid proves that the characteristic function of the associated quantifier is, in fact, additive. This is enough to show that phrases of the form *some N [sg]* belong to the class of additive expressions.

9 Negative concord

Independent evidence for the linguistic relevance of the distinction between anti-additive and antimultiplicative expressions comes from the phenomenon of negative concord. Negative concord is the general term for the many cases where multiple occurrences of morphologically negative elements express a single semantic negation (Labov 1972, Ladusaw 1992, Van der Wouden and Zwarts 1994, Van der Wouden 1994). A typical example is the French sentence in (42a) which contains a distinguished negative element (*ne*) in addition to the anti-additive noun phrase *personne* 'no one'.

French

(42) a. Personne n' a vu les enfants
No one not has seen the children
'No one saw the children'

b. * Peu d'enfants n' ont vu Jean
Few children not have seen Jean
'Few children saw Jean'

c. Peu d'enfants ont vu Simone
Few children have seen Simone
'Few children saw Simone'

As (42b) shows, we cannot replace *personne* by the monotone decreasing noun phrase *peu d'enfants* 'few children'. Instead, the particle *ne* must be removed, which give us the grammatical sentence (42). This shows that the form of concord known as negative doubling is restricted to anti-additive expressions in French.[13] Even antimultiplicative noun phrases of the form *pas tout le monde* 'not everyone' are unable to licence the occurrence of the negative particle *ne*, as illustrated by the contrast in (43).

(43) a. Pas tout le monde a vu Simone
Not everyone has seen Simone
'Not everyone saw Simone'

[13]The only exceptions appear to be monotone decreasing expressions like *presque personne* which elicit negative doubling, judging from the grammatical sentence *Presque personne n'a rien vu*.

b. * Pas tout le monde n' a vu Jean
Not everyone not has seen Jean
'Not everyone saw Jean'

Afrikaans, which also exhibits negative doubling, allows such cases. Ponelis (1985, 372) gives the following examples of an antimultiplicative noun phrase licensing the negative particle *nie*.

Afrikaans

(44) a. Nie alle bestuurders sal dit in die stadsverkeer waag nie
Not all drivers will this in the city trafic dare not
'Not all drivers will dare this in city traffic'

b. Nie almal sal dit in die stadsverkeer waag nie
Not everyone will this in the city traffic dare not
'Not everyone will dare this in city traffic'

As observed by Van der Wouden (1994), this suggests that both antimultiplicative and anti-additive expressions may license *nie* in Afrikaans.

10 A hierarchy of negations

The preceding discussion has made it clear that there can be substantial differences between the behavior of the determiner, on the one hand, and that of the corresponding noun phrase, on the other. Characteristic examples are the expressions *every* and *all*. When we compare the semantical properties of these elements with those of noun phrases of the forms *every N* and *all N*, it turns out that we are dealing with anti-additive expressions in the first case, but multiplicative expressions in the second. Things are different, however, when, in place of *every* and *all*, we consider the universal negations *no*, *not a* and *not one*, for now it is not only the determiner, but also the corresponding noun phrase which is anti-additive in nature. Even more surprising is the behavior of expressions as *both*, *each*, *the n*, *neither* and *none of the n*. In all of these cases, the determiner is clearly non monotonic. On the other hand, noun phrases of the forms *both N*, *each N* and *the n N* exhibit a multiplicative character, as opposed to those of the forms *neither N* and *none of the n N*, which instead are anti-additive.

Table 5: Comparison of the logical behavior of determiners with that of the corresponding noun phrases

	Determiner	Noun phrase
at least n	monotone increasing	monotone increasing
some/sm [sg]	additive	additive
some/sm [pl]	monotone increasing	monotone increasing
no	anti-additive	anti-additive
at most n	monotone decreasing	monotone decreasing
not a single	anti-additive	anti-additive
not every	additive	antimultiplicative
not all	additive	antimultiplicative
every	anti-additive	multiplicative
all	anti-additive	multiplicative
some but not all	monotone increasing	nonmonotonic
nearly all	nonmonotonic	monotone increasing
both	nonmonotonic	multiplicative
each	nonmonotonic	multiplicative
many	nonmonotonic	monotone increasing
the n	nonmonotonic	monotone increasing
most	nonmonotonic	monotone increasing
almost no	nonmonotonic	monotone decreasing
not each	nonmonotonic	antimultiplicative
neither	nonmonotonic	anti-additive
few	nonmonotonic	monotone decreasing
none of the n	nonmonotonic	anti-additive
(precisely) n	nonmonotonic	nonmonotonic

This description of the different possibilities is still not complete, for it is also possible that neither the determiner nor the corresponding noun phrase is multiplicative or anti-additive in nature. Expressions which belong to this class include *at least n*, *at most n*, *not all*, *some but not all*, *many*, *few*, *most* and *(precisely) n*. *Not all* is an additive determiner which produces an antimultiplicative noun phrase. *At least one*, on the other hand, is an additive determiner which gives rise to an additive noun phrase. Following our earlier policy, a number of these results have been collected in table 5. On the basis of this survey, it is easy to see that the logical behavior of the determiner is wholly independent of that of the corresponding noun phrase. To forestall any misunderstandings, it should be pointed out that multiplicative and additive functions are also monotone increasing. In an analogous way, one easily establishes that the classes of anti-additive and antimultiplicative functions are both subsets of the class of monotone decreasing functions.

In the same way that a determiner can be associated with a function carrying sets of individuals into collections of such sets, a sentential connective can be assigned a function from the algebra of truth values, *2*, to the power set algebra $P(2)$. It is not uncommon, for example, to find an expression like *and* being portrayed as a function which maps sentences into so-called adsentences. In accordance with this, the element in question will be associated semantically with a function that assigns the singleton set $\{1\}$ to the truth value *1*, and the empty set to the truth value *0*. What this means is that a sentence of the form *The dog barks and the cat meows* must be regarded as true just in case both of its component sentences are true. It requires little reflection to realize that the connectives *or*, *if* and *without* can be treated in an analogous manner. The following definition provides the necessary details.

Definition 11

$And(0) = \emptyset$	$Or(0) = \{1\}$
$If(0) = \{0,1\}$	$Without(0) = \{1\}$
$And(1) = \{1\}$	$Or(1) = \{0,1\}$
$If(1) = \{1\}$	$Without(1) = \emptyset$

From this description it is immediately clear that the connectives *and* and *or* are additive and multiplicative in nature; *if* and *without*, on the other hand, are anti-additive and antimultiplicative. These and similar findings have been collected in table 6. On the basis of this survey, it is easy to see that there can be substantial differences between the logical behavior of the connective and that of the corresponding adsentence. Characteristic examples are the expressions *if* and *without*. When we compare the semantical properties of these elements with those of adsentences of the forms *if S* and *without S*, it turns out that we are dealing with anti-additive and antimultiplicative expressions in the first case, but additive and multiplicative expressions in the second. Things are different, however, when, instead of *if* and *without*, we consider the elements *and* and *or*, for in this case it is not only the adsentence, but also the corresponding connective which is additive and multiplicative in nature.

We are now in a position to answer the question that has guided us throughout. As the functional characterization in (21) shows, the logical behavior of monotone decreasing expressions is governed by one half of the first law of De Morgan and one half of the second law of De Morgan. This means that the class of monotone decreasing elements shows a close resemblance to what Hazen (1992) refers to as *subminimal negation*. This is an order-inverting operation which satisfies the rule of contraposition in (45).

(45) $(x \rightarrow y) \rightarrow (\neg y \rightarrow \neg x)$

Table 6: Comparison of the logical behavior of connectives with that of the corresponding adsentences

	Connective	Adsentence
and	additive/multiplicative	additive/multiplicative
or	additive/multiplicative	additive/multiplicative
if	anti-additive/antimultiplicative	additive/multiplicative
without	anti-additive/antimultiplicative	additive/multiplicative

Downward monotonic expressions exhibit an analogous pattern, as is clear from Definition 4. By reference to Definitions 5 and 7 it can likewise be shown that the logical behavior of anti-additive expressions is determined by the first law of De Morgan as a whole and one half of the second law of De Morgan. In this respect, anti-additive phrases behave in much the same way as the operation of minimal negation in nonclassical logic. The class of antimultiplicative elements, on the other hand, satisfies the second law of De Morgan as whole and one half of the first law of De Morgan. Interestingly enough, there does not seem to be a corresponding logical operation. Following (Dunn 1993), any operation which satisfies both laws of De Morgan will be referred to as *De Morgan negation*. A good example of an expression which belong to this category is the particle *not*. The connectives *if* and *without* also exhibit the characteristic properties of De Morgan negation. In other words, we seem to have discovered a hierarchy of negative elements of increasing strength, which can be portrayed as the diamond in figure 1. This completes our exposition.

Figure 7

De Morgan negation
(anti-additive and antimultiplicative)

Minimal negation
(anti-additive)

GAP
(antimultiplicative)

Subminimal negation
(monotone decreasing)

References

Barwise, J.: On branching quantifiers in English. *Journal of Philosophical Logic* 8 (1979), 47–80.

Barwise, J., & Cooper, R.: Generalized quantifiers and natural language. *Linguistics and Philosophy* 4 (1981), 159–219.

Birkhoff, G.: *Lattice Theory*. Providence (American Mathematical Society), (1940, 1948, 1967).

Chellas, B.F.: *Modal logic: an introduction*. Cambridge 1980.

Dunn, J.M.: Perp and Star: Two Treatments of Negation. In: *Philosophical Perspectives* (*Philosophy of Language and Logic*) 7 (1993), ed. by J. Tomberlin, 331–357.

Hazen, A.: Subminimal Negation. Philosophy Department Preprint 1/92 University of Melbourne, 1992.

Keenan, E.L., & Faltz, L.M.: *Boolean semantics for natural language*. Dordrecht 1985.

Labov, W.: Negative attraction and negative concord in English grammar, *Language* 48 (1972), 773–818.

Ladusaw, W.A.: *Polarity sensitivity as inherent scope relations*. PhD thesis, University of Texas at Austin, 1979.

Ladusaw, W.A.: Expressing negation. In: *SALT II: Proceedings of the conference on Semantics and Linguistic Theory 2*, ed. by Ch. Barker & D. Dowty, Columbus/Ohio (The Ohio State University) 1992, 237–259.

Laka Mugarza, I.: *Negation in syntax: On the nature of functional categories and projections*. PhD thesis (MIT working papers in linguistics), MIT, Cambridge/MA. 1990.

Ponelis, F.F.: *Afrikaanse sintaksis*. Pretoria 1985[2].

Quine, W.V.O.: *Methods of Logic*. London 1952.

Sikorski, R.: *Boolean Algebras*, Berlin 1969[3].

Stoll, R.: *Sets, Logic, and Axiomatic Theories*. San Francisco 1974[2].

van Benthem, J.: *Essays in logical semantics*. Dordrecht 1986.

van der Wouden, T.: *Negative Contexts*. PhD thesis, University of Groningen, 1994.

van der Wouden, T., & Zwarts, F.: A semantic analysis of negative concord. In: *SALT III: Proceedings of the third conference on semantics and linguistic theory*, ed. by U. Lahiri & A.Z. Wyner, Ithaca/NY (Cornell University, Dep. of Modern Languages and Linguistics) 1993, 202–219.

Zanuttini, R.: Two strategies for negation: Evidence from Romance. In: *Proceedings of ESCOL V*, ed. by J. Powers & K. de Jong, The Ohio State University, Columbus, 1989, 535–46.

Zanuttini, R.: *Syntactic Properties of Sentential Negation: A Comparative Study of Romance Languages*. PhD thesis, University of Pennsylvania, 1991.

Zwarts, F.: The syntax and semantics of negative polarity. Ms., University of Groningen, 1993.

Zwarts, F.: Three types of polarity. Ms. Groningen, to appear in: *Plural Quantification*, ed. by F. Hamm & E. Hinrichs, Dordrecht 1995.

Part III

Negation in Artificial Intelligence

The Role of Negation in Nonmonotonic Logic and Defeasible Reasoning

Gerhard Schurz

ABSTRACT. §1 of this paper introduces into nonmonotonic reasoning (NMR) from the viewpoint of negation by default. §2 contains a classification of the most important NMR systems in terms of syntactic representation and axiomatization. One leaf of this classification is defeasible reasoning (DR), which is the focus of §§3-4. §3 reconstructs Donald Nute's strict defeasible logic as a natural deduction (or sequent) calculus (DR1), including several examples of proof trees, linear proofs and metalogical theorems (about Cut, Cautious Monotonicity, etc.). One characteristic feature of DR is its notion of the procedural negation $n(A)$ of a formula A, with the intended meaning that $n(A)$ should be derivable from a knowledge base **K** iff A is not derivable from **K**. DR systems which satisfy this biconditional are called n-complete. DR1 is not n-complete, because it cannot handle circularities. Based on the idea of keeping track on branches, an extended DR system (DR2) is constructed and proved to be n-complete in §4. Its n-completeness rests essentially on the fact that its underlying classical first order fragment is decidable. §5 discusses some fundamental connections between n-completeness, semantical completeness and decidability. DR systems which extend the full first order logic can be made n-complete only by introducing procedural negation by default.

1 Negation by Default in Nonmonotonic Logic

A great part of human reasoning is based on *uncertain laws* like

(1) Normally, birds can fly. formalized: $\text{Bird}(x) \Rightarrow \text{CanFly}(x)$

($\Rightarrow$ for uncertain implication). Uncertain laws $Ax \Rightarrow Bx$ have two characteristic *epistemic* features: (i) we know they are not strict, but have *exceptions* — 'abnormal' cases where Aa is true but Ba is false; and (ii) we do not known the corresponding statistical probabilities $p(Bx/Ax)$, even not vaguely. (i) implies that detachment "$Ax \Rightarrow Bx,\ Aa \mid\!\sim Ba$" is not truth-preserving and hence not deductively valid. (ii) implies that probability calculations are not applicable to uncertain laws.

That uncertain laws play an important role in common sense reasoning as well as in science was detected by philosophers of science already decades ago. Scriven (1959) called them *normic* laws. Consider examples like:

(2) Normally, a match being struck will light.
Normally, frustration causes aggression.
Normally, an economic crisis will increase radicalism.

Philosophers at that time did not make much sense of normic statements.[1] Scriven called them "truisms" and did not consider them as proper scientific laws. This led him to the claim that in history as well as in practical life, we have explanations without covering laws. The dominant attitude in the philosophy of science of the 60ies and 70ies was that there are only two strategies by which a normic law can be transformed into a truely scientific law. Either the normic law is completed such that it becomes a truely deterministic law. This presupposes a complete list of perturbing factors describing the possible exceptions which are required to be absent. For instance "if a match is struck, and it is not wet, and there is enough oxygen, and the temperature is high enough, and ..., then it will light". In most of all practical cases, this strategy is impossible on grounds of complexity – there are too many of rare but possible exceptional cases. In recent AI literature, this has been called the *qualification* problem (cf. Georgeff 1987, Prendinger/Schurz 199+). The second strategy proposed by Hempel (1965) was to replace the normic law by a probabilistic law. As mentioned above, usually we do not know the probability value, but even if we would know, this would not bring us very far, because knowledge of many other probability values is needed to enable sound probabilistic reasoning. For instance, if one wants to calculate the chances that a match which has been struck (Ax) will light (Bx) and one knows in addition that the match comes from a blue matchbox (Cx) of the shop round the corner (Dx), then the probability which has to be known is not (only) $p(Bx/Ax)$ but $p(Bx/Ax \wedge Cx \wedge Dx)$.[2] Also the second strategy fails on complexity grounds.

So we have to stick with the above representation of uncertain laws. They can be classified as qualitative (i.e., nonnumeric) inductive laws. The main steps to establish a "logic" of uncertain laws came from AI in the 80ies and was called nonmonotonic logic (NML), or nonmotonic reasoning (NMR). Note that according to philosophical terminology, NML should be better treated as a branch of inductive rather than deductive logic, because it is not truth-preserving. Meanwhile, many systems of NMR have been developed, for instance: nonmonotonic logic (McDermott/Doyle 1980), autoepistemic logic (Moore 1985), default reasoning (Reiter 1980, Poole 1988), defeasible reasoning (Nute 1992, Pollock 1994, Schurz 1995a,c), default logic based on conditionals (Delgrande 1988), on belief revision (Makinson/Gärdenfors 1991), or on ϵ-semantics (Pearl 1988, Schurz 1994b). These systems differ

[1] Adams (1966, 1975) is an outstanding exception.
[2] This is the requirement of total evidence going back to Carnap (1962, 211, 563).

in several important respects (see §2), but all of them share the following basic idea: one is allowed to detach the instantiated consequent Ba from an uncertain law $Ax \Rightarrow Bx$ and the instantiated antecedent Aa *only* as long as *nothing else* is derivable from one's *knowledge* which implies that Ba is false. Let $\vdash$ denote monotonic (deductive) and $\mid\!\sim$ denote nonmonotonic inference, and assume, for instance, that our knowledge base **K** contains that Tweety is a bird and that normally, birds can fly. Then we may detach that Tweety can fly:

(3) $\{\text{Bird}(x) \Rightarrow \text{CanFly}(x), \text{Bird}(\text{tweety})\} \mid\!\sim \text{CanFly}(\text{tweety})$

We assume by *default* that Tweety is a *normal* and not an exceptional bird, since our knowledge base does not imply the opposite. Call a statement expressing an exception an exception statement. Then we may express the basic idea of NML as follows: *negations of exception statements are assumed by default* – as long as an exception statement E is not derivable from **K**, we assume by default that $\neg E$.

Assume that we know in addition that penguins (strictly) can't fly. Then the statement *Penguin(tweety)* would be a typical exception statement w.r.t. (with respect to) the Bird-CanFly-Law. As long as **K** does not imply *Penguin(tweety)* we assume by default that $\neg$*Penguin(tweety)* holds. Hence we are allowed to detach $CanFly(tweety)$:

(4) $\{\text{Bird}(x) \Rightarrow \text{CanFly}(x), \text{Penguin}(x) \rightarrow \neg\, \text{CanFly}(x), \text{Bird}(\text{tweety})\}$
$\mid\!\sim \text{CanFly}(\text{tweety})$

However, as soon as we know that Tweety is a penguin we are no longer allowed to detach this consequence, because our knowledge base now deductively entails the opposite:

(5) $\{\text{Bird}(x) \Rightarrow \text{CanFly}(x), \text{Penguin}(x) \rightarrow \neg\, \text{CanFly}(x),$ $\vdash \neg\, \text{CanFly}(\text{tweety})$
$\text{Bird}(\text{tweety}), \text{Penguin}(\text{tweety})$ $\not\mid\!\sim \text{CanFly}(\text{tweety})$

We see why inference from uncertain laws is *nonmonotonic:* because additional knowledge may make previously derived consequences underivable. The so-called (meta-)rule of *monotonicity* $\Gamma \vdash A \Rightarrow \Gamma \cup \Delta \vdash A$ (where Γ, Δ are sets of sentences) becomes invalid in nonmonotonic logic. In defeasible reasoning, one also says that the Bird-Can-Fly law is *defeated*, and calls uncertain laws *defeasible laws.*

Call a knowledge base *conflicting* if the simultaneous application of detachment to all uncertain laws contained in it would lead to an inconsistency. To determine how to reason from conflicting knowledge is the most difficult question in NMR. One case is clear. If an uncertain law $Ax \Rightarrow Bx$ with instantiated antecedent Aa stands in conflict with a fact $\neg Ba$ or with a deterministic law $Cx \rightarrow \neg Bx$ with instantiated antecedent Ca, then the uncertain law will always be defeated, i.e. detachment from it will be blocked.

The difficult case is when two or more uncertain laws are in conflict, as e.g. in the knowledge base $\{Ax \Rightarrow Bx, Cx \Rightarrow \neg Bx, Aa, Ca\}$. The answers given by various systems of NMR are so different that a profound conceptual foundation of NML is certainly an urgent task.

Instead of attempting to solve this task here, let me try to give some orientation. As always in logic, this task involves two main problems.

The *first* problem is to give a semantic definition of the *correctness* of a nonmonotonic inference, and based on it, a *justification* of NML. In deductive logic, correctness coincides with *truth-preservation:* if the premises are true, then the conclusion is *certainly* true. But what is – or better: *should* be – *preserved* in a nonmonotonic inference? I see two main approaches. One is based on similarity relations between *possible worlds* – roughly speaking, the inference "$p \Rightarrow q, p \mid\sim q$" preserves truth in all possible worlds which are 'normal' w.r.t. p and q (cf. Kraus et al. 1990, Delgrande 1988). This approach is powerful as an abstract mathematical semantics, but less satisfactory for the purpose of a practical justification, because – this is the very point of uncertain reasoning – we never know whether *our* world is 'normal' or not. Hence this semantic definition does not tell us what a nonmonotonic inference preserves in *our* world. In this respect, the more important approach is *probability semantics.* It starts from the assumption that a necessary condition for an uncertain law $Ax \Rightarrow Bx$ to be true is that the probability of Bx given Ax is high – without specifying 'how high' it has to be. The probabilistic definition of the correctness of nonmonotonic inference requires, roughly speaking, that if the premises are true, then the conclusion has to be true in a 'sufficiently' high percentage of cases (cf. Adams 1975, ch. 2, Pearl 1988, ch. 10.2, Goldszmidt et. al. 1990, Schurz 1994b). Though this paper is not addressed to probability semantics (cf. Schurz 1994b), I will return to this question in §5.

The *second problem* is to give an appropriate syntactic representation and axiomatization. Most work done in NML has been devoted to this second problem. In the next section I try to give a classification of the main kinds of NMR in terms of differences in syntactic representation and axiomatization. One leaf of this classification – my favoured leaf – is defeasible reasoning, which is investigated in §§3–5.

2 Syntactic Representation and Axiomatization: An Overview

Let $A, B, \ldots$ range over formulas and $\Gamma, \Delta, \ldots$ over sets of them. Almost all systems of NML accept the following three rules:[3]

(R) $\frac{A \in \Gamma}{\Gamma \mid\sim A}$ *Reflexivity*

(C) $\frac{\Gamma \mid\sim A, \Gamma \cup \{A\} \mid\sim B}{\Gamma \mid\sim B}$ *Cut*

(SC) $\frac{\Gamma \vdash A}{\Gamma \mid\sim A}$ *Supraclassicality*

As is well-known, (R) and (C) together guarantee that the corresponding nonmonotonic consequence operation $\mathbf{C}_{\mid\sim}$ is a fixed point operation, i.e. satisfies $\Delta \in \mathbf{C}(\Delta)$ and $\mathbf{C}(\Delta) = \mathbf{C}(\mathbf{C}(\Delta))$. The exact way these axioms work depends on the underlying formal language. I see the following three major branching points in styles of formal representations.

(1.) Uncertain laws are (1.1) implicit or (1.2) explicit. In the systems of Makinson/Gärdenfors (1991) and Kraus et al. (1990), the uncertain laws are not explicit part of the language. It is assumed that the nonmonotonic inference relation is relativized to a given background of uncertain laws, also called 'defaults'. Therefore, the language in these approaches is either ordinary propositional language $\mathcal{L}0$ or first order language $\mathcal{L}1$. Nonmonotonic inferences are formulated as follows: "Bird(tweety) $\mid\sim$ Can-Fly(tweety)", or in $\mathcal{L}0$: "$p \mid\sim q$". – In all the other systems of NMR (known to me), the uncertain laws are explicit parts of the language. All these approaches have to deal with an *extended* language, extending $\mathcal{L}0$ or $\mathcal{L}1$ by some new primitive symbols. In defeasible reasoning, e.g., this new primitive is the defeasible impication sign $\Rightarrow$. Nonmonotonic inferences are then formulated in the following form: "Bird(x) $\Rightarrow$ Can-Fly(x), Bird(tweety) $\mid\sim$ Can-Fly(tweety)", or in $\mathcal{L}0$: "$p \Rightarrow q, p \mid\sim q$".

This difference in representation causes an important axiomatic difference. The axiom of structurality says that the inference relation is closed under substitutions (σ) for propositonal variables or predicates (cf. Schurz (1995b) on this notion):

(ST) $\frac{\Gamma \mid\sim A}{\sigma\Gamma \mid\sim \sigma A}$ *Structurality*

(ST) holds in the systems of type (1.2), but fails to hold in the systems of type (1.1).[4] Since I share the traditional view that logic is a matter of form and hence should be closed under substitution, I opt for systems of type (1.2).

[3]Cf. Kraus et al. (1990): (SC) is not mentioned there, but it is entailed by (LE) and (RW), see below.

[4]By the way, that (ST) can't hold for type (1.1) systems in $\mathcal{L}0$ follows already from the fact that propositional logic is the strongest consistent structural logic in $\mathcal{L}0$ and that (SC) holds in type (1.1) systems.

(1.2) Uncertain laws as (1.2.1) statements or as (1.2.2) rules. McDermott/Doyle (1980) formalize uncertain laws as statements involving material implication and a possibility operator: "Bird(x) $\wedge \Diamond$ Can-Fly(x) $\rightarrow$ Can-Fly (x)". Thereby, $\Diamond B(x)$ is derivable from the knowledge base **K** if $\neg B$ is not derivable from it. A further development of this kind of NMR is Moore's autoepistemic reasoning (Moore 1985). Alternatively, Reiter (1980) reconstructs uncertain laws as default rules of the form "Bird(x): M Can-Fly(x) / Can-Fly(x)", with the same condition for "M" ("might") as for "$\Diamond$" above. Konolige (1988) shows that both representations are intertranslatable. Hence I think the difference between type (1.2.1) and type (1.2.2) systems is superficial but not fundamental. The same remark applies to later developments. Systems of type (1.2.1) are Delgrande (1988), Pearl (1988, ch. 11) and Schurz (1995a); systems of type (1.2.2) are Poole (1988) and Nute (1988). The reason why I opt for systems of type (1.2.1) is twofold: first I think it is more natural to consider uncertain laws as statements, and second, I want to reserve the name "rule" for the inference rules expressed in the metalanguage (I think this helps to avoid confusions). We have simplified the things a little bit. Both McDermott/Doyle and Reiter admit a more general version of uncertain laws, of the form $p \wedge \Diamond r \rightarrow q$ (or $p : Mr\,/\,q$, respectively), where the formula in the scope of $\Diamond$ may be different from the conclusion. Laws of the form $p \wedge \Diamond q \rightarrow q$ are called *normal defaults* (Reiter 1980, 1985). Given one intends a probabilistic justification of NMR, then I think a restriction to normal defaults is justified. Such a restriction is important for the next branching point.

(1.2.i) Blocking Clause (1.2.i.1) implicit in inference rules or (1.2.i.2) explicit as part of the uncertain law. This branching occurs on both leafs of the previous branches, hence $i \in \{1, 2\}$. With the "blocking clause" I mean the clause specifying when detachment from an uncertain law is allowed, and when it is 'blocked' (or defeated). In the systems of McDermott/Doyle (1980), Reiter (1980), Moore (1985) and Poole (1988), the blocking clause is part of the uncertain law itself – in form of the possibility clause. For this reason they may be called *autoepistemic* – although Moore (1985), who invented this name, reserved it for systems of type 1.2.1. In the systems of Delgrande (1988), Pearl (1988, ch. 11) and in the systems of defeasible reasoning (DR), the blocking clause is not part of the uncertain law but is implicitly encoded in the (metalogical) inference rules of the system. The reason is that all the latter systems restrict themselves to normal default laws. Here, the possibility clause of McDermott/Doyle and Reiter becomes redundant, because it is determined by the rest of the uncertain law. In other words, a defeasible law "$p \Rightarrow q$" corresponds to a McDermott/Doyle law of the form $p \wedge \Diamond q \rightarrow q$ or to a Reiter rule of the form "$p : Mq/q$". The function of the possibility clause is taken over by the DR inference rules

which allow the detachment of q from $\{p \Rightarrow q, p\}$ only if $\neg q$ does not follow from the knowledge base. – On the reasons mentioned above I opt for systems of type (1.2.i.1). Provided one restricts to normal default laws, I think that translations between type (1.2.i.1) and type (1.2.i.2) systems should be easily possible, although I am not aware of such translations (except my own one in Schurz (1994b)).

(1.2.i.j) Unique(1.2.i.j.1) versus multiple (1.2.i.j.2) extensions. This is the most important branching point in NMR. According to the idea of Reiter,[5] one may infer nonmonotonic consequences from a consistent knowledge base by successive applications of detachment from uncertain laws and the classical inference steps as long as the set of consequences remains consistent. Given that the default laws are 'normal', this procedure will always reach some fixed point or 'extension'; however, it may end up in several different extensions, because this kind of nonmonotonic inference procedure is sensitive to the *ordering* in which formulas are detached. Here are some well-known examples.

(6) $\{\text{Mammal}(x) \Rightarrow \neg\text{Can-Fly}(x), \text{Bat}(x) \Rightarrow \text{Can-Fly}(x),$ $\text{Mammals}(x) \rightarrow \text{Bat}(x), \text{Mammal}(\text{dracula}), \text{Bat}(\text{dracula})\}$

(7) $\{\text{Adult}(x) \Rightarrow \text{Employed}(x), \text{Student}(x) \Rightarrow \neg\text{Employed}(x),$ $\text{Student}(x) \Rightarrow \text{Adult}(x), \text{Adult}(\text{peter}), \text{Student}(\text{peter})\}$

(8) $\{\text{Republican}(x) \Rightarrow \neg\text{Pacifist}(x), \text{Quaker}(x) \Rightarrow \text{Pacifist}(x),$ $\text{Republican}(\text{nixon}), \text{Quaker}(\text{nixon})\}$

In (6), if we first apply detachment to the mammal-law we infer that dracula can't fly, which blocks further detachment from the bat-law because of the consistency requirement. If we first apply detachment to the bat-law we infer that dracula can fly. So we end up in two possible extensions. Similary for the examples (7) and (8). While Reiter sticks with the existence of multiple extensions, most of the other authors have tried to develop methods which single out some *unique* preferred extension. This option is crucial for a further axiom:

(CM) $\dfrac{\Gamma \mid\!\sim A, \Gamma \mid\!\sim B}{\Gamma \cup \{A\} \mid\!\sim B}$ *Cautious Monotonicity* [6]

Together with Cut this axiom implies that if $\Gamma \mid\!\sim A$ and $\Gamma \cup \{A\} \mid\!\sim B$ then also $\Gamma \mid\!\sim B$ and $\Gamma \cup \{B\} \mid\!\sim A$; so the derivation process will be *cumulative*, i.e. independent from the ordering in which formulas are derived, and thus,

[5] See Reiter (1980, §3.1); cf. also Poole (1988) and Schurz (1994b, §2).

[6] Cf. Kraus et al. (1990) and Makinson/Gärdenfors (1991, 197). The axiom is called "restricted monotonicity" by Gabbay (1985, 447) and "triangularity" by Pearl (1988, 468).

will converge to the unique extension $\mathbf{C}(\Delta)$. On this reason, "cut" and "cautious monotonicity" together are also called "cumulativity" (after Makinson 1989). Hence, (CM) will not hold in type (1.2.i.j.2) systems allowing multiple extensions, but only in type (1.2.i.j.1) systems. Different suggestions have been made to single out a preferred extension. The *intersection approach*, suggested by McDermott/Doyle (1980), takes the intersection of all extensions. The priority approach, enhanced by Poole (1988), Brewka (1990), (1991, ch. 5) and also contained in Nute (1992, §5), assumes priorities between the uncertain laws; if a conflict arises the law with higher priority fires and the other one is blocked. A special kind of the *priority* approach is the specifity approach, favoured, e.g., by Delgrande (1988) and by the systems of DR. A law antecedent $A(x)$ is called *more specific* than another one $B(x)$ if we know that all or at least almost all A's are B's but not vice versa, i.e. if $A(x) \rightarrow B(x)$ or $A(x) \Rightarrow B(x)$ is entailed by our knowledge base without that the reverse direction is entailed by it (the first case is called *strict* specifity and the second case *defeasible* specifity; cf. Nute (1992 §§7–8). The specifity approach says that if two uncertain laws are in conflict, then only the law with the more specific antecedent 'fires', i.e. is open to detachment, while the other law is defeated. This entails that in a case where no one of the antecedents is more specific than the other, both laws are defeated and no detachment obtains. Example (6) is a case of strict specifity, and we may infer that Dracula can fly; example (7) is a case of defeasible specifity and we may infer that Peter is unemployed; finally (8) is a case where both laws defeat each other and nothing can be inferred about Nixon being a pacifist or not. The main reason why I opt for the specifity approach is that it is the only one which gets probabilistically justified (cf. Pearl 1988, ch. 11, Schurz 1994b).

The four axioms presented so far have some well-known consequences:

(LE) $\dfrac{\{A\} \dashv\vdash \{B\}, \{A\} \mid\!\sim C}{\{B\} \mid\!\sim C}$ *Left Logical Equivalence*

(RW) $\dfrac{\Gamma \mid\!\sim A, \{A\} \vdash B}{\Gamma \mid\!\sim B}$ *Right Weakening*

(CON) $\dfrac{\Gamma \mid\!\sim A, \Gamma \mid\!\sim B}{\Gamma \mid\!\sim A \wedge B}$ *Conjunction*

In the presence of (R), (C) and (CM), {(LE),(RW)} and (SC) are interderivable. Several authors aggree that (R), (C), (CM) and (SC) are the basis of reasonable nonmonotonic inference relations.[7]

[7]According to Kraus et al. (1990, 179), (R), (C), (CM), (LE) and (RW) are the basic axioms for cumulative consequence relations. (For Gabbay (1985, 455), (R), (C) and (CM) are basic requirements; he also requires (SC) (1985, 446).

The last branching point in our classification concerns the type of inference rules. Inference rules of nonmonotonic logics must always have some underivability preconditions among their premises. Hence they always have the form

$$(9)\quad \frac{\Gamma \mid\!\sim A, \Gamma \not\mid\!\sim B (\ldots)}{\Gamma \mid\!\sim f(A,B)}$$

where $f(A, B)$ is a formula determined by A and B. The question is how to handle these underivability preconditions. The route taken by most nonmonotonic systems is via consistency checks (cf. McDermott/Doyle, 1980, Reiter 1980, Poole 1988, Delgrande 1988). Another possible route is the characteristics of DR in Nute's style to which we turn in the next section. It is based on an explicit axiomatization of $\not\mid\!\sim$ (simultaneous with the axiomatization of $\mid\!\sim$) with help of a special negation operator, the n-operator meaning that a certain formula is not derivable. Defeasible reasoning has the advantage that it can be reconstructed in a style which is close to traditional monotonic logics. Although Nute himself presents his systems in a very different style, solely based on proof trees, I will show in the next section how his system can be reconstructed as a natural deduction (or sequent) calculus. This is not of purely esthetic purpose but will enable simple proofs of certain metalogical theorems.

3 Nute's strict defeasible logic

Nute's defeasible logic (Nute 1992), in short DL, is very close to PROLOG. Its predecessor (Nute 1988) was a PROLOG-implementation of defeasible reasoning, so-called *d-Prolog*. The language of DL, $\mathcal{L}$(DL), contains the following primitive symbols of $\mathcal{L}1$: (i) indiviual constants, variables and relation symbols (but not function symbols); (ii) strict (material) implication $\rightarrow$; (iii) classical negation $\neg$, (iv) as in PROLOG, n-ary conjunctions are represented a sets; (v) universal quantifiers are implicit in the understanding of laws. In addition, it contains the following operators extending $\mathcal{L}1$: $\Rightarrow$ for defeasible implication, the d-operator for defeasible derivability, and the n-operator for demonstrative non-derivability, which is called *derivational negation* and corresponds to PROLOG's negation by (finite) failure. Except for (implicit) conjunction, iteration is not allowed for any of these operators.

A *literal* is an atomic formula or its (classical) negation. From now on, $A, B, \ldots$ denote *closed literals* and $\mathbf{A}$, $\mathbf{B}$, ... finite sets of them; $\alpha, \beta, \ldots$ denote open literals and $\boldsymbol{\alpha}, \boldsymbol{\beta}, \ldots$ finite sets of them. A *knowledge base* is a pair $\mathbf{K} = \langle \mathbf{L}, \mathbf{F} \rangle$, where the set of *facts* $\mathbf{F}$ is a finite set of closed literals, and the set of *laws* $\mathbf{L}$ is a finite set of *strict* laws of the form $\boldsymbol{\alpha} \rightarrow \beta$ or *defeasible* laws of the form $\boldsymbol{\alpha} \Rightarrow \beta$. An *extended* closed literal is a formula of the form A, $n(A)$, $d(A)$ and $n(d(A))$; to save brackets we abbreviate $nd(A) := n(d(A))$.

X, Y, ... range over extended closed literals and **X**, **Y**, ... over finite sets of them. We have only one inference relation $\vdash$; the difference between strict (monotonic) and defeasible (nonmonotonic) inference is indicated by the d-operator.[8] Formulas of the form $n(A)$ or $nd(A)$ are called *n-shape formulas*. The inference relation is defined between knowledge bases and extended literals. The meaning of "$\vdash$" is as follows: $\mathbf{K} \vdash A$ stands for "A is strictly derivable from **K**", $\mathbf{K} \vdash n(A)$ for "A is demonstratively not strictly derivable from **K**", $\mathbf{K} \vdash d(A)$ for "A is defeasibly derivable from **K**", and $\mathbf{K} \vdash nd(A)$ for "A is demonstratively not defeasible derivable from **K**". For *sets* (conjunctions) of closed literals, we use the following abbreviations: $\mathbf{K} \vdash d(\mathbf{A})$ abbreviates $\mathbf{K} \vdash d(A)$ for *all* $A \in \mathbf{A}$, $\mathbf{K} \vdash n(\mathbf{A})$ abbreviates $\mathbf{K} \vdash n(A)$ for *some* $A \in \mathbf{A}$, and $\mathbf{K} \vdash nd(\mathbf{A})$ abbreviates $\mathbf{K} \vdash nd(A)$ for *some* $A \in \mathbf{A}$. The domain of possible instantiations of the laws in **L** is restricted to constants which occur either in **K** or in the proof goal X; hence it is finite. We let $i, j, ...$ range over functions mapping variables into constants. These function are called *unifiers*. $UF[\psi/\phi]$ denotes the set of all unifiers restricted to the variables occurring ψ and to the constants occuring either in **K** or in ϕ; thereby ψ and ϕ may be any formulas or formula sets of $\mathcal{L}$(DL). $\boldsymbol{\alpha} i$ stands for the i-instantiation of $\boldsymbol{\alpha}$, i.e. the result of replacing each variable z in $\boldsymbol{\alpha}$ by $i(z)$; similarly $\boldsymbol{\alpha} i := \{\beta i | \beta \in \boldsymbol{\alpha}\}$.

We call our basic system **DR1**. It differs from Nute's strict DL in some superficial respects. Nute describes his system in terms of proof trees, and he consideres the additional operators $\Rightarrow, d, n$ and nd as part of the meta-language (he speaks of "rules" insteasd of "laws").[9] While Nute presents his system mainly for propositonal logic, I immediately proceed to the first order representation, because I think the *meaning* of defeasible laws is only clear if they are represented with variables.[10] For sake of space only the basic part of Nute's system is presented: we skip his "might-defeaters" (§1) and also the complications in order to handle "preemting defeaters" (§6). I will present the rules in two versions, called the *quantifier* version and the *match* version. The quantifier version corresponds to Nute's own formulation, the match version is easier comprehensible and facilitates the proofs of several theorems (which are missing in Nute 1992). Here are the rules for strict derivability.[11] Though strict derivation in the given fragment of $\mathcal{L}1$ is trivial

[8]In my (Schurz 1995a) I did not include this d-operator, instead, I distinguished between $\vdash$ and $\mid\!\sim$. But this causes complications for the proper formulation of the axiom of Cut; therefore I prefer now the above representation.

[9]Nute writes Ep^+, p^- and $E(p^-)$ instead of $d(p)$, $n(p)$ and $nd(p)$, resp.

[10]If I say that something is *normally* the case, then I must assume that this "something" occurs *repeatedly*, hence it must contain a variable which may be instantiated in different ways.

[11](M^+1) plus (M^+2) – conjoined as indicated in remark (2.) below – correspond to Nute's rule M^+ in §3; (M^-) corresponds to Nute's (M^-). Since we have stated the rules

(because it is decidable anyway), the study of these rules may enlighten several characteristics of DL systems.

DR1, strict rules, quantifier version:

$$(\mathrm{M}^+1):\quad \frac{A \in \mathbf{F}}{\langle \mathbf{L}, \mathbf{F}\rangle \vdash A} \qquad (\mathrm{M}^+2):\quad \frac{\begin{array}{l}\boldsymbol{\alpha} \to \beta \in \mathbf{L}\\ i \in UF[\boldsymbol{\alpha}, \beta/A] \textit{ and } \beta i = A\\ \langle \mathbf{L}, \mathbf{F}\rangle \vdash \boldsymbol{\alpha} i\end{array}}{\langle \mathbf{L}, \mathbf{F}\rangle \vdash A}$$

$$(\mathrm{M}^-)\quad \frac{\begin{array}{l}A \notin \mathbf{F}\\ \forall(\boldsymbol{\alpha} \to \beta) \in \mathbf{L}\, \forall j \in UF[\boldsymbol{\alpha}, \beta/A]:\\ \qquad \textit{if } \beta j = A \textit{ then } \langle \mathbf{L}, \mathbf{F}\rangle \vdash n(\boldsymbol{\alpha} j)\end{array}}{\langle \mathbf{L}, \mathbf{F}\rangle \vdash n(A)}$$

In words: A is strictly derivable if A is a fact (M^+1) or A i-instantiates the consequent of a strict law and all i-instantiated antecedent conjuncts of this law are strictly derivable (M^+2); A is demonstratively not strictly derivable if A is not a fact and for each strict law having A as an i-instantiation of its consequent, at least one of its i-instantiated antecedents conjuncts is demonstratively not strictly derivable (M^-). The assumptions of *finite domain* and *finite* **K** are essential for yielding decidable rules for n-shape formulas. (Thereby, I think I have corrected a mistake in Nute's system by restricting unifiers to finite sets of variables.[12]) The finite **K** assumption is obvious. The finite domain assumption is necessitated by laws which contain *anonymous* variables in the antecedent, i.e., variables which do not occur in the consequent, for example in the law $Fxy \to Gx$ (equivalent to $\forall x(\exists y Fyx \to Gx)$). In order to prove $n(Ga)$ via (M^-) one has to prove $n(Fi(x)i(y))$ for every unifier i such that $i(x) = a$; hence one has to prove $n(Fua)$ for every constant u in **K**. This rule would not be decidable if the domain were infinite. Metalogical questions of the finite domain assumption are discussed in §5. – At hand at the M-rules we can observe some important properties of DR calculi.

1.) There are positive rules (for derivability) and negative rules (for demonstrative non-derivability). The negative rules differ from usual (monotonic) natural deduction calculi insofar some of their preconditions involve a universal quantification over all facts, laws and unifiers. Note that "$A \notin \mathbf{F}$" is an implicit universal quantification: $\forall F \in \mathbf{F} : F \neq A$. Dually, the positive rules involve implicit existential quantification ($A \in \mathbf{F}$ and $\boldsymbol{\alpha} \to \beta \in \mathbf{L}$).

---

not in the propositional but in the first order version we have modified them according to Nute's example $\mathrm{QD}^+_<$ in §9 (instead of an explicit translation of his propositional rules to first order logic Nute just gives one example.)

[12]If the unifiers are defined on the infinite domain of all variables (as in Nute's system), then even if the range of unifiers is restricted to finitely many constants one has still infinitely many unifiers; and this makes the rules undecidable.

2.) Each kind of rule (e.g. M^+-type) covers *start* rules and *iterative* rules. Start rules with verified preconditions terminate the derivation process – they yield the leafs in the corresponding proof tree. (M^+1) is a start rule (facts are called "call terms" in PROLOG), and (M^+2) an iterative rule. Also the negative rule contains implicitly a start rule, namely (M_s^-): "$A \notin \mathbf{F}$ *and* $\forall(\boldsymbol{\alpha} \rightarrow \beta) \in \mathbf{L}\, \forall j \in UF[\boldsymbol{\alpha}, \beta / A] : \beta j \neq A \,/\, \langle \mathbf{L}, \mathbf{F}\rangle \vdash n(A)$." If the preconditions of M_s^- are verified, the strict derivation of an n-shape formula terminates. Only because there exists no reasonable way to split (M^-) up into (M_s^-) and a second conjunct, (M^-) it presented as one single rule. – Vice versa, we may always join several rules for the same kind of goal into one single rule by connecting their preconditions via a disjunction. If we do this for the M^+-rules, we obtain (M^+): "$A \in \mathbf{F}$ *or* $\exists(\boldsymbol{\alpha} \rightarrow \beta) \in \mathbf{L}\, \exists i \in UF[\boldsymbol{\alpha}, \beta \,/\, A] : \beta i = A$ *and* $\langle \mathbf{L}, \mathbf{F}\rangle \vdash \boldsymbol{\alpha} i \,/\, \langle \mathbf{L}, \mathbf{F}\rangle \vdash A$." We call this rule the *joint* M^+-rule (similar for the D-rules below).

3.) The negative duals of the positive rules for one kind of goal G can be obtained by the following procedure called *negate*: (i) negate the precondition and the conclusion of the joint rule for G, (ii) apply the logical transformations for obtaining negative duals (de Morgan and $\forall$-$\exists$-transformation), thereby driving "not" in front of the derivability statements, and (iii) replace each phrase of the form "not $\vdash X$" by "$\vdash \underline{n}(X)$". Thereby, $\underline{n}(X)$ is called the n-complement of an extended literal X and is defined as follows: $\underline{n}(A) = n(A)$, $\underline{n}(n(A)) = A$, $\underline{n}(d(A)) = nd(A)$, and $\underline{n}(nd(A)) = d(A)$. (This definition of *negate* is applicable also to the rules for defeasible derivability below.) The operator $\underline{n}$ is a definitional extension of derivational negation (cf. §§4-5) which can be applied iteratively and satisfies $\underline{n}(\underline{n}(X)) = X$. We make the following observation (proofs will be given later):

Observation 3.1 Applying *negate* to the joint M^+-rule gives the joint M^--rule and vice versa.

4.) PROLOG's inference eninge is *backward chaining:* it tries to find a proof for a given goal X by scanning through all possibilities of proving X, thereby passing from instantiated law consequents to instantiated law antecedents. Basic for this search method is the notion of the *search tree* for a given proof goal X from a given base $\mathbf{K}$. It is an *And-Or-tree*, since a goal may be a conjunction of subgoals, and it may be proved in different ways (connected by 'or'). The nodes of the tree consist of extended literals, hence all matching conditions are omitted. The arcs lead from instantiated law consequents downward to instantiated law antecedents, or they lead from an and/or node to its conjuncts/disjuncts. Nodes which are verified by a start rule are associated with "+" (true), nodes which are neither facts nor match with some law are associated with "–"("fail"). "+" and "–" are propagated upwards the tree according to the obvious rules for and- and or-nodes. As an example, consider the knowledge base:

(10) $\{\{\text{Parent}(x,y), \text{Parent}(y,z)\} \rightarrow \text{Grandparent}(x,z)$,
Parent(a,b), Parent(b,c), Parent(a,d)}

The *search tree* for Grandparent(a,d) is given in Fig. 1. Unconnected arcs represent the Ors, connected arcs represent Ands.

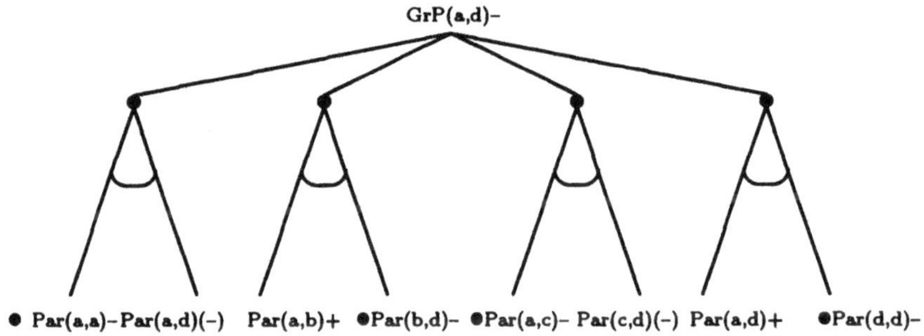

Fig. 1: *Search tree for Grandparent(a,d) - Example (10)*

PROLOG scans through the search tree in a *depth first* and *left to right* manner. Important subtrees of the search tree are the following. The *termination tree* for a goal X consists of that part of the search tree for X through which the search mechanism has actually scanned before *terminating* either with "true" or with "fail." A *proof tree* for a goal X is an And-subtree of the termination tree for X which has "+" on all of its leafs. A *failure tree* for a goal X is an Or-subtree of the termination tree which has "–" on all of its leafs. In the example of Fig. 1, the failure tree of "GrP(a,d)" is the Or-subtree consisting of the nodes indicated by a dot at the left, and the termination tree is that subtree in which all leafs having their value (+/–) in brackets are omitted – these leafs are not needed in the computation of the goal value. Termination, proof and failure tree are always finite, while search trees may be infinite (cf. §4). – It is easy to verify the following observations:

Observation 3.2 The search tree for a positive goal A and that for its n-complement $n(A)$ are *dual* in the sense of being structurally isomorphic, except that formulas and their n-complements, And- and Or-connections of arcs, and "+" and "–" are exchanged.

Observation 3.3 The proof-tree of an n-shape formula $n(A)$ is identical with the dual of A's failure tree.

Observation 3.4 The n-operator corresponds exactly to PROLOG's negation by (finite) failure.

Observations 3.2 and 3.3 follow from remark 3.) about the procedure *negate* and the definitions of search, proof and failure trees. They will be important for §4. For example, the proof tree for n(GrP(a,d)) is illustrated in Fig. 2.

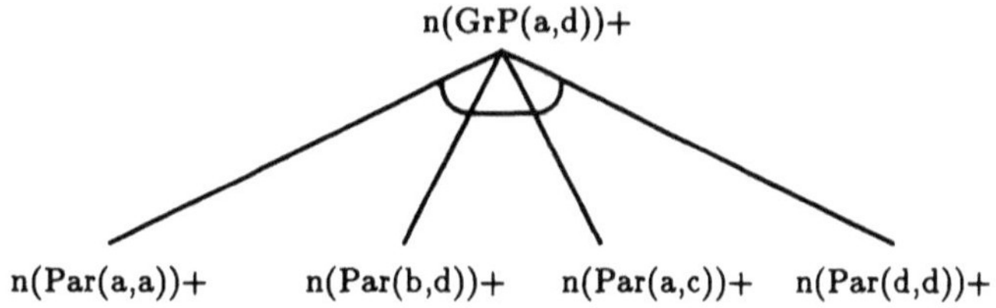

Fig. 2: *Proof tree for n(GrP(a,d)) – Example (10), cf. Fig. 1*

It is exactly the dual of the above failure tree for GrP(a,d) (the subtree indicated by dots). Observation 3.4 follows from Observation 3.3. For, PROLOG's definition "not A :- A, !, fail; true" implies that the attempt to prove "not A" returns "true" exactly if the search for A yields a failure tree for A; it returns no if the search for A leads to a proof tree for A. The remaining possibility (which is the topic of §4) is that scanning through the search tree leads into an infinite regress or loop, in which case PROLOG tells you a memory overflow. Hence, our rules for n-shape formulas are an *axiomatization of* PROLOG*'s negation by (finite) failure*, telling the exact preconditions for proving "not A" – something which is not obvious from PROLOG's definition of "not".

5.) The quantifier version of the rules has the disadvantage that there is no way to break the quantified premises into parts, which brings difficulties for comprehension as well as for metalogical targets. We present them now in an equivalent version which is close to PROLOG's *matching* procedure. A law with consequent β 'matches' a proof goal B if $B = \beta i$ for some i. Let $M(B, \mathbf{Ls})$ ("M" for "match") stand for the set of all instantiated antecedents sets of strict laws in $\mathbf{L}$ matching B; more precisely:

$$M(B, \mathbf{Ls}) = \{\mathbf{A} \mid \exists\ \boldsymbol{\alpha} \rightarrow \beta \in \mathbf{L}\, \exists i \in UF(\boldsymbol{\alpha}, \beta/B) : \beta i = B \textit{ and } \boldsymbol{\alpha} i = \mathbf{A}\}$$

Similarly, $M(B, \mathbf{Ld})$ and $M(B, \mathbf{L})$ stand for the set all instantiated antecedent sets of all defeasible laws or all laws, respectively. With this auxiliary notion, the M-rules become this:

DR1, strict rules, match version:

(M+1): $\dfrac{A \in \mathbf{F}}{\langle \mathbf{L}, \mathbf{F} \rangle \vdash A}$ (M+2): $\dfrac{\mathbf{B} \in M(A, \mathbf{Ls}) \quad \langle \mathbf{L}, \mathbf{F} \rangle \vdash \mathbf{B}}{\langle \mathbf{L}, \mathbf{F} \rangle \vdash A}$

(M−): $\dfrac{A \notin \mathbf{F} \quad M(A, \mathbf{Ls}) = \{\mathbf{B}_1, ..., \mathbf{B}_m\} \quad \text{For all } 1 \leq i \leq m : \langle \mathbf{L}, \mathbf{F} \rangle \vdash n(\mathbf{B}_i)}{\langle \mathbf{L}, \mathbf{F} \rangle \vdash n(A)}$

Instead of quantifying over strict laws, we list all elements of $M(-, \mathbf{Ls})$. The equivalence of the quantifier and the match version is obvious. Here is an example of a linear proof.[13] It is always a good tactic to state first all match-conditions as premises. As an abbreviation we introduce the notation $M(\alpha, \mathbf{Ls}) = \emptyset$ meaning that no instantiation of α matches with some strict law.

Linear proof of n(GrP(a,d)), corresponding to the proof tree in Fig. 2:

1)	$\mathbf{F}$ = {Par(a,b), Par(b,c), Par(a,d)}	Premise
2)	M(Par(x,y), $\mathbf{Ls}$) = $\emptyset$	Premise
3)	M(GrP(a,d), $\mathbf{Ls}$) = {{Par(a,a),Par(a,d)}, {Par(a,b),Par(b,d)}, {Par(a,c),Par(c,d)}, {Par(a,d),Par(d,d)} }	Premise
4)	$\langle \mathbf{F}, \mathbf{L} \rangle \vdash n$(Par(a,a))	By (M−) from 1,2
5)	$\langle \mathbf{F}, \mathbf{L} \rangle \vdash n$({Par(a,a),Par(a,d)})	By def. from 4
6)	$\langle \mathbf{F}, \mathbf{L} \rangle \vdash n$(Par(b,d))	By (M−) from 1,2
7)	$\langle \mathbf{F}, \mathbf{L} \rangle \vdash n$({Par(a,b),Par(b,d)})	By def. from 6
8)	$\langle \mathbf{F}, \mathbf{L} \rangle \vdash n$(Par(a,c))	By (M−) from 1,2
9)	$\langle \mathbf{F}, \mathbf{L} \rangle \vdash n$({Par(a,c),Par(c,d)})	By def. from 8
10)	$\langle \mathbf{F}, \mathbf{L} \rangle \vdash n$(Par(d,d))	By (M−) from 1,2

[13]Note that the derivation rules in the match version do not completely correspond to PROLOG's inbuilt inference algorithm; the exception are anonymous variables. In the example of Fig. 1, PROLOG does not construct all instantiations of {P(a,x), P(x,d)} with constants in **K**, then trying to prove these instantiations, one after the other. Rather, it tries to match the leftmost conjunct P(a,x) either with some fact or with some consequent of a law. If it succeeds with unifier i, it then tries to prove the second conjunct instantiated with i. So, PROLOG will never attempt to prove the instantiations Par(a,a) and Par(a,c) – they are excluded by the matching operation, which is much more economically. An axiomatization which ***procedurally*** corresponds to PROLOG is possible by extending the inference relation to sets of open literals (implicitly understood as existentially quantified conjunctions).

11)	$\langle \mathbf{F}, \mathbf{L} \rangle \vdash n(\{\text{Par(a,d)},\text{Par(d,d)}\})$	By def. from 10
12)	$\langle \mathbf{F}, \mathbf{L} \rangle \vdash n(\text{GrP(a,d)})$	By (M^-) from 1,3,5,7,9,11

We turn to the rules for defeasible derivability. The complement operation "–" on literals is defined as $-At := \neg At$ and $-\neg At := At$ (for At an atomic formula). For sake of space, we present only the match version – how to translate it into the quantifier version is obvious.[14]

DR1, defeasible rules, match version:

(D^+1):
$$\frac{\langle \mathbf{L}, \mathbf{F} \rangle \vdash A}{\langle \mathbf{L}, \mathbf{F} \rangle \vdash d(A)}$$

(D^+2):
$$\frac{\begin{array}{l} \mathbf{B} \in M(A, \mathbf{Ls}) \\ \langle \mathbf{L}, \mathbf{F} \rangle \vdash d(\mathbf{B}) \\ \langle \mathbf{L}, \mathbf{F} \rangle \vdash n(-A) \end{array}}{\langle \mathbf{L}, \mathbf{F} \rangle \vdash d(A)}$$

(D^+3):
$$\frac{\begin{array}{l} \mathbf{B} \in M(A, \mathbf{Ls}) \\ \langle \mathbf{L}, \mathbf{F} \rangle \vdash d(\mathbf{B}) \\ \langle \mathbf{L}, \mathbf{F} \rangle \vdash n(-A) \\ M(-A, \mathbf{Ls}) = \{\mathbf{C}_1, ..., \mathbf{C}_k\} \\ \text{For all } 1 \le i \le k : \langle \mathbf{L}, \mathbf{F} \rangle \vdash nd(\mathbf{C}_i) \\ M(-A, \mathbf{Ld}) = \{\mathbf{D}_1, ..., \mathbf{D}_m\} \\ \text{For all } 1 \le i \le m\text{:} \quad \textit{Either } \langle \mathbf{L}, \mathbf{F} \rangle \vdash nd(\mathbf{D}_i) \\ \qquad\qquad\qquad\qquad \textit{or: } \langle \mathbf{L}, \mathbf{B} \rangle \vdash d(\mathbf{D}_i) \textit{ and } \langle \mathbf{L}, \mathbf{D}_i \rangle \vdash nd(\mathbf{B}) \end{array}}{\langle \mathbf{L}, \mathbf{F} \rangle \vdash d(A)}$$

(D^-1):
$$\frac{\begin{array}{l} \langle \mathbf{L}, \mathbf{F} \rangle \vdash n(A) \\ \langle \mathbf{L}, \mathbf{F} \rangle \vdash -A \end{array}}{\langle \mathbf{L}, \mathbf{F} \rangle \vdash nd(A)}$$

(D^-2):
$$\frac{\begin{array}{l} \langle \mathbf{L}, \mathbf{F} \rangle \vdash n(A) \\ M(A, \mathbf{Ls}) = \{\mathbf{B}_1, ..., \mathbf{B}_k\} \\ \text{For all } 1 \le i \le k : \langle \mathbf{L}, \mathbf{F} \rangle \vdash nd(\mathbf{B}_i) \\ M(A, \mathbf{Ld}) = \{\mathbf{C}_1, ..., \mathbf{C}_m\} \\ \text{For all } 1 \le i \le m : \textit{Either: } \langle \mathbf{L}, \mathbf{F} \rangle \vdash nd(\mathbf{C}_i) \\ \textit{or: } \mathbf{D} \in M(-A, \mathbf{Ls}) \textit{ and } \langle \mathbf{L}, \mathbf{F} \rangle \vdash d(\mathbf{D}) \\ \textit{or: } \mathbf{E} \in M(-A, \mathbf{Ld}) \textit{ and } \langle \mathbf{L}, \mathbf{F} \rangle \vdash d(\mathbf{E}) \\ \qquad \textit{and } [\langle \mathbf{L}, \mathbf{C}_i \rangle \vdash nd(\mathbf{E}) \textit{ or } \langle \mathbf{L}, \mathbf{E} \rangle \vdash d(\mathbf{C}_i)] \end{array}}{\langle \mathbf{L}, \mathbf{F} \rangle \vdash nd(A)}$$

In words: (D^+1) says that whatever is strictly derivable is also defeasibly derivable. (D^+2) tells us that A is defeasibly derivable if A is the instantiated consequent of a strict law and all of its instantiated antecedent conjuncts are defeasibly derivable, provided the complement of A is demonstratively

[14] Nute writes "¬" instead of "–". (D^+1) corresponds to Nute's (E^+) and (D^-1) to Nute's E^- – both are found in his §3. Nute calls these two rules together with the M-rules the *monotonic kernel* of defeasible logic, because the inference relation defined by them is monotonic. (D^+2) corresponds to the rule S^+ in his §4; (D^+3) and (D^-2) correspond to his rules to $D^+_\rightarrow$ and $D^-_\rightarrow$ in §8 (which extends the same rules of §7) – except that we don't have might-defeaters.

not strictly derivable. Hence strict laws with defeasibly derivable antecedents are defeated only if the complement of the instantiated consequent is strictly derivable. (D^+3) says that A is defeasibly derivable if it is the instantiated consequent of a defeasible law and all the instantiated antecedent conjuncts of this law are defeasibly derivable, provided the complement of A is demonstratively not strictly derivable and the law is not defeated, where the latter conditions means this: (i) for each complementary strict law (that is, strict law having the complement of A as instantiated consequent), at least one of its instantiated antecedent conjuncts is demonstratively not defeasibly derivable, and (ii) for each complementary defeasible law, either at least one antecedent conjunct is not defeasibly derivable or the first law is more specific than the second one, where this latter condition means that the instantiated antecedent of the second law is defeasibly derivable from the instantiated antecedent of the first law via the laws in **L**, but not vice versa.[15] (D^-1) says that A is demonstratively not defeasibly derivable if A is demonstratively not strictly derivable and A's complement is strictly derivable. Finally, (D^-2) tells us that A is demonstratively not defeasibly derivable if A is demonstratively not strictly derivable and for all strict laws having A as instantiated consequent, at least one instantiated antecedent conjunct is demonstratively not defeasibly derivable, and for all defeasible laws having A as instantiated consequent, either at least one instantiated antecedent conjunct is demonstratively not defeasibly derivable, or the defeasible law is defeated by another law, which means that either there exists a complementary strict law with all instantiated antecedents conjuncts being defeasibly derivable, or there exists a complementary defeasible law with all instantiated antecedent conjuncts being defeasibly derivable which is not less specific than the first defeasible law.

An easy exercise verifies that application of the procedure *negate* to the joint D^+-rule leads to the joint D^--rule and vice versa.[16] Let us demonstrate how defeasible derivability works at hand of the following example:

---

[15]Unlike Nute, we need not restrict to laws with nonempty antecedents because we deal within predicate logic, and Nute's problem (in §8) arises only in propositional logic.

[16]Parallel to this so-called 'strict' system, Nute defines a 'semi-strict' system. Given two conflicting strict laws $Ax \rightarrow Bx$ and $Cx \rightarrow \neg Bx$ with defeasibly derivable instantiated antecedents $d(Aa)$ and $d(Ca)$, the strict systems infers both $B(a)$ and $\neg B(a)$, provided neither $\neg B(a)$ nor $B(a)$ are strictly derivable. In contrast, the semi-strict system infers neither $B(a)$ nor $\neg B(a)$ (because here the rule D^+2 contains an additional defeat clause). I prefer the strict system because it follows from probability theory that the instantiated consequents of strict laws are at least as probable as their instantiated antecedents and hence should be inferred. If this leads to contradictions then the culprit of this is not the strict rule D^+1; this rule only makes the contradiction explicit. We will see in §5 that Nute's system indeed fails to satisfy the axiom of consistency preservation, by passing to the semi-strict version this failure can't be avoided.

(11) $\{Bx \Rightarrow Fx, Ax \rightarrow Fx, Px \rightarrow Bx, Px \Rightarrow \neg Fx, Pt, Bt\}$

Read: Bx – Bird(x), Fx – Can-Fly(x), Ax – Aeroplane(x), Px – Penguin(x), t – tweety

The termination tree for $\langle \mathbf{L}, \mathbf{F} \rangle \vdash d(\neg Ft)$ is presented in Fig. 3. It is not enough that the nodes contain formulas (as in the trees for strict derivability), because in testing the precondition of being more specific, formulas have to be derived from knowledge bases which are different from the given knowledge base $\langle \mathbf{L}, \mathbf{F} \rangle$, e.g. from $\langle \mathbf{L}, \mathbf{A} \rangle$ (etc.). So our nodes have to contain derivability claims of the form $\langle \mathbf{L}, \mathbf{A} \rangle \vdash X$. Since the set of laws remains unchanged through the search, we abbreviate these claims as $\mathbf{A} : X$ without mentioning $\mathbf{L}$. Because of limitations of space we develop the tree now from left to right. Double vertical lines stand for And-connnections, single vertical lines for Or-connections; single arcs are written horicontally. We also state the linear proof.

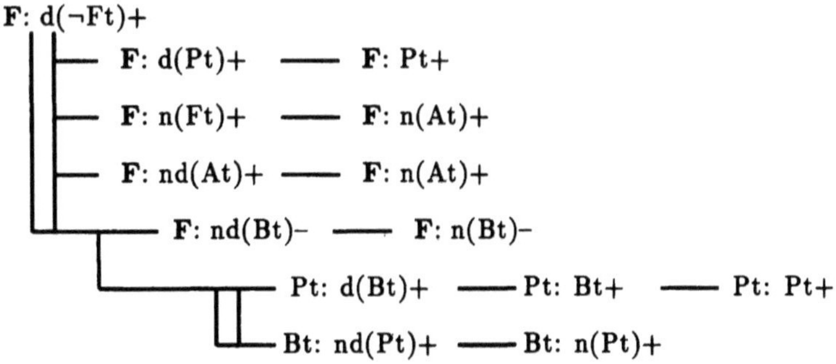

Fig. 3: *Termination tree for* **F**: *d(¬Ft) – Example (11)*

Linear proof for **F**: *d(¬Ft) – Example (11):*

1) $\mathbf{F} = \{\text{Pt, Bt}\}$ Premise
2) $M(\neg\text{Ft}, \mathbf{Ls}) = \{\{\text{Pt}\}\}$ Premise
3) $M(\text{Ft}, \mathbf{Ls}) = \{\{\text{At}\}\}$ Premise
4) $M(\text{Ft}, \mathbf{Ld}) = \{\{\text{Bt}\}\}$ Premise
5) $M(\text{Bt}, \mathbf{Ls}) = \{\{\text{Pt}\}\}$ Premise
6) $M(\text{At}, \mathbf{L}) = \emptyset$ Premise
7) $M(\text{Pt}, \mathbf{L}) = \emptyset$ Premise
8) **F**: Pt By $M^{+}1$ from 1
9) **F**: d(Pt) By $D^{+}1$ from 8
10) **F**: n(At) By M^{-} from (1), (6)

11) **F**: n(Ft)	By M^- from (1), (3), (10)
12) **F**: nd(A(t))	By D^-2 from (6), (10)
13) Pt: Pt	By M^+1
14) Pt: Bt	By M^+2 from 13, 5
15) Pt: d(Bt)	By D^+1 from 14
16) Bt: n(Pt)	By M^- from 7
17) Bt: nd(Pt)	By D^-2 from 16, 7
18) **F**: $d(\neg$Ft)	By D^+3 from 2, 9, 11, 3, 12, 4, 15, 17

We turn to the four basic axioms (of §2), (R), (C), (CM) and (SC). The proper formulation of (SC) is "$\mathbf{K} \vdash A \,/\, \mathbf{K} \vdash d(A)$"; this axiom is verified by the rule (D^+1). In order to prove (R), (C) and (CM), we have to modify the system **DR1** in an inessential respect. So far the inference relation $\Gamma \vdash \psi$ of **DR1** is defined only if X is an extended literal, but not a law, and if Γ is a "knowledge base", i.e. contains laws and literals, but not extended literals. In order to verify (R) we must allow for laws to appear in the conclusion. In order to verify (C): "$\mathbf{K} \vdash X$, $\mathbf{K} \cup \{X\} \vdash Y \Rightarrow \mathbf{K} \vdash Y$" we must add an extended literal X to the premise set **K** and hence allow the premises to consist of arbitary formulas of $\mathcal{L}$(DL). Let $\phi_1, \phi_2, ..., \psi_1, \ldots$ range over arbitrary formulas of $\mathcal{L}$(DL) and $\Gamma_1, \Gamma_2, ..., \Delta_1, \ldots$ over arbitrary sets of them. Each $\phi \in \Gamma$ is either a strict or a defeasible law or a closed literal or an extended (closed) literal. Γ is called an *extended* knowledge system; the extended literals of Γ are called the *hypotheses* of Γ. Hence, extended knowledge systems consist of knowledge bases plus arbitrary hypotheses. $\mathbf{L}_\Gamma$ stands for the set of laws in Γ (similarly for $\mathbf{Ls}_\Gamma$ and $\mathbf{Ld}_\Gamma$). We call the extended basic system **DR1$^+$**. Its rules are obtained from the rules of **DR1** by the following simple modifications:

(1.) The new rule (M^+1) has the form $\phi \in \Gamma / \Gamma \vdash \phi$. The new rule ($M^-1$) has the form "$A \notin \Gamma$, $M(A, \mathbf{Ld}_\Gamma) = \{\mathbf{B}_1, \ldots, \mathbf{B}_n\}$, for all $1 \leq i \leq m : \Gamma \vdash n(\mathbf{B}_i) \,/\, \Gamma \vdash n(A)$." — (2.) In all other rules we replace "$\langle \mathbf{L}, \mathbf{F} \rangle$" by "$\Gamma$", and "**L**" by "$\mathbf{L}_\Gamma$" where **L** occurs in contexts different from "$\langle \mathbf{L}, \mathbf{F} \rangle$" (in match-conditions or in specifity conditions), and similarly for **Ls** and **Ld**.

Theorem 3.1 DR1$^+$ satisfies (R), (C), (CM), (SC).

Proof: (SC) was proved above. (R) follows from the new rule (M^+1). For the rest, note the following

Lemma 3.1 If $\Gamma \vdash \psi$, then $\mathbf{L}_{\Gamma \cup \{\psi\}} = \mathbf{L}_\Gamma$.

Lemma 3.1 holds because the only way to derive a law from Γ is via M^+1.

For (C): Assume (i) $\Gamma \vdash \psi_1$ and (ii) $\Gamma, \psi_1 \vdash \psi_2$ (as usual "Γ, ψ" abbreviates "$\Gamma \cup \{\psi\}$"). We prove by induction on the length of the linear proof of $\Gamma, \psi_1 \vdash \psi_2$ that for all sequents of the form $\Gamma, \psi_1 \vdash \psi$ occurring in this proof, also $\Gamma \vdash \psi$ is derivable. Instead of going through all the rules separately, we give a *general argument.* Assume $\Gamma, \psi_1 \vdash \psi$ has been derived by some of our rules, call it RU, from certain premises. We show that then also the modified premises where "$\Gamma \cup \{\psi_1\}$" is replaced by "Γ" are derivable. This implies that $\Gamma \vdash \psi$ can be derived by the *same* rule RU from these modified premises – with one exception where $\Gamma \vdash \psi$ follows from assumption (i). All premises of all of our rules have one of the following forms: (a) $\Gamma \cup \psi_1 \vdash \phi$. Then we conclude $\Gamma \vdash \phi$ by IH (ind. hyp.). (b) $M(A, \mathbf{L}_{\Gamma \cup \{\psi_1\}}) = \{\mathbf{B}_1, ..., \mathbf{B}_m\}$. Then $M(A, \mathbf{L}_\Gamma) = \{\mathbf{B}_1, ..., \mathbf{M}_m\}$ follows by Lemma 3.1. The same holds if **Ls** or **Ld** stands in place of **L**. (c) The condition $\psi \in \Gamma \cup \{\psi_1\}$ in M^+1. Then either $\psi \in \Gamma$, whence we conclude $\Gamma \vdash \psi$ by M^+1. Or $\psi = \psi_1$, in which case $\Gamma \vdash \psi$ follows from assumption (i). (d) The condition $A \notin \Gamma \cup \{\psi_1\}$ in M^-, where $\psi = n(A)$. This implies $A \notin \Gamma$. (e) The conditions $\langle \mathbf{L}_{\Gamma \cup \{\psi_1\}}, \mathbf{A} \rangle \vdash d(\mathbf{B})$ and $\langle \mathbf{L}_{\Gamma \cup \{\psi_1\}}, \mathbf{A} \rangle \vdash nd(\mathbf{B})$ for suitable **A** and **B** in D^+3 and D^-2. By Lemma 3.1, $\mathbf{L}_{\Gamma \cup \{\psi_1\}} = \mathbf{L}_\Gamma$, so these premises and hence their derivations keep the same as in the proof of $\Gamma, \psi_1 \vdash \psi_2$. So in all cases, $\Gamma \vdash \psi$ follows from the modified premises by the same rule RU. By induction, $\Gamma \vdash \psi_2$ is $\mathbf{DR1}^+$-derivable.

For (CM): By a similar general argument. We assume $\Gamma \vdash \psi_1$ and $\Gamma \vdash \psi_2$ and prove that $\Gamma, \psi_1 \vdash \psi_2$ by induction on the length of the proof of $\Gamma \vdash \psi_2$. Again we prove that for each sequent of the form $\Gamma \vdash \psi$ occurring in this proof, $\Gamma, \psi_1 \vdash \psi$ is derivable. Assume $\Gamma \vdash \psi$ has been derived from premises by one of our rules, say RU. We show that then also the modified premises where "Γ" is replaced by "$\Gamma \cup \{\psi_1\}$" are derivable, whence also $\Gamma \cup \{\psi_1\} \vdash \psi$ is derivable by RU. (a) If a premise has the form $\Gamma \vdash \phi$, we conclude $\Gamma \cup \{\psi_1\} \vdash \phi$ by IH. (b) If a premise is a match-condition, we argue as in (b) above. (c) The premise $\psi \in \Gamma$ in (M^+1) implies the modified premise $\psi \in \Gamma \cup \{\psi_1\}$. (d) The premise is the condition $A \notin \Gamma$ in (M^-), and $\psi = n(A)$. Assume in the opposite that $A \in \Gamma \cup \{\psi_1\}$. Then $A = \psi_1$, whence $\Gamma \vdash A$ by assumption (i). Because $A \notin \Gamma$, $\Gamma \vdash A$ can only be derived via (M^+2), which implies that $\mathbf{B} \in M(A, \mathbf{Ls}_\Gamma)$ and $\Gamma \vdash \mathbf{B}$ must hold. But this contradicts the fact that $\Gamma \vdash n(A)$ has been derived via (M^-). So $A \notin \Gamma \cup \{\psi_1\}$. (e) If the premise its a specifity condition, we argue as in (e) above. Q.E.D.

The distinction between strictly and defeasibly derivable formulas within our language is essential for the validity of (C) and (CM). Without this distinction these axioms would fail, because our rules are sensitive to whether the consequences have been derived strictly or defeasibly. To give an example, assume $\mathbf{K} = \{Ax \Rightarrow Bx, Bx \rightarrow Cx, Dx \rightarrow \neg Cx, Aa, Da\}$. Then (i) $\mathbf{K} \vdash d(Ba)$, (ii) $\mathbf{K} \nvdash Ca$, (iii) $\mathbf{K} \cup \{d(Ba)\} \nvdash Ca$, but (iv) $\mathbf{K} \cup \{Ba\} \vdash Ca$. So, if we add Ba instead of $d(Ba)$ to **K**, Ca becomes derivable, which would be a violation of Cut if we had not distinguished between Bd and $d(Ba)$.[17]

4 The problem of negation completeness

First some general remarks on the relation between *classical* and *derivational negation*. For classical negation, the concepts of consistency and completeness are these:

$$\mathbf{K} \text{ is } \neg\text{-consistent iff: } \mathbf{K} \vdash A \Rightarrow \textit{not}\, \mathbf{K} \vdash -A \quad (\text{for all } A)$$

$$\mathbf{K} \text{ is } \neg\text{-complete iff: } \mathbf{K} \vdash A \Leftarrow \textit{not}\, \mathbf{K} \vdash -A \quad (\text{for all A})$$

Let us also generally arrange that the conjunction of *x-consistency* and *x-completeness* (x is a specification, e.g. "$\neg$") is called *x-adequateness*. Obviously, $\neg$-consistency coincides with classical or *strict* consistency: *not* **K** $\vdash A, \neg A$ for any A); $\neg$-completeness coincides with the notion of completeness as defined for first order theories: **K** $\vdash A$ or **K** $\vdash \neg A$ for all A. Knowledge bases should be $\neg$-consistent, but they need not be $\neg$-complete, since this would presuppose complete knowledge about that part of the real world which is described by the language.

Consider the corresponding notions for the derivational negation n. While $\neg$-consistency and $\neg$-completeness are properties of **K**, n-consistency and n-completeness are properties of the underlying inference relation $\vdash$. Hence, in defining these properties we have to quantify over **K**.

$$\vdash \text{ is } n\text{-consistent iff: } \mathbf{K} \vdash X \Rightarrow \textit{not}\, \mathbf{K} \vdash \underline{n}(X) \quad (\text{for all } X \text{ and } \mathbf{K})$$

$$\vdash \text{ is } n\text{-complete iff: } \mathbf{K} \vdash X \Leftarrow \textit{not}\, \mathbf{K} \vdash \underline{n}(X) \quad (\text{for all } X \text{ and } \mathbf{K})$$

The relation between classical and derivational negation can be explained in terms of $n/\neg$- consistency and completeness as follows:

$$\text{If } \mathbf{K} \text{ is } \neg\text{-consistent and } \vdash \text{ is n-complete, then: } \mathbf{K} \vdash \neg A \Rightarrow \mathbf{K} \vdash \underline{n}(A)$$

$$\text{If } \mathbf{K} \text{ is } \neg\text{-complete and } \vdash \text{ is } n\text{-consistent, then: } \mathbf{K} \vdash \neg A \Leftarrow \mathbf{K} \vdash \underline{n}(A)$$

$$\text{If } \mathbf{K} \text{ is } \neg\text{-adequate and } \vdash \text{ is } n\text{-adequate, then: } \mathbf{K} \vdash \neg A \Leftrightarrow \mathbf{K} \vdash \underline{n}(A)$$

Thus, negation by finite failure coincides with classical negation only if **K** is $\neg$-adequate – which is nothing but the so-called *closed world asumption.*

Both n-consistency *and* n-completeness is a presupposition of the intuitive adequateness of defeasible logics, because the intended meaning of $n(A)$ is just the nonderivability of A. Nute (1992) proves that all of his systems are n-consistent. Hence, X and $\underline{n}(X)$ will never be derivable from any **K**,

[17] In my (Schurz 1995a) I made the diagnosis that the basic systems fails to satisfy (RW) – because here I did not distinguish between A- and $d(A)$-formulas.

even not if **K** is (strictly) inconsistent: if $\mathbf{K} \vdash A, \neg A$ then neither $n(A)$ nor $n(\neg A)$ is derivable. (The situation changes drastically if we consider *extended* knowledge systems Γ; they may be n-inconsistent even if the underlying inference relation is n-consistent.)

It is easy to see that the basic system is not n-complete. For example, if **K** contains *circular laws*, as in the example

(12) $\mathbf{K} = \{Ax \Rightarrow Bx, Bx \Rightarrow Ax\}$

then neither $n(Aa)$ nor $n(Ba)$ are derivable from **K** in **DR1**, although also neither Aa nor Ba are derivable. Any search procedure for the goal **F**: $d(Aa)$ – whether in PROLOG's depth first left-to-right fashion or otherwise – will immediately run into an infinite loop: deriving **F**: $d(Aa)$ presupposes deriving **F**: $d(Ba)$ which presupposes deriving **F**: $d(Aa)$, etc. The same happens for the dual goal **F**: $nd(Aa)$. In both cases the search tree contains an infinite branch **F**: $d(Aa)$ – **F** : $d(Ba)$ – **F** : $d(Aa)$ ... or **F**: $nd(Aa)$ – **F** : $nd(Ba)$ – **F** : $nd(Aa)$ ..., respectively.– The *search trees* for **F**: $d(Aa)$ and for **F**: $nd(Aa)$ are given in Fig. 4.

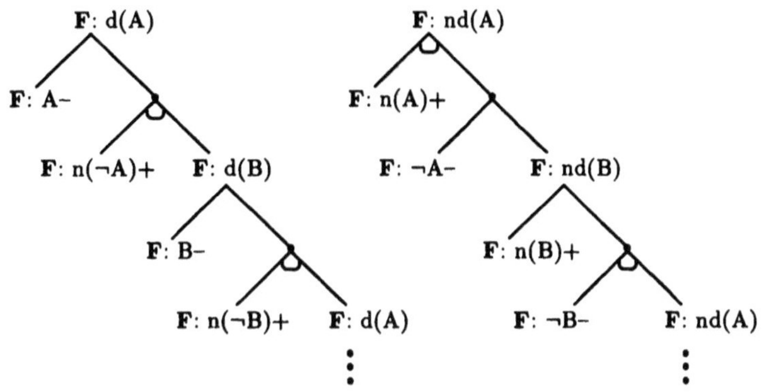

Fig. 4: *Search trees for* **F**: *d(A) and* **F**: *nd(A) in* **DR1** – *Example (12)*

This failure may lead to intuitively wrong results, as in the example

(13) $\mathbf{K} = \{Ax \Rightarrow Bx, Bx \Rightarrow Ax, Ca, Cx \Rightarrow Dx, Ax \Rightarrow \neg Dx\}$

Intuitively, **F**: $d(Da)$ should be derivable because the antecedent Aa of the conflicting law $Ax \Rightarrow Dx$ is not defeasibly derivable, but since **F**: $nd(Aa)$ is not derivable because of the circularity, the basic system **DR1** fails to derive **F**: $d(Da)$. It often happens that the laws of an expert system are circular, in particular if it contains *causal* as well as noncausal *symptom* laws. For instance, coplic spots indicate measles, which in turn causally imply the symptom coplic spots (cf. Schurz 1991). Of course, circularity of strict laws causes the same problems.

Circular laws are not the only cause of inferential loops. A second cause are circularities between subproofs concerning specifity. Here is one example:

(14) $\mathbf{K} = \{\alpha \Rightarrow \delta, \gamma \Rightarrow \neg\delta, \alpha \Rightarrow \beta, \gamma \Rightarrow \neg\beta, \alpha \Rightarrow \gamma, \beta \Rightarrow \neg\gamma, \alpha i, \beta i, \gamma i\}$

In order to derive $d(\delta i)$ from $\mathbf{K}$, one has to derive that αi is more specific than γi, because of the law $\gamma \Rightarrow \neg\delta$ and the fact γi; hence one must derive αi: $d(\gamma i)$ via the law $\alpha \Rightarrow \gamma$ [and also, γi: $nd(\alpha i)$]. Because of the law $\beta \Rightarrow \neg\gamma$ and the fact βi this presupposes to derive that αi is more specific than βi, hence to derive $\alpha i : d(\beta i)$ via the law $\alpha \Rightarrow \beta$ [plus $\beta i : nd(\alpha i)$]. Because of the law $\gamma \Rightarrow \neg\beta$ and the fact γi, this in turn presupposes to derive that αi is more specific than γi, i.e. $\alpha i : d(\gamma i)$. Hence we have the following infinite circular branch of the proof tree: $\mathbf{F}$: $d(\delta i) - \alpha i : d(\gamma i) - \alpha i : d(\beta i) - \alpha i : d(\gamma i) -$ etc.

If DR should be reliable, it has to overcome the circularity problem. It can be solved by *keeping track of the branch* $\langle G_n, ..., G_1\rangle$ which leads from the current subgoal G_n to the goal G_1 at the top of the search tree. The goals G_i are sequents abbreviated as $\mathbf{A} : X$, meaning that $\langle \mathbf{L}, \mathbf{A}\rangle \vdash X$. Each G_i is a derivational precondition for G_{i-1} (i.e. a daughter of G_{i-1} in the search tree for G_1). We number the branch members from right to the left because we formalize branches with help of PROLOG's convenient notation of (finite) lists. $[H|T]$ denotes the list (sequence) with H (head) as its first member and T (tail) as the list of the following members. Similarly, $[H_1, H_2|T]$ is the list with H_1 as its first, H_2 as its second member and T as the list of the other members. $[\,]$ is the empty list. Hence, $[a, b] = [a|[b]] = [a, b|[\,]]$ and $[a] = [a|[\,]]$. $\Pi_1, \Pi_2, ...$ are variable ranging over branches, i.e. lists of sequents of the form $\mathbf{A}$: X, where $\mathbf{A}$ are sets of literals and X is an extended literal. The head of a branch is always the current subgoal, its last member is the top goal. $\vdash [\mathbf{F} : X \mid \Pi]$ means that X is derivable from $\langle \mathbf{L}, \mathbf{F}\rangle$ as the last member of the branch $[\mathbf{F} : X \mid \Pi]$. A branch is called *circular* if it contains one and the same sequent twice. Whenever the attempt to prove the current subgoal G_n as the last member of the noncircular branch $[G_n \mid \Pi]$ returns a new goal G_{n+1}, we extend the branch, obtaining $[G_{n+1}, G \mid \Pi]$ and check whether the extended branch is still non-circular. If yes, we go on. If no, then G_{n+1} occurs twice on the branch. In this case we conclude *fail* if G_{n+1} is a positive goal and *true* if it is a negative goal. Explanation of details follows soon. First have a look on the rules of the new DR system, **DR2**, which keeps track of branches. We present them in the match version. We write *circ*Π if Π is circular and *straight*Π if Π is noncircular. Note that we cannot use our set-theoretic abbreviations $\mathbf{K} \vdash \mathbf{A}$ or $\mathbf{K} \vdash n(\mathbf{A})$ any longer since different goals in the consequence set may have different branches.

DR2 (match version):

(Init)
$$\frac{\vdash [\mathbf{F} : X]}{\langle \mathbf{L}, \mathbf{F}\rangle \vdash X} \quad \text{for } X \in \{A, n(A)\}$$

(M⁺1):
$$\frac{\begin{array}{l} A \in \mathbf{F} \\ \mathit{straight}[\mathbf{F} : A \mid \Pi] \end{array}}{\vdash [\mathbf{F} : A \mid \Pi]}$$

(M⁺2):
$$\frac{\begin{array}{l} \mathbf{B} \in M(A, \mathbf{Ls}) \\ \forall B \in \mathbf{B} : \vdash [\mathbf{F} : B, \mathbf{F} : A \mid \Pi] \\ \mathit{straight}[\mathbf{F} : A \mid \Pi] \end{array}}{\vdash [\mathbf{F} : A \mid \Pi]}$$

(M⁻1):
$$\frac{\mathit{circ}[\mathbf{F} : n(A) \mid \Pi]}{\vdash [\mathbf{F} : n(A) \mid \Pi]}$$

(M⁻2):
$$\frac{\begin{array}{l} A \in \mathbf{F} \\ M(A, \mathbf{Ls}) = \{\mathbf{B}_1, \ldots, \mathbf{B}_n\} \\ \text{For all } 1 \le i \le n : \exists B \in \mathbf{B}_i : \vdash [\mathbf{F} : n(B), n(A) \mid \Pi] \end{array}}{\vdash [\mathbf{F} : n(A) \mid \Pi]}$$

(D⁺1):
$$\frac{\begin{array}{l} \mathit{straight}[\mathbf{F} : d(A) \mid \Pi] \\ \vdash [\mathbf{F} : A] \end{array}}{\vdash [\mathbf{F} : d(A) \mid \Pi]}$$

(D⁻1): like (M⁻1), with $nd(A)$ instead of $n(A)$

(D⁺2):
$$\frac{\begin{array}{l} \mathit{straight}[\mathbf{F} : d(A) \mid \Pi] \\ \vdash [\mathbf{F} : n(-A)] \\ \mathbf{B} \in M(A, \mathbf{Ls}) \\ \text{For all } B \in \mathbf{B} : \vdash [\mathbf{F} : d(B), \mathbf{F} : d(A) \mid \Pi] \end{array}}{\vdash [\mathbf{F} : d(A) \mid \Pi]}$$

(D⁻2):
$$\frac{\begin{array}{l} \vdash [\mathbf{F} : n(A)] \\ \vdash [\mathbf{F} : -A] \end{array}}{\vdash [\mathbf{F} : nd(A) \mid \Pi]}$$

(D⁺3):
$$\frac{\begin{array}{l} \mathit{straight}[\mathbf{F} : d(A) \mid \Pi] \\ \vdash [\mathbf{F} : n(-A)] \\ \mathbf{B} \in M(A, \mathbf{Ld}) \\ \text{For all } B \in \mathbf{B} : \vdash [\mathbf{F} : d(B), \mathbf{F} : d(A) \mid \Pi] \\ M(-A, \mathbf{Ls}) = \{\mathbf{C}_1, \ldots, \mathbf{C}_k\} \\ \text{For all } 1 \le i \le k : \exists C \in \mathbf{C}_i : \vdash [\mathbf{F} : nd(C), \mathbf{F} : d(A) \mid \Pi] \\ M(-A, \mathbf{Ld}) = \{\mathbf{D}_1, \ldots, \mathbf{D}_m\} \\ \text{For all } 1 \le i \le m : \mathit{Either}\ \exists D \in \mathbf{D}_i : \vdash [\mathbf{F} : nd(D), \mathbf{F} : d(A) \mid \Pi] \\ \qquad\qquad \mathit{or} : \quad \forall D \in \mathbf{D}_i : \vdash [\mathbf{B} : d(D), \mathbf{F} : d(A) \mid \Pi]\ \mathit{and} \\ \qquad\qquad\qquad\quad \exists B \in \mathbf{B} : \vdash [\mathbf{D}_i : nd(B), \mathbf{F} : d(A) \mid \Pi] \end{array}}{\vdash [\mathbf{F} : d(A) \mid \Pi]}$$

(D⁻3):
$$\frac{\begin{array}{l} \vdash [\mathbf{F} : n(A)] \\ M(A, \mathbf{Ls}) = \{\mathbf{B}_1, \ldots, \mathbf{B}_k\} \\ \text{For all } 1 \le i \le k : \exists B \in \mathbf{B}_i : \vdash [\mathbf{F} : nd(B), \mathbf{F} : nd(A) \mid \Pi] \\ M(A, \mathbf{Ld}) = \{\mathbf{C}_1, \ldots, \mathbf{C}_m\} \\ \text{For all } 1 \le i \le m : \mathit{Either}\ \exists C \in \mathbf{C}_i : \vdash [\mathbf{F} : nd(C), \mathbf{F} : nd(A) \mid \Pi] \\ \qquad \mathit{or} : \quad \mathbf{D} \in M(-A, \mathbf{Ls})\ \mathit{and}\ \forall D \in \mathbf{D} : \vdash [\mathbf{F} : d(D), \mathbf{F} : nd(A) \mid \Pi] \\ \qquad \mathit{or} : \quad \mathbf{E} \in M(-A, \mathbf{Ld})\ \mathit{and}\ \forall E \in \mathbf{E} : \vdash [\mathbf{F} : d(E), \mathbf{F} : nd(A) \mid \Pi] \\ \qquad\qquad \mathit{and} : \quad \exists E \in \mathbf{E} : \vdash [\mathbf{C}_i : nd(E), \mathbf{F} : nd(A) \mid \Pi] \\ \qquad\qquad\qquad\quad \mathit{or}\ \forall C \in \mathbf{C}_i : \vdash [\mathbf{E} : d(C), \mathbf{F} : nd(A) \mid \Pi] \end{array}}{\vdash [\mathbf{F} : nd(A) \mid \Pi]}$$

The rule Init leads derivability of X simpliciter back to derivability within the singleton branch $[X]$. To explain how the "straight" and "circ" conditions work we intruce notion of the *complement* $\underline{n}(\Pi)$ of a branch Π, defined by $\underline{n}([G_n, ..., G_1]) = [\underline{n}(G_n), ..., \underline{n}(G_1)]$, where the goals are sequents and the complement of a sequent is defined by $\underline{n}(\Gamma : X) = \Gamma : \underline{n}(X)$. All positive rules, i.e. rules for positive goals P contain a check whether the current branch $[\mathbf{F} : P \mid \Pi]$ is straight. Assume $[\mathbf{F} : P \mid \Pi]$ is a branch for the top goal G which has become circular, i.e. $\mathbf{F}$: P occurs twice on it. Then derivability of the positive goal P within this branch fails. We want to conclude in this case that $\mathbf{F}$: $n(P)$ is derivable within the corresponding branch of the dual search tree for the complementary goal $\underline{n}(G)$. We know from Observations 3.2 and 3.3 of §3 that this branch is the complement of the current branch – $[\mathbf{F} : n(P) \mid \underline{n}(\Pi)]$. Since Π is circular iff $\underline{n}(\Pi)$ is circular, it suffices to introduce the general rule ($D^{-}1$) [since Π varies over arbitrary branches we may replace $\underline{n}(\Pi)$ by Π in all negative rules]. — If $\mathbf{F}_1 : Y_1$ is a derivational precondition for $\mathbf{F} : X$, the branch $[\mathbf{F} : X \mid \Pi]$ becomes extended at its head with $\mathbf{F}_1 : Y_1$. — In the D-rules we *cut off* the branch whenever we pass from some precondition about strict derivability to a conclusion about defeasible derivability. The reason we *may* do this is that strict derivability statements do never have preconditions about defeasible derivability. Hence given that $[\mathbf{F} : d(A) \mid \Pi]$ is straight, also $[\mathbf{F} : A, \mathbf{F} : d(A) \mid \Pi]$ must be straight, because Π can't contain any strict derivability statements (since this would mean that $\mathbf{F} : d(A)$ is a precondition for a strict derivability claim). The reason why we *want* to cut off is that it reduces complexity.

To see how a PROLOG implementation of **DR2** works, consider again the example with circular laws:

(15) $\mathbf{K} = \{Ax \Rightarrow Bx, Bx \Rightarrow Ax\}$

We present the termination tree in the left to right fashion. For simplicity, we let the nodes just be branches, omitting the derivability signs and the Init step. The termination trees for the goals $\mathbf{F}$: $d(Aa)$ and $\mathbf{F}$: $nd(Aa)$ of example (15) are given in Fig. 5 and Fig. 6.

Linear proof of $\langle \mathbf{L}, \mathbf{F} \rangle \vdash nd(A)$ *in* **DR2** – *Example (15):*

1)	$\mathbf{F} = \emptyset$	Prem
2)	$M(Aa, \mathbf{Ls}) = \emptyset$	Prem
3)	$M(Aa, \mathbf{Ld}) = \{\{Ba\}\}$	Prem
4)	$M(Ba, \mathbf{Ls}) = \emptyset$	Prem
5)	$M(Ba, \mathbf{Ld}) = \{\{Aa\}\}$	Prem
6)	$\vdash [\mathbf{F} : nd(Aa), \mathbf{F} : nd(Ba), \mathbf{F} : nd(Aa)]$	By $D^{-}1$
7)	$\vdash [\mathbf{F} : n(Ba)]$	By $M^{-}2$ from (1) and (4)
8)	$\vdash [\mathbf{F} : nd(Ba), \mathbf{F} : nd(Aa)]$	By ($D^{-}3$) from (4), (5), (6), (7)

9) $\vdash [\mathbf{F} : n(Aa)]$ By $M^{-}2$ from (1) and (2)

10) $\vdash [\mathbf{F} : nd(Aa)]$ By $D^{-}3$ from (2), (3), (8), (9)

11) $\langle \mathbf{L}, \mathbf{F} \rangle \vdash nd(Aa)$ By Init from (10)

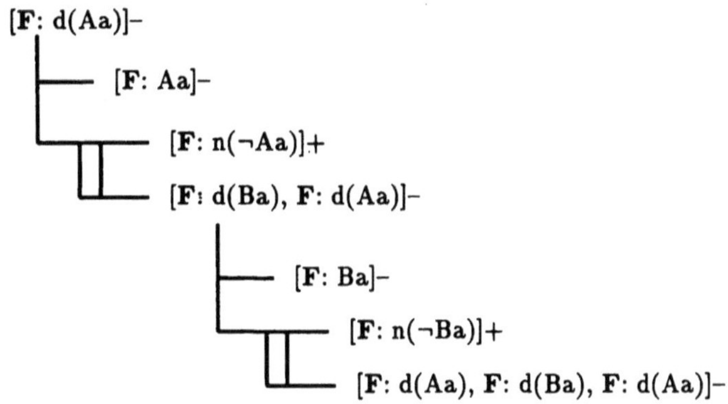

Fig. 5: *Termination tree for* **F***: d(Aa) in* **DR2** *– Example (15)*

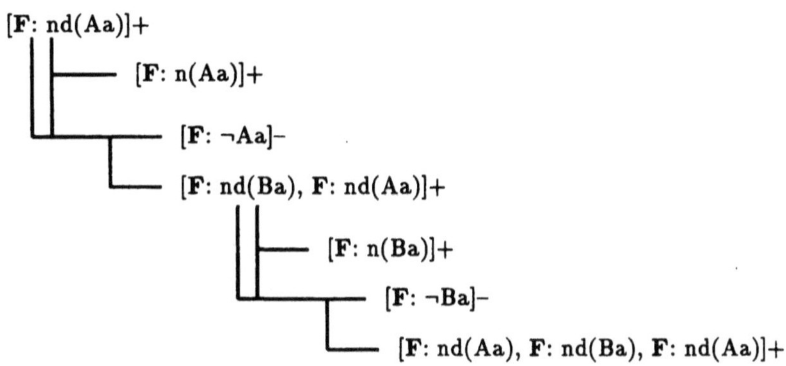

Fig. 6: *Termination tree for* **F***: nd(Aa) in* **DR2** *– Example (15)*

Now consider the extended example:

(16) $\mathbf{K} = \{Ax \Rightarrow Bx, Bx \Rightarrow Ax, Cx \Rightarrow Ax, Ca\}$

Termination trees for the goals **F**: d(Aa) and **F**: nd(Aa) (in left to right fashion) are given in Fig. 7 and Fig. 8.

Comparing example (15) with that of (16) we recognize that in the termination tree of **F**: $d(Aa)$ of (16) the top goal is an Or-node which now contains

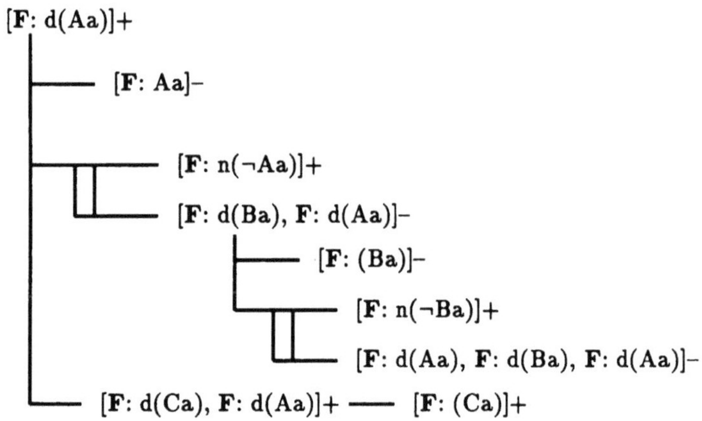

Fig. 7: *Termination tree for* **F***: d(Aa) – Example (16)*

the additional daugther $[\mathbf{F} : d(Ca), \mathbf{F} : d(Aa)]+$. Hence $\mathbf{F} : d(Aa)$ succeeds in (16) while it fails in (15). Dually for the termination trees of $\mathbf{F} : nd(Aa)$ – it fails in (16) but succeeds in (15). As a final example, consider how **DR2** masters our example with circular laws:

(17) $\mathbf{K} = \{Ax \Rightarrow Bx, Bx \Rightarrow Ax, Ca, Cx \Rightarrow Dx, Ax \Rightarrow \neg Dx\}$

The proof tree for $[\mathbf{F} : d(Da)]$ is given in Fig. 9.

We turn to some general theorems. First we verify that **DR2** *extends* **DR1**. The proof of Theorem 4.1 rests on a well-known fact about "detourless" proofs.

Theorem 4.1 $\langle \mathbf{L}, \mathbf{F} \rangle \vdash_{\mathbf{DR1}} X$ implies $\langle \mathbf{L}, \mathbf{F} \rangle \vdash_{\mathbf{DR2}} X$ (for all **L**, **F** and X).

Proof: Take a proof tree for the goal G in **DR1**. If it contains a sequent G' more than once, we eliminate the whole subtree between the first and the last occurrence of G'. Doing this for all G's occurring more then once, we obtain a proof tree for G without circular branches. The result will be a noncircular proof for G. By induction on the complexity of the noncircular proof tree it is quickly shown that also $\langle \mathbf{L}, \mathbf{F} \rangle \vdash_{\mathbf{DR2}} X$ holds. *Sketch:* assume, **F**: X is derived from $\mathbf{F}_1 : Y_1, ..., \mathbf{F}_n : Y_n$ with some rule RU of **DR1**. By IH, $\vdash_{\mathbf{DR2}} [\mathbf{F}_i : Y_i, \mathbf{F} : X \mid \Pi]$ holds for all $1 \leq i \leq n$. Since $[\mathbf{F} : X \mid \Pi]$ is straight, this implies that $\vdash [\mathbf{F} : X \mid \Pi]$ by the corresponding rule RU of **DR2**. Q.E.D.

Next we verify that Observation 3.3 about *negate* still holds for the extended rules. Of course we must extend the *negate* procedure to branches: we replace *not* $\vdash \Pi$ by $\vdash \underline{n}(\Pi)$ and *not straight*(Π) by $circ(\underline{n}(\Pi))$ (recall that $circ(\Pi)$ *iff* $circ(\underline{n}(\Pi))$).

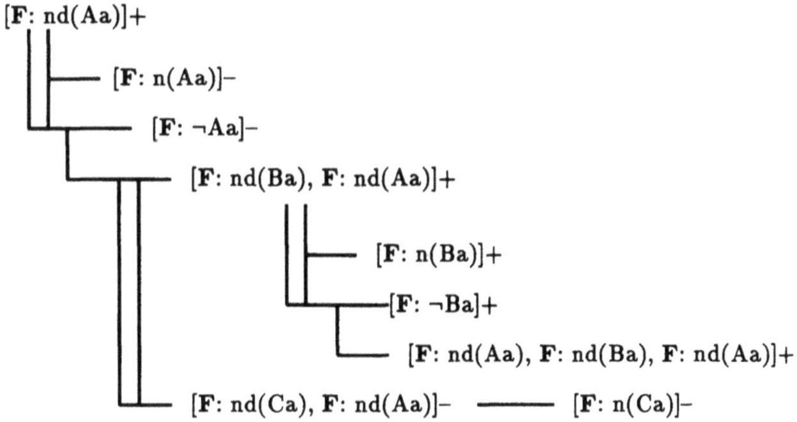

Fig. 8: *Termination tree for* **F**: *d(Aa) – Example (16)*

Theorem 4.2 Application of "negate" to the positive joint M-rule/D-rule of **DR2** leads to the negative joint M-rule/D-rule of **DR2**, respectively; and vice versa.

Proof: The disjunction of the preconditions of (M^+1,2) complemented with quantifiers is: ($A \in \mathbf{F}$ and $straight[\mathbf{F} : A \mid \Pi]$) or ($\exists \mathbf{B}_i \in M(A, \mathbf{Ls}) \forall B \in \mathbf{B}_i(\vdash [\mathbf{F} : B, \mathbf{F} : A \mid \Pi])$ and $straight[\mathbf{F} : A \mid \Pi]$). Application of *negate* leads to: ($A \notin \mathbf{F}$ or $circ[\mathbf{F} : n(A) \mid \underline{n}(\Pi)]$) and ($\forall \mathbf{B}_i \in M(A, \mathbf{Ls}) \exists B \in \mathbf{B}_i(\vdash [\mathbf{F} : n(B), \mathbf{F} : n(A) \mid \underline{n}(\Pi)])$ or $circ[\mathbf{F} : n(A) \mid \underline{n}(\Pi)]$). Writing "$ij$" for the jth disjunct of the ith conjunct, we abbreviate this expression as (11 or 12) & (21 or 12), since 22 = 12. This is equivalent with 12 or (11 & 21). 12 is the precondition of the rule (M^-1) and 11 & 21 the precondition of rule (M^-2); since $\underline{n}(\Pi)$ is a variable branch we can write Π instead of $\underline{n}(\Pi)$.

Applying *negate* to the disjunction of the D^+-preconditions and writing Π instead of $\underline{n}(\Pi)$ leads to the following conjunction of disjunctions: ($circ[\mathbf{F} : nd(A) \mid \Pi]$ or $[\mathbf{F} : n(A)]$) & ($circ[\mathbf{F} : nd(A) \mid \Pi]$ or $\forall \mathbf{B} \in M(A, \mathbf{Ls}) \exists B \in \mathbf{B}(\vdash [\mathbf{F} : nd(B), \mathbf{F} : nd(A) \mid \Pi])$ or $\vdash [\mathbf{F} : -A]$) & ($circ[\mathbf{F} : nd(A) \mid \Pi]$ or $\vdash [\mathbf{F} : -A]$ or $\forall \mathbf{B} \in M(A, \mathbf{Ld})\,(\exists B \in \mathbf{B}(\vdash [\mathbf{F} : nd(B), \mathbf{F} : nd(A) \mid \Pi])$ or $\exists \mathbf{C} \in M(-A, \mathbf{Ls}) \forall C \in \mathbf{C}(\vdash [\mathbf{F} : d(C), \mathbf{F} : nd(A) \mid \Pi])$ or $\exists \mathbf{D} \in M(-A, \mathbf{Ld})(\forall D \in \mathbf{D}(\vdash [\mathbf{F} : d(D), \mathbf{F} : nd(A) \mid \Pi])$ & $(\exists D \in \mathbf{D}(\vdash [\mathbf{B} : nd(D), \mathbf{F} : nd(A) \mid \Pi])$ or $\forall B \in \mathbf{B}(\vdash [\mathbf{D} : d(B), \mathbf{F} : nd(A) \mid \Pi])))))$. We abbreviate this as: (11 or 12) & (11 or 22 or 23) & (11 or 23 or 33), for note that 21 = 31 = 11 and 32 = 23. This is equivalent with 11 or (12 & 23) or (12 &22 & 33), which are exactly the preconditions of our three rules D^-1, D^-2 and D^-3.

- [**F**: d(Da)]+
  - [**F**: d(Ca), **F**: d(Da)]+ —— [**F**: Ca]+
  - [**F**: n(¬Ca)]+
  - [**F**: nd(Aa), **F**: d(Da)]+
    - [**F**: n(Aa)]+
    - [**F**: nd(Ba)], **F**: nd(Aa), **F**: d(Da)]+
      - [**F**: n(Ba)]+
      - [**F**: nd(Aa), **F**: nd(Ba), **F**: nd(Aa), **F**: d(Da)]+

Fig. 9: *Termination tree for* **F** : *d(Da) - Example (17)*

Instead of proving the other direction similarly, one may established it from the first direction as follows. Let $\mathbf{P}^+$ and $\mathbf{P}^-$ be the joint preconditions of the positive and negative rules (for M or for D), respectively. It is easy to see that *negate*(*negate*($\mathbf{P}^+$)) = $\mathbf{P}^+$ — for, driving "not" into the formula (cf. 3.(ii) of §3) two times leads back to the original formula except that its atoms have"not not" in front of them, and the twofold replacement of the derivability conditions "not $\vdash \Pi$" by "$\vdash \underline{n}(\Pi)$" leads to "$\vdash \underline{n}(\underline{n}(\Pi))$" = "$\vdash \Pi$". Since we know from the above that *negate*($\mathbf{P}^+$) = $\mathbf{P}^-$, it follows that *negate*($\mathbf{P}^-$) = $\mathbf{P}^+$. - Q.E.D.

With help of Theorem 4.2 it is easy to prove *n*-consistency; we need not proceed by induction doing all rules separately (as Nute does), but can do it in one step.

Theorem 4.3 **DR2** is n-consistent, i.e. $\langle \mathbf{L}, \mathbf{F} \rangle \vdash_{\mathbf{DR2}} X \Rightarrow not\langle \mathbf{L}, \mathbf{F} \rangle \vdash_{\mathbf{DR2}} \underline{n}(X)$.

Proof: Assume in the opposite, $\underline{n}(X)$ is derived from premise set Prem_1 either by the joint M^--rule or by the joint D^--rule. By assumption, X has been derived from premise set Prem_2 by the joint M^+- or D^+-rule, respectively. Let Prem_1^* be the result of replacing every precondition "$\vdash \Pi$" in Prem_1 by "*not* $\vdash \underline{n}(\Pi)$". By IH, Prem_1 implies Prem_1^* (because there are only unnegated derivability conditions in Prem_1). From Theorem 4.2 we know that Prem_1^* is equivalent with not(Prem_2) (the procedure leading from Prem_1 via Prem_1^* to Prem_2 is the reverse of *negate*). This means that Prem_2 can't be true and hence that $\langle \mathbf{L}, \mathbf{F} \rangle \vdash_{\mathbf{DR2}} X$ can't hold; contradiction. - Q.E.D.

It is not difficult to prove that **DR2** is n-complete. The proof rests on the finite domain assumption which implies a limitation of the size of any **DR2** search tree.

Theorem 4.4 DR2 is n-complete, i.e. $\langle \mathbf{L}, \mathbf{F}\rangle \not\vdash_{\mathbf{DR2}} X \Rightarrow \langle \mathbf{L}, \mathbf{F}\rangle \vdash_{\mathbf{DR2}} \underline{n}(X)$.

Proof: Consider the full search tree for $\mathbf{F}$: X in **DR2**, except that its nodes are just sequents (i.e. we replace each list $[H \mid T]$ just by its head). Because no extended literals occurs more than once on a branch of this tree, and the number of different extended literals is finite, the depth of this search tree must be finite. Moreover, each node has only finitely many daughters. Therefore the entire search tree is finite. Hence, the full search tree contains a failure tree for $\mathbf{F}$: X. From Theorem 4.2 and Observation 3.3 of §3 we know that the dual of this failure tree is a proof tree for $\mathbf{F}$: $\underline{n}(X)$. – Q.E.D.

5 Generalizations of DR2: An Outlook

5.1 n-completeness, semantic completeness and decidability

This section discusses some fundamental connections between n-completeness, decidability and semantical completeness properties.

Observation 5.1 If $\vdash$ is recursively axiomatized and n-adequate, then $\vdash$ is decidable.

Proof: For any $\mathbf{K}$ and A (or $d(A)$): If $\mathbf{K} \vdash A$ holds, then it will appear in the recursive enumeration of inferences after some finite time. If $\mathbf{K} \not\vdash A$, then $\mathbf{K} \vdash n(A)$ by n-complenetess, so $\mathbf{K} \vdash n(A)$ will appear in the r. e. after some finite time, from which we can conclude by n-consistency that $\mathbf{K} \not\vdash A$. Hence, we have an algorithm which decides $\mathbf{K} \vdash A$ after some finite time. – Q.E.D.

The notion of semantic completeness presupposes a semantic notion of *valid* inference, which we possess only for the strict subsystem **M**. So let us discuss this question for M. $\mathbf{K} \Vdash A$ means whenever (i.e., in all truth valuations where) all premises in $\mathbf{K}$ are true, A is true. The semantical properties are defined as follows:

$\vdash$ is *s*(emantically)-correct if: (i) $\vdash A \Rightarrow \Vdash A$ and (ii) $\vdash n(A) \Rightarrow \not\Vdash A$

$\vdash$ is *s*(emantically)-complete if: (i) $\vdash A \Leftarrow \Vdash A$ and (ii) $\vdash n(A) \Leftarrow \not\Vdash A$

Observation 5.2 If $\vdash$ is n-complete, then: $\vdash$ is s-correct $\Rightarrow \vdash$ is s-complete.

Proof: Assume $\Vdash A$. So $\not\vdash n(A)$ by s-correctness (ii), hence $\vdash A$ by n-completeness (for, $\underline{n}(n(A)) = A$). Assume $\not\Vdash A$. So $\not\vdash A$ by s-correctness (i), hence $\vdash n(A)$ by n-complenetess. – Q.E.D.

Observation 5.3 If $\vdash$ is n-consistent, then: $\vdash$ is s-correct $\Leftarrow \vdash$ is s-complete.

Proof: Assume $\vdash A$. So $\not\vdash n(A)$ by n-consistency, hence $\Vdash\!\!-A$ by s-completeness (ii). Assume $\vdash n(A)$. So $\not\vdash A$ by n-consistency, hence $|\!\not\vdash\!\!-A$ by s-completeness (i). – Q.E.D.

Observation 5.4 If $\vdash$ is n-adequate, then: $\vdash$ is s-correct $\Leftrightarrow \vdash$ is s-complete.

Proof: From Observations 5.3 and 5.4.

In other words: for n-adequate DR-systems the semantical notions of correctness and completeness coincide. This is *typical* for DR systems. With this in mind, let us now turn to Observation 5.1. It implies that the system **DR2** is decidable, and so is its monotonic subsystem, call it **M2**. It is well-known that classical inferences within the restricted $\mathcal{L}1$-fragment of $\mathcal{L}$(DR) – those involving only purely universal or singular statements without function symbols – are decidable. With this restrictions it is possible to have a DR system with a monotonic subsystem which is is n-adequate and still s-adequate, i.e. which coincides with classical inference. But if we go to the general first order case decidability gets lost. Therefore, the monotonic subsystem of an n-adequate DR-system must either be n-incomplete or else s-incorrect. This seems to constitute a *dilemma* for any nonmonotonic logic. We know from the previous examples that n-incompleteness leads to intuitively wrong results. But if n-completeness goes along with s-incorrectness, we have intuitively wrong results anyway. (We say 'intuitively wrong' and not 'incorrect' because so far we have not established a semantics for defeasible inference.) A proposal to solve this dilemma is made in §5.3. First we have a closer look to extensions of **M2**.

5.2 Extensions of the monotonic subsystem of **DR2**

Nute's monotonic subsystem is obviously s-inadequate. For instance, the system does not master contraposition. This may lead to intuitively wrong results, as in the example $\mathbf{K} = \{Ax \rightarrow Bx, Cx \Rightarrow Dx, \{Cx, \neg Ax\} \Rightarrow \neg Dx, Ca, \neg Ba\}$ [Cx - x is a bird, Dx - x can fly, Ax - x is alive, Bx - x is breathing]. Since $\neg Aa$ follows logically from $\mathbf{K}$, $d(\neg Da)$ should be derivable; however, our DR-system derives $d(Da)$ instead because of its unability to derive $\neg Aa$.

As long as we remain in the restricted $\mathcal{L}1$-fragment of $\mathcal{L}$(DR) there are some well-known algorithms for deciding classical inference - *resolution*, for example. So let us just assume that we have defined $\vdash$ in some way which is s-adequate and still n-adequate. Nothing is changed by that assumption in our D-rules. However, as soon as we allow existential quantifiers or function symbols to enter our language, the situation becomes undecidable. If we apply Skolemization we obtain knowledge bases which are as before except that facts, strict and defeasible laws may now contain arbitrary function symbols. As is well-known, this implies that the finite domain assumption breaks down, for now we may instantiate with arbitrarily complex terms, e.g. $f(a)$, $f(f(a))$, ... etc. As a result, we may have search trees containing branches which are infinite without being circular. For instance, if $\mathbf{K} = \{Af(x) \rightarrow Ax\}$, then the search tree for the goal Aa will produce the infinite branch $Aa - Af(a) - Af(f(a)) - \ldots$ etc.

5.3 Negation by default on the procedural level

In the general first order case, the only possibility to guarantee n-adequacy is to implement negation by default on the procedural level (or metalevel) by *restricting the search depth.* Whenever it is not possible to derive a goal X from $\mathbf{K}$ within a reasonable bound of search depth, we conclude per default that $\mathbf{K} \vdash X$ fails if X is a positive goal, and that $\mathbf{K} \vdash X$ succeeds if X is a negative goal. The implementation of this idea in **DR2** is simple because nodes are lists and the search depth of a node is just its length. Search depth restricted inference relations are unusual in logics; suggestions have been made by Elkan (1990). Can arguments be given for search depth restrictions which are more than ad hoc restrictions following the slogan "better a false result than no result"?

In the first order case, we can have either n-adequateness or s-adequateness but not both. What is the better choice? In monotonic logic, the answer seems to be s-adequateness, because it is the very *intention* of deductive logic that its inferences are *certain.* To realize this intention the inference relation has to be s-correct. If for a particular problem the inference engine runs into an infinite branch of the search tree, it is better to state *no* result than to conclude by default that the goal is not provable, because such a conclusion would not be certain. In nonmonotonic logic, however, the situation is very different. Recall the remarks about my favoured probability semantics for NMR. Of course, NMR leads to failures in predictions, but these failures occur in NMR *anyway.* Since the *intention* of NMR is only to *preserve truth in most cases, but not to preserve it in all cases*, the introduction of a procedural default negation would be justified as long as we can show that *in most of the cases where the search depth of a goal exceeds the depth bound, the goal is in fact unprovable.* Hence we would need a statistical test of the depth bound restriction of the nonmonotonic inference engine (indeed, *given* the deductive concept of logic this would be a rather absurd enterprise). Such a test could be performed in a language fragment which is decidable. How to define a reasonable heuristics for depth restrictions which satisfies the above statistical requirement is completeley beyond this paper.

5.4 Indirect conflict and collective defeat

System **DR2** does not master two further problems: indirect conflict and collective defeat. Indirect conflict is given whenever two defeasible laws with instantiated antecedents have instantiated consequents which are not complementary but yet inconsistent with each other, given the strict laws of **L**. An example is:

(18) $\mathbf{K} = \{Ax \Rightarrow Bx, Cx \Rightarrow Dx, Bx \rightarrow Ex, Dx \rightarrow \neg Ex, Aa, Ca\}$
Read: Ax - x is a quaker, Bx - x is pacifist, Cx - x is conservative,
Dx - x is an adherent of nuclear deterrance,
Ex - x votes for cutting down military budget, a - Nixon.

It would be reasonable to remain sceptical and infer nothing about Nixon, but the systems **DR1** and **DR2** infer $d(Ba)$ and $d(Da)$ and then $d(Ea)$ and $d(\neg Ea)$. As a consequence, these systems violate a further important axiom, that of consistency preservation. **K** is *defeasibly consistent* if for no literal A, both $\mathbf{K} \vdash d(A)$ and $\mathbf{K} \vdash d(-A)$ holds. Then the axiom of consistency preservation (CP) says:

(CP): If **K** is strictly consistent, then **K** is defeasibly consistent.

Collective defeat (so called after Pollock 1987, 493ff.) is given when more than two defeasible laws with instantiated antecedents have consequents which are jointly inconsistent, given the strict laws of **L**. Consider the following example:

(19) $\mathbf{K} = \{Ax \Rightarrow Bx, Ax \Rightarrow Cx, \{Bx, Cx\} \rightarrow Dx, Ex \Rightarrow \neg Dx, Aa, Ea\}$
Read: Ax evidence for being married (Bx)
and for being a priest (Cx),
Dx - being a protestant, Ex - being an Italian.

The conjunction of the consequents of the first two defeasible laws ($\{Ba, Ca\}$) strictly implies the negation of the consequent of the third one ($\neg Da$). Intuitively, neither $\{Ba, Ca\}$ nor $\neg Da$ should be derivable, and since we have no reason to prefer Ba over Ca, neither Ba nor Ca should be derivable. **DR1** and **DR2**, however, derive from (19) Ba, Ca and from that Da, which is inadequate. A slight variation of (19) is given when the fourth law is not defeasible but strict. A well known example of collective defeat of this sort is the *lottery paradox* (cf. Poole 1991). In (Schurz 1995a) I have suggested extensions of **DR1** which handle indirect conflict and collective defeat, on the cost of more complex definitions of the inference relations. An elaboration of this idea is beyond the scope of this paper.

References

Adams, E.W.: Probability and the Logic of Conditionals. In: *Aspects of Inductive Logic*, ed. by J. Hintikka & P. Suppes, Amsterdam 1966, 265–316.

Adams, E.W.: *The Logic of Conditionals.* Dordrecht 1975.

Bacchus, F. et al: Statistical Foundations for Default Reasoning. In: *Proc. 13th International Joint AI Conference* Vol. I 1993, 563–569.

Brewka, G.: Preferred Subtheories: An Extended Logical Framework for Default Reasoning. In: *Proc. 11th International Joint AI Conference* Vol. 2 1990.

Brewka, G.: *Nonmonotonic Reasoning. Logical Foundations of Commonsense.* Cambridge 1991.

Carnap, R.: *Logical Foundations of Probability.* Chicago 1962².

Delgrande, J.P.: An Approach to Default Reasoning Based on a First-Order Conditional Logic: Revised Report. *Artificial Intelligence* 36 (1988), 63–90.

Elkan, C.: Incremental, Approximate Planning. In: *Proc. 8th National Conference on Artificial Intelligence (AAAI-90)* 1990, 145–150.

Gabbay, D.M.: Theoretical Foundations for Non-Monotonic Reasoning in Expert Systems. In: *Proceedings NATO Advanced Study Institute on Logics and Models of Concurrent Systems*, ed. by K.R. Apt, Berlin 1985, 439–57.

Georgeff, M.P.: Many Agents are Better than One. In: *Proceedings of the 1987 Workshop on the Frame Problem in Artificial Intelligence*, ed. by F.M. Brown, Lawrence/Kansas 1987, 57–75.

Goldszmidt, M. et al: A Maximum Entropy Approach to Nonmonotonic Reasoning. In: *Proc. 8th National Conference on Artificial Intelligence (AAAI-90)* 1990, 646–652.

Hempel, C.G.: *Aspects of Scientific Explanation*, New York 1965.

Konolige, K.: On the Relation between Default and Autoepistemic Logic. *Artificial Intelligence* 35 (1988), 343–382.

Kraus, S., Lehmann, D., & Magidor, M.: Nonmonotonic Reasoning, Preferential Models and Cumulative Logics. *Artificial Intelligence* 44 (1990), 167–207.

Makinson, D.: General Theory of Cumulative Inference. In: *Proceedings of the 2nd International Workshop on Non-Monotonic Reasoning*, Berlin 1989.

Makinson, D., & Gärdenfors, P.: Relations between the Logic of Theory Change and Nonmonotonic Logic. In: *The Logic of Theory Change*, ed. by A. Fuhrmann, & M. Morreau, Berlin 1991, 85–205.

McDermott D., & Doyle, J.: Non-Monotonic Logic I. *Artificial Intelligence* 13 (1980), 41–72.

Moore, R.C.: Semantic Considerations on Nonmonotonic Logic. *Artificial Intelligence* 25 (1985), 75–94.

Nute, D.: Defeasible Reasoning: A Philosophical Analysis in Prolog. In: *Aspects of Artificial Intelligence*, ed. by J. Fetzer, Dordrecht 1988.

Nute, D.: Basic Defeasible Logic. In: *Intentional Logics for Programming*, ed. by L. Farinas-del-Cerro & M. Pentonnen, Oxford 1992, 125–154.

Pearl, J.: *Probabilistic Reasoning in Intelligent Systems.* Santa Mateo 1988.
Pollock, J.L.: Defeasible Reasoning. *Cognitive Science* 11 (1987), 481–518.
Pollock, J L.: Justification and Defeat. *Artificial Intelligence* 67 (1994), 377–408.
Poole, D.: A Logical Framework for Default Reasoning. *Artificial Intelligence* 36 (1988), 27–47.
Poole, D.: The Effect of Knowledge on Belief: Conditioning, Specifity and the Lottery Paradox in Default Reasoning. *Artificial Intelligence* 49 (1991), 281–307.
Prendinger, H., & Schurz, G.: Reasoning About Action and Change. A Dynamic Logic Approach. *submitted for publication* 199+.
Reiter, R.: A Logic for Default Reasoning. *Artificial Intelligence* 13 (1980), 1–132.
Rescher, N.: *Hypothetical Reasoning.* Amsterdam 1964.
Schurz, G., ed.: *Erklären und Verstehen in der Wissenschaft.* Vienna–Munich (paperback 1990), 235–298.
Schurz, G.: Erklärungsmodelle in der Wissenschaftstheorie und in der Künstlichen Intelligenz. In: *Erklärung im Gespräch – Erklärung im Mensch-Maschine-Dialog*, ed. by H. Stoyan, Berlin 1991, 1–42.
Schurz, G.: Nonmonotonic Reasoning and Changes of Belief. In: *Scientific Belief and Religious Belief*, ed. by P. Weingartner, Dordrecht 1994a, 47–64.
Schurz, G.: Probabilistic Justification of Default Reasoning Based on Minimal Assumptions. *KI-94: Advances in Artificial Intelligence*, ed. by B. Nebel & L. Dreschler-Fischer, Berlin 1994b, 248–259.
Schurz, G.: Defeasible Reasoning Based on Constructive and Cumulative Rules. In: *Philosophy and the Cognitive Sciences. (Proceedings of the 16th Int. Wittgenstein Symposium)*, ed. by R. Casati, B. Smith & G. White, Vienna 1995a, 297–310.
Schurz, G.: Most General First Order Theorems are Not Recursively Enumerable. *Theoretical Computer Science* 148 (1995b).
Schurz, G.: Theories and their Applications – a Case of Nonmonotonic Reasoning. In: *Proceedings of the 1994 Conference of the IUHPS*, ed. by R. Woycicki, Amsterdam 1995c.
Scriven, M.: Truisms as Grounds for Historical Explanations. In: *Theories of History*, ed. by P. Gardiner, New York 1959, 443–468.

Belnap's Epistemic States and Negation-as-Failure

Gerd Wagner

ABSTRACT. Using the general concept of a *knowledge system* proposed in (Wagner 1994a) we show how Belnap's system of *epistemic states* (Belnap 1977) can be extended by adding a second negation in the style of Negation-as-Failure. We call the resulting nonmonotonic system with two kinds of negation *Disjunctive Fact Bases.* While Belnap's system is the basis for defining standard 4-, 3-, and 2-valued logic, Disjunctive Fact Bases capture the nonmonotonic reasoning based on minimal, resp. paraminimal models. In order to relate both kinds of negation to the law of the excluded middle and to classical logic we formalize Körner's distinction between exact and inexact predicates (Körner 1966). In the system of Disjunctive Fact Bases, it is possible to simplify the representation of exact predicates (and reduce the complexity created by them) by means of an appropriate form of the Closed-World Assumption (CWA). Thus, we finally distinguish between three kinds of predicates: inexact predicates, exact predicates subject to the CWA, and exact predicates not subject to the CWA.

1 Introduction

In this paper, we shall address issues of disjunctive information processing, resp. disjunctive knowledge representation and reasoning, such as the interplay between disjunctive and negative information (concerning explicit negative information, negation-as-failure and the Closed-World Assumption), and the notion of inclusive versus exclusive disjunctive information. These issues have first been discussed in the field of disjunctive logic programming (see, e.g. (Minker 1982, Ross/Topor 1988, Sakama 1989, Gelfond/Lifschitz 1991, Lobo/Minker/Rajasekar 1992)). However, there is no generally acknowledged semantics for disjunctive logic programs, and many proposals are not based on clear logical principles but seem to be rather ad-hoc. It appears that the non-rule-related issues of disjunctive information processing are complicated by the presence of deduction rules, and a semantical account for the interplay between disjunctive, negative and deductive knowledge is hard to find. Therefore, we believe it is methodologically preferable to settle these issues in a simpler framework without rules, and this is the approach taken here.

The concept of a knowledge representation and reasoning system, or shorter: *knowledge system (KS)*, consists essentially of two main components: an inference and an update operation manipulating knowledge bases as abstract objects,[1] together with a set of formal properties these operations may have. In general, there are no specific restrictions on the internal structure of a knowledge base. It appears, however, that a computational design can be achieved by 'compiling' incoming information into some normal form rather than leaving it in the form of arbitrarily complex formulas. This is the case, for instance, in Belnap's KS which can be considered as a paradigm for knowledge systems.

The concept of a KS constitutes a useful framework for the classification and comparison of various computational systems and formalisms like, e.g., relational and deductive databases, logic programs and other rule-based systems. It is more general than that of a logic (i.e. a consequence relation). A standard logic can be viewed as a special kind of KS. On the other hand, by defining the inference and update operations procedurally, KSs can serve as the basis for the operational definition of logics. For instance, by appropriate settings of certain 'parameters' in the system of disjunctive fact bases one can obtain three-valued and classical logic, in addition to Belnap's four-valued logic.

Both relational and deductive database systems can be considered as computational paradigms of real world knowledge systems. They implement a form of nonmonotonic reasoning caused by the use of negation-as-failure refering to default-implicit negative information. On the other hand, relational and deductive databases, as well as normal logic programs, are not capable of representing and processing explicit negative information. This shortcoming has led to the extension of logic programming by adding a second negation (in addition to negation-as-failure) as proposed independently in (Gelfond/Lifschitz 1990, Gelfond/Lifschitz 1991) and in (Pearce/Wagner 1990, Wagner 1991). We call the general concept of an operator expressing default-implicit negative information in the style of negation-as-failure *weak negation*, and denote it by '$-$', while the concept of an operator expressing explicit negative information will be called *strong negation*, denoted by '$\sim$'. Our concept of a *vivid knowledge system (VKS)* is a two-fold generalization:

1. it extends already known logics, such as Belnap's 4-valued or Nelson's constructive logic,[2] by adding *weak* negation, and

[1]This distinction was already proposed in (Levesque 1984) where the resp. operations were called *ASK* and *TELL*.

[2]See (Belnap 1977), resp. (Almukdad/Nelson 1984).

2. it extends already known knowledge systems, such as relational or deductive database systems, by adding *strong* negation.

In the framework of a VKS, a specific meaning is assigned to the *Closed-World Assumption*: if the Closed-World Assumption holds for a predicate, its weak negation implies its strong negation, in other words, an atomic sentence formed with such a predicate is already false if it is false by default.

The structure of the paper is as follows: below we shall give a brief introduction into partial logics with two kinds of negation; in section 2 we shall present some basic concepts of the theory of knowledge representation proposed in (Wagner 1994a), in section 3 we formulate Belnap's epistemic states in terms of this theory, in section 4 the system of disjunctive fact bases is defined, and in section 5 the distinction between exact and inexact predicates is formalized, and its relation to the Closed-World Assumption is discussed.

1.1 Preliminaries

A *signature* $\sigma = \langle Rel, ExRel, Const \rangle$ consists of a set of relation symbols *Rel*, a set $ExRel \subseteq Rel$ of exact relation symbols, and a set of constant symbols *Const*.[3]

We consider the following logical functors: conjunction ($\wedge$), disjunction ($\vee$), strong negation ($\sim$), weak negation (alias negation-as-failure, denoted by $-$), exclusive disjunction ($|$), and the truth constant 1; relation symbols are denoted by $p, q, r, \ldots$; constant symbols by $c, d, \ldots$; and variables by $x, y, \ldots$. Quantifiers, $\exists$ and $\forall$, are only incidentally considered. If $\mathcal{F}$ is a set of logical functors, $L(\sigma; \mathcal{F})$ denotes the corresponding set of wellformed formulas. $L(\sigma) = L(\sigma; -, \sim, \wedge, |, \vee)$ is the smallest set containing the atomic formulas of σ, and being closed with respect to the following condition: if $F, G \in L(\sigma)$, then $\{-F, \sim F, F \wedge G, F \vee G, F|G\} \subseteq L(\sigma)$.

With respect to a signature σ we define the following sublanguages: $\mathrm{At}(\sigma) = L(\sigma; \emptyset)$, the set of all atomic sentences (also called *atoms*); $\mathrm{Lit}(\sigma) = L(\sigma; \sim)$, the set of all *literals*; and $\mathrm{XLit}(\sigma) = \mathrm{Lit}(\sigma) \cup \{-l : l \in \mathrm{Lit}(\sigma)\}$, the set of all *extended literals*. We shall frequently omit the reference to a specific signature, and simply write L instead of $L(\sigma)$. We introduce the following convention: when L is a set of sentences, L^x denotes the corresponding set of formulas.

[3]For the sake of simplicity we shall not consider functional terms but only variables and constants; therefore, we define signatures without function symbols leading to a finite Herbrand universe.

An *atom* $a \in$ At is called *proper*, if $a \neq 1$. We use $a, b, \ldots, l, k, \ldots, e, f, \ldots$, and $F, G, H, \ldots$ as metavariables for atoms, literals, extended literals and well-formed formulas, respectively.

With each negation a complement operation for the resp. type of literal is associated: $\tilde{a} = \sim a$ and $\widetilde{\sim a} = a$, $\bar{l} = -l$ and $\overline{-l} = l$. These complements are also defined for sets of resp. literals $L \subseteq$ Lit, and $E \subseteq$ XLit: $\widetilde{L} = \{\tilde{l} : l \in L\}$, resp. $\overline{E} = \{\bar{e} : e \in E\}$. We distinguish between the positive and negative elements of $E \subseteq$ XLit by writing $E^+ := E \cap \text{Lit}$ and $E^- := \{l : -l \in E\}$.

If X is a set of sets, then Fin(X) denotes its restriction to finite elements. If Y is an ordered set, then Min(Y) denotes the set of all minimal elements of Y, i.e. $\text{Min}(Y) = \{X \in Y \mid \neg\exists X' \in Y : X' < X\}$.

Let $L \subseteq L(\sigma)$ be a nonempty language. An operation $C : 2^L \to 2^L$ is called an *inference operation*. The corresponding *inference relation* $\vdash$ is defined by $X \vdash F$ iff $F \in C(X)$. An inference operation (relation) is called a *consequence* operation (relation) if it satisfies Inclusion (Reflexivity), Idempotence (Transitivity), and Monotony.

A *model-theoretic system* $\langle L, \boldsymbol{I}, \models \rangle$ is determined by a language L, a set $\boldsymbol{I}$ whose elements are called *interpretations* and a *model relation* $\models \subseteq \boldsymbol{I} \times L$ between interpretations and formulas. With every model-theoretic system $\langle L, \boldsymbol{I}, \models \rangle$, we can associate a model operator Mod_I, a consequence operation C^I, and a consequence relation $\models_I$ in the following way. Let $X \subseteq L$, then the associated model operator is defined as $\text{Mod}_I(X) = \{\mathcal{I} \in \boldsymbol{I} : \mathcal{I} \models X\}$, where $\mathcal{I} \models X$ iff for every $F \in X : \mathcal{I} \models F$. The associated consequence operation is defined by $C^I(X) = \{F \in L : \text{Mod}_I(X) \subseteq \text{Mod}_I(F)\}$, and finally $X \models_I F$ iff $F \in C^I(X)$.

An inference operation C is called *correct*, resp. *complete*, with respect to the model-theoretic system $\langle L, \boldsymbol{I}, \models \rangle$ iff $C(X) \subseteq C^I(X)$, resp. $C(X) = C^I(X)$.

1.2 Partial Logics with Two Kinds of Negation

Definition 1 (Interpretation) *Let $\sigma = \langle Rel, ExRel, Const \rangle$ be a signature. A partial Herbrand σ-interpretation $\mathcal{I}$ consists of:*

1. *A set U, the universe or domain of $\mathcal{I}$ which is equal to the set of constant symbols, $U = Const$;*

2. *an assignment $c^{\mathcal{I}} = c$ to every constant symbol $c \in Const$;*

3. *an assignment of a pair $\langle r^{\mathcal{I}}, \tilde{r}^{\mathcal{I}} \rangle$ to every relation symbol $r \in Rel$ such that*

$$r^{\mathcal{I}} \cup \tilde{r}^{\mathcal{I}} \subseteq U^{a(r)},$$

and in the special case of an exact relation symbol $r \in ExRel$,

$$r^{\mathcal{I}} \cup \tilde{r}^{\mathcal{I}} = U^{a(r)},$$

where $a(r)$ denotes the arity of r.

In the sequel we shall also simply say 'interpretation' instead of 'partial Herbrand interpretation'.

The class of all partial Herbrand σ-interpretations is denoted by $\boldsymbol{I}_4(\sigma)$. We define the classes of *coherent*, of *total*, and of total coherent (or *2-valued*) interpretations by

$$\boldsymbol{I}_c(\sigma) = \{\mathcal{I} \in \boldsymbol{I}_4(\sigma) : r^{\mathcal{I}} \cap \tilde{r}^{\mathcal{I}} = \emptyset \text{ for all } r \in Rel\}$$
$$\boldsymbol{I}_t(\sigma) = \{\mathcal{I} \in \boldsymbol{I}_4(\sigma) : r^{\mathcal{I}} \cup \tilde{r}^{\mathcal{I}} = U^{a(r)} \text{ for all } r \in Rel\}$$
$$\boldsymbol{I}_2(\sigma) = \boldsymbol{I}_c(\sigma) \cap \boldsymbol{I}_t(\sigma)$$

The model relation $\models \subseteq \boldsymbol{I}_4(\sigma) \times L(\sigma; -, \sim, \wedge, \vee, |, \exists, \forall)$ between an interpretation and a sentence is defined inductively as follows.

Definition 2 (Model Relation)

1. $\mathcal{I} \models r(c_1, \ldots, c_m)$ *iff* $\langle c_1, \ldots, c_m\rangle \in r^{\mathcal{I}}$.
   $\mathcal{I} \models \sim r(c_1, \ldots, c_m)$ *iff* $\langle c_1, \ldots, c_m\rangle \in \tilde{r}^{\mathcal{I}}$.
2. $\mathcal{I} \models F \wedge G$ *iff* $\mathcal{I} \models F$ *and* $\mathcal{I} \models G$.
3. $\mathcal{I} \models F \vee G$ *iff* $\mathcal{I} \models F$ *or* $\mathcal{I} \models G$.
4. $\mathcal{I} \models \exists x F(x)$ *iff* $\mathcal{I} \models F(c)$ *for some* $c \in Const$.
5. $\mathcal{I} \models \forall x F(x)$ *iff* $\mathcal{I} \models F(c)$ *for all* $c \in Const$.
6. $\mathcal{I} \models -F$ *iff* $\mathcal{I} \not\models F$.

All other cases of compound formulas are handled by the following DeMorgan-style rewrite rules:

$$\sim(F \wedge G) \longrightarrow \sim F \vee \sim G \qquad \sim(F \vee G) \longrightarrow \sim F \wedge \sim G$$
$$\sim \exists x F(X) \longrightarrow \forall x \sim F(x) \qquad \sim \forall x F(X) \longrightarrow \exists x \sim F(x)$$
$$\sim\sim F \longrightarrow F \qquad \sim -F \longrightarrow F$$

and the definition for exclusive disjunction:

$$F|G \longrightarrow (F \wedge -G) \vee (G \wedge -F)$$

in the sense that for every rewrite rule $LHS \longrightarrow RHS$, we define $\mathcal{I} \models LHS$ iff $\mathcal{I} \models RHS$.

Mod_* denotes the model operator associated with the system $\langle L(\sigma), \boldsymbol{I}_*, \models\rangle$, and $\models_*$ denotes the resp. consequence relation, for $* = 4, c, t, 2$.

Definition 3 (Diagram) *The* diagram *of* $\mathcal{I} \in \boldsymbol{I}_4(\sigma)$ *is defined as* $D_{\mathcal{I}} = \{l \in \mathrm{Lit}(\sigma) : \mathcal{I} \models l\}$.

Observation 1 *Partial Herbrand interpretations can be identified with their diagrams, i.e. there is a one-to-one correspondence between* $\boldsymbol{I}_4(\sigma)$ *and* $2^{\mathrm{Lit}(\sigma)}$.

Definition 4 (Extension) *Let* $\mathcal{I}$ *and* $\mathcal{I}'$ *be two interpretations. We say that* $\mathcal{I}'$ extends $\mathcal{I}$, *symbolically* $\mathcal{I} \leq \mathcal{I}'$, *if* $D_{\mathcal{I}} \subseteq D_{\mathcal{I}'}$.

Definition 5 (Minimal Models) *For* $F \in L(\sigma) \supseteq X$, *and* $* = 4, c$, *we define* $\mathrm{MinMod}_*(X) = \mathrm{Min}(\mathrm{Mod}_*(X))$, *and* $X \models^m_* F$ *iff* $\mathrm{MinMod}_*(X) \subseteq \mathrm{Mod}_*(F)$.

Observation 2 *An interpretation is the unique minimal model of its diagram: for* $* = 4, c$, *and for every* $\mathcal{I} \in \boldsymbol{I}_*(\sigma)$, $\mathrm{MinMod}_*(D_{\mathcal{I}}) = \{\mathcal{I}\}$, *and consequently,* $\mathcal{I} \models F$ *iff* $D_{\mathcal{I}} \models^m_* F$.

Formulas of partial logic (with two kinds of negation) can be normalized in the same manner as those of classical logic. For this purpose, we introduce DNS(F), the *disjunctive normal set* of a formula F, which is defined as follows:

$$\begin{aligned}
\mathrm{DNS}(1) &= \{\emptyset\} \\
\mathrm{DNS}(e) &= \{\{e\}\} \\
\mathrm{DNS}(F \wedge G) &= \{E \cup D : E \in \mathrm{DNS}(F),\ D \in \mathrm{DNS}(G)\} \\
\mathrm{DNS}(F \vee G) &= \mathrm{DNS}(F) \cup \mathrm{DNS}(G)
\end{aligned}$$

All other cases of compound formulas can be handled by the above and the following DeMorgan-style rewrite rules:

$$\begin{array}{rclrcl}
-(F \vee G) & \longrightarrow & -F \wedge -G & -(F \wedge G) & \longrightarrow & -F \vee -G \\
-{\sim}(F \vee G) & \longrightarrow & -{\sim}F \vee -{\sim}G & -{\sim}(F \wedge G) & \longrightarrow & -{\sim}F \wedge -{\sim}G \\
-{\sim}{\sim}F & \longrightarrow & -F & --F & \longrightarrow & F \\
-{\sim}-F & \longrightarrow & -F & & &
\end{array}$$

The disjunctive normal form of a formula $G \in L(-, \sim, \wedge, \vee)$ is obtained as

$$\mathrm{DNF}(G) = \bigvee_{E \in \mathrm{DNS}(G)} \bigwedge E$$

Observation 3 *Let* $F \in L(\sim, \wedge, \vee)$. *Then,*

1. $\mathrm{MinMod}_4(F) \subseteq \mathrm{DNS}(F)$.

2. $\mathrm{DNS}(F) \subseteq \mathrm{Mod}_4(F)$.

2 Knowledge Systems

Before presenting the formal definitions, we start with a semi-formal discusssion of the basic concepts to be introduced, notably: knowledge base, query, inference, answer, information ordering, input and update.

In general, a knowledge base (KB) can consist of any kind of data structures capable of representing knowledge, e.g. a set, or multiset, or sequence, of (logical) expressions, or a directed graph, etc. For the sake of simplicity, we shall assume that a KB is a set of expressions from a representation language. Only certain formulas may make sense for representing knowledge, that is, there will be a specific representation language L_{Repr}, and a KB will be a (usually finite) collection of elements of L_{Repr}, possibly constrained in some way determined by the set L_{KB} of all admissible KBs: $\mathrm{KB} \in L_{\mathrm{KB}} \subseteq 2^{L_{\mathrm{Repr}}}$. Likewise, since not every formula may be appropriate as a sensible query, the set of admissible queries is specified by L_{Query}.

The basic scenario of a knowledge system (KS) consists of two operations: an inference operation processing queries posed to the KB, and an update operation processing inputs entered by users or by other (e.g. sensoric) information suppliers. A KS restricts the admissible inputs to elements of a specific input language L_{Input}, and an update is performed by processing the input formula in an appropriate way in order to assimilate its information content into the KB. Since it appears reasonable to require that any information entered to a KB can be queried afterwards, we shall assume that $L_{\mathrm{Input}} \subseteq L_{\mathrm{Query}}$.

Definition 6 (Knowledge System) *An abstract knowledge system* $\boldsymbol{K}$ *is a quintuple:*[4]

$$\boldsymbol{K} = \langle L_{\mathrm{KB}}, \vdash, L_{\mathrm{Query}}, \mathsf{Upd}, L_{\mathrm{Input}} \rangle$$

where the inference relation $\vdash \subseteq L_{\mathrm{KB}} \times L_{\mathrm{Query}}$, *together with the update operation* $\mathsf{Upd} : L_{\mathrm{KB}} \times L_{\mathrm{Input}} \rightarrow L_{\mathrm{KB}}$, *satisfy for any* $X \in L_{\mathrm{KB}}$,

(KS1) $X \vdash 1$, *and* $\mathsf{Upd}(X, 1) \doteq X$.

(KS2) $L_{\mathrm{Input}} \subseteq L_{\mathrm{Query}}$.

(KS3) $\mathsf{Upd}(X, F) \vdash F$, *for any* $F \in L_{\mathrm{Input}}$ *which is consistent with* X.[5]

---

[4]The formulation of a KS in terms of query and input processing was already implicitly present in Belnap's (1977) view of a KS. In (Levesque 1984) it was proposed as a 'functional approach to knowledge representation'. In (Wagner 1994a) the concept of knowledge systems was further extended and used as an integrating framework for knowledge representation and logic programming.

[5]According to some notion of consistency associated with the abstract knowledge system. E.g., one might want to exclude contradictory pieces of information from this reflexivity principle: $\mathsf{Upd}(\{\sim p\}, p) \not\vdash p$. We shall not discuss this issue in the present paper, however.

If elements of L_{KB} are finite sets (resp. structures), $\boldsymbol{K}$ is called *finitary*. In the sequel, we shall sometimes simply write 'KB' in formal expressions standing for an arbitrary knowledge base $X \in L_{\text{KB}}$. An inference operation C is defined as usual:

$$C(\text{KB}) = \{F \in L_{\mathsf{Query}} : \text{KB} \vdash F\}$$

Also, an answer operation taking a knowledge base and an open query formula, and providing the corresponding set of valid answers, can be defined in terms of the inference relation.

Definition 7 (Answer Operation) *If KB is* definite,[6] *an answer to an open query formula $F(x)$ is a tuple, and the set of all answers is a relation:*

$$\mathsf{Ans}(\text{KB}, F(x)) = \{t : \text{KB} \vdash F(t)\}$$

where x is a variable (tuple) and t a constant (tuple). In general, however, an answer may be indefinite, i.e. a minimal set of possible answer substitutions corresponding to a minimal disjunctive consequence:

$$\mathsf{Ans}(\text{KB}, F(x)) = \text{Min}(\{T : \text{KB} \vdash \bigvee\{F(t) : t \in T\}\})$$

Not all open query formulas can be answered sensibly. We therefore require that queries are *evaluable*.[7] Answers to evaluable queries on the basis of definite KBs may be computed by means of relational algebra operations, such as projection, selection, set difference, union, and join. For instance,

$$\mathsf{Ans}(\text{KB}, F(x,y) \wedge G(y,z)) = \mathsf{Ans}(\text{KB}, F(x,y)) \bowtie \mathsf{Ans}(\text{KB}, G(y,z))$$

In many cases, it is useful to be able to update by a set of inputs and we 'overload' the symbol Upd to denote also this more general update operation

$$\mathsf{Upd} : L_{\text{KB}} \times 2^{L_{\mathsf{Input}}} \longrightarrow L_{\text{KB}}$$

which has to be defined in such a way that for any finite $A \subseteq L_{\mathsf{Input}}$, $\mathsf{Upd}(\text{KB}, A) = \mathsf{Upd}(\text{KB}, \bigwedge A)$. We sometimes write $\text{KB} + F$ as an abbreviation of $\mathsf{Upd}(\text{KB}, F)$, resp. $\text{KB} - F$ as an abbreviation of $\mathsf{Upd}(\text{KB}, -F)$.

Example 1 (Fact Bases) *A KB consisting of ground literals (viewed as positive and negative facts) is called a* fact base. *For instance, the fact base*

$$X_1 = \{r(S), r(P), s(S), {\sim} s(L), {\sim} s(T), m(P,L), m(T,S)\}$$

may represent the information that Susan and Peter are residents, Susan is a smoker, Linda and Tom are nonsmokers, Peter is married with Linda, and Tom is married with Susan.

[6] KB is called *definite* if $\text{KB} \vdash F \vee G$ implies that either $\text{KB} \vdash F$, or $\text{KB} \vdash G$, where $\vdash$ is a constructive inference relation (see below).

[7] See (van Gelder/Topor 1991) for the notion of *evaluable*, resp. *domain-independent*, formulas.

As a kind of natural deduction from positive and negative facts an inference relation $\vdash$ *between a fact base* $X \subseteq \mathrm{Lit}$ *and a sentence is defined in the following way:*[8]

$$
\begin{array}{ll}
(\vdash a) & X \vdash a \ \textit{if} \ a \in X \\
(\vdash \sim a) & X \vdash \sim a \ \textit{if} \ \sim a \in X \\
(\vdash -l) & X \vdash -l \ \textit{if} \ l \notin X \\
(\vdash \wedge) & X \vdash F \wedge G \ \textit{if} \ X \vdash F \ \& \ X \vdash G \\
(\vdash \vee) & X \vdash F \vee G \ \textit{if} \ X \vdash F \ \textit{or} \ X \vdash G \\
(\vdash |) & X \vdash F|G \ \textit{if} \ X \vdash F \wedge -G \ \textit{or} \ X \vdash G \wedge -F \\
(\vdash \exists) & X \vdash \exists x F(x) \ \textit{if} \ X \vdash F(c) \ \textit{for some constant} \ c
\end{array}
$$

For instance, one might ask X_1 *"who is married with a non-resident?",*

$$\mathsf{Ans}(X_1, \exists y[m(x,y) \wedge -r(y)]) = \{\langle P \rangle\}$$

answered by "Peter", or "who is married with a nonsmoker?",

$$\mathsf{Ans}(X_1, \exists y[m(x,y) \wedge \sim s(y)]) = \{\langle P \rangle\}$$

also answered by "Peter". The only updates we consider are insertions, $\mathsf{Upd}(X,l) := X \cup \{l\}$, *of literals. For* $K \subseteq \mathrm{Lit}$, *we have* $\mathsf{Upd}(X,K) = X \cup K$. *The knowledge system of fact bases is then defined as*

$$\boldsymbol{F} := \langle 2^{\mathrm{Lit}}, \vdash, L(-,\sim,\wedge,\vee,|,\exists), \mathsf{Upd}, \mathrm{Lit} \rangle$$

The restricted system where weak negation and exclusive disjunction are excluded from the query language, $L_{\mathsf{Query}} = L(\sim,\wedge,\vee)$, *is denoted by* $\boldsymbol{F}^+$.

It is easy to check that KS1, KS2 and KS3 hold.

Observation 4 *The inference relation of* $\boldsymbol{F}$ *captures model-based reasoning, resp. minimal reasoning on the basis of definite knowledge, in partial logic. Every fact base corresponds to the diagram of a partial Herbrand model, and for every* $F \in L(-,\sim,\wedge,\vee,|,\exists)$ *and every* $\mathcal{I} \in \boldsymbol{I}_4$, *it holds that* $D_{\mathcal{I}} \vdash F$ *iff* $\mathcal{I} \models F$, *and consequently, for every* $X \subseteq \mathrm{Lit}$,

$$X \vdash F \quad \textit{iff} \quad X \models_4^m F$$

where $\models_4^m$ *denotes entailment based on 4-valued minimal models (see definition 5).*

Whenever we deal with both kinds of negation, the strong negation $\sim$ is the principal negation operator (expressing explicit falsity), and the weak negation $-$ is an auxiliary negation operator (used, e.g., to express the CWA, see below). Sometimes we shall also make use of the symbol $\neg$ standing either for classical negation (which differs from both $\sim$ and $-$), or for an arbitrary negation.

[8]This inductive definition is completed by the DeMorgan-style rewrite rules listed in section 1.2.

2.1 Regular Knowledge Systems

In order to compare knowledge bases in terms of their information content we assume that there is an *information*, or *knowledge ordering* $\leq$ between KBs such that

$$\mathrm{KB}_1 \leq \mathrm{KB}_2 \quad \textit{if}\ \mathrm{KB}_2\ \textit{contains at least as much information as}\ \mathrm{KB}_1.$$

The information ordering should be defined in terms of the structural components of knowledge bases and not in terms of higher-level notions (like derivability).[9] The informationally *empty KB* will be denoted by 0. By definition, $0 \leq X$ for all $X \in L_{\mathrm{KB}}$, i.e. 0 is the least element of $\langle L_{\mathrm{KB}}, \leq \rangle$.

In general, more information does not mean more consequences. In other words: answers are not necessarily preserved under growth of information. Queries, for which this is the case, are called *persistent.*

Definition 8 (Persistent Queries) *A closed, resp. open, query formula F is called* persistent *if* $\forall X_1, X_2 \in L_{\mathrm{KB}} : X_1 \vdash F \;\Rightarrow\; X_2 \vdash F$, *resp.* $\mathsf{Ans}(X_1, F) \subseteq \mathsf{Ans}(X_2, F)$, *whenever* $X_1 \leq X_2$. *If all $F \in L_{\mathrm{Query}}$ are persistent, the KS and its inference relation $\vdash$ are called persistent. The set of all persistent query formulas is denoted by L_{PersQ}. An operator of the query language is called persistent, if every query formed with it and with persistent subformulas is again persistent.*

Definition 9 (Ampliative Inputs) *An input formula F is called (i)* ampliative[10] *if* $\mathrm{KB} \leq \mathsf{Upd}(\mathrm{KB}, F)$, *or (ii)* reductive *if* $\mathrm{KB} \geq \mathsf{Upd}(\mathrm{KB}, F)$. *A KS and its update operation* Upd *are called* ampliative, *if all inputs $F \in L_{\mathrm{Input}}$ are ampliative. The set of all ampliative input formulas is denoted by L_{Ampl}.*

In this paper, we shall only consider ampliative updating.

A certain subset $L_{\mathrm{Unit}} \subseteq L_{\mathrm{Input}}$ designates those elementary expressions which will be called *information units*, e.g. atoms, literals, or weighted (resp. labelled, or annotated) atoms, and the like. An information unit represents an elementary piece of information with a positive information content. A knowledge base may contain contradictory pieces of information, and we assume that all inconsistent information units contained in $X \in L_{\mathrm{KB}}$ are collected by $\mathrm{Inc}(X) \subseteq L_{\mathrm{Unit}}$.

Definition 10 (Regular KS) *A knowledge system* $\boldsymbol{K}$ *is called* regular, *if there is a preorder $\langle L_{\mathrm{KB}}, \leq, 0 \rangle$ with least element 0, a designated set $L_{\mathrm{Unit}} \subseteq L_{\mathrm{Input}}$, and an operation* $\mathrm{Inc} : L_{\mathrm{KB}} \rightarrow 2^{L_{\mathrm{Unit}}}$, *such that*

[9] The usual way to compare the information content of two KBs in standard logic, namely by means of checking the inclusion of consequences: $\mathrm{KB}_1 \leq \mathrm{KB}_2$ if $C(\mathrm{KB}_1) \subseteq C(\mathrm{KB}_2)$, does not work in a nonmonotonic setting.

[10] The name is adopted from (Belnap 1977).

(KS4) $X \leq X + u$, *for any* $X \in L_{\mathrm{KB}}$, *and for any* $u \in L_{\mathrm{Unit}}$, *such that* $u \notin \mathrm{Inc}(X)$, *and* $\mathrm{Inc}(X + u) \subseteq \mathrm{Inc}(X)$.

(KS5) *For all* $X_1, X_2 \in L_{\mathrm{KB}}$, $X_1 \leq X_2$ *iff* $\forall F \in L_{\mathrm{AmplI}} \forall G \in L_{\mathrm{PersQ}} : X_1 + F \vdash G \Rightarrow X_2 + F \vdash G$

(KS6) *Ampliative inputs are persistent queries:* $L_{\mathrm{AmplI}} = L_{\mathrm{PersQ}} \cap L_{\mathrm{Input}}$.

A regular KS will be represented as a 9-tuple

$$\langle 0, \leq, L_{\mathrm{KB}}, \vdash, L_{\mathrm{Query}}, \mathsf{Upd}, L_{\mathrm{Input}}, \mathrm{Inc}, L_{\mathrm{Unit}} \rangle$$

Definition 11 (Consistency)

1. KB *is called* consistent, *if* $\mathrm{Inc}(\mathrm{KB}) = \emptyset$.
2. $F \in L_{\mathrm{Input}}$ *is called* consistent, *if* $\mathrm{Inc}(0 + F) = \emptyset$.
3. $F \in L_{\mathrm{Input}}$ *is called* consistent within $X \in L_{\mathrm{KB}}$, *if*
   - *3.1* F *is consistent, and*
   - *3.2* $\mathrm{Inc}(X + F) \subseteq \mathrm{Inc}(X)$ *(requiring that* F *does not increase the inconsistency of* X*), and*
   - *3.3 for every* $u \in \mathrm{Inc}(X)$, $0 + F - u \vdash F$ *(requiring that the information of* F *is not already inconsistent in* X*).*

Example 2 (Standard Logics) *A standard logic (such as classical, or intuitionistic logic), given by a language* L *and a consequence relation* $\vdash \subseteq 2^L \times L$, *resp. by the corrresponding consequence operation* C, *can be viewed as an infinitary knowledge system*

$$\langle \emptyset, \subseteq, \{X \in 2^L : X = C(X)\}, \vdash, L, \mathsf{Upd}, L, \mathrm{Inc}, L \rangle$$

where 1) a KB is a deductively closed set of formulas, 2) the knowledge ordering is set inclusion, 3) update by F *is the addition of* F *and subsequent closure, i.e.* $\mathsf{Upd}(X, F) = C(X \cup \{F\})$,[11] *4) the query, unit and input languages are all equal to* L, *and 5)* $\mathrm{Inc}(X) = \emptyset$ *if* $X \neq L$, *and* $\mathrm{Inc}(X) = X$ *otherwise. KS1–KS6 hold, more or less, trivially. Notice, however, that it is not clear whether a standard logic corresponds to a sensible finitary knowledge system, because set inclusion is no longer an adequate knowledge ordering if KBs are not deductively closed (KS5 is violated).*

Example 3 (Fact Bases) *The knowledge system of fact bases is also regular:*

$$\boldsymbol{F} := \langle \emptyset, \subseteq, 2^{\mathrm{Lit}}, \vdash, L(-, \sim, \wedge, \vee, |, \exists), \mathsf{Upd}, \mathrm{Lit}, \mathrm{Inc}, \mathrm{Lit} \rangle$$

[11] Notice that this corresponds to the AGM *expansion* of 'belief sets', see (Gärdenfors 1988).

where $\mathrm{Inc}(X) = X \cap \widetilde{X}$. We have to show that KS4–KS6 hold. Proof: it is obvious that KS4 holds. Since $L_{\mathsf{AmpI}} = \mathrm{Lit}$, and $L_{\mathsf{PersQ}} = L(\sim, \wedge, \vee)$, KS5 follows by straightforward induction on the complexity of queries (it corresponds to the *permanence principle* of partial logic). KS6 is again obvious. □

2.2 Formal Properties of Knowledge Systems

2.2.1 Basic Properties of Inference and Update

Definition 12 (Constructive Inference) *Let $L \subseteq \mathrm{Lit}$ be an arbitrary set of literals, and $\lceil L \rceil := \mathsf{Upd}(0, L)$. Then $\vdash$ is called* constructive *if it satisfies (i) constructible truth, and (ii) constructible falsity, i.e. both of the following conditions: for any $F, G \in L_{\mathsf{Query}}$: (i) $\lceil L \rceil \vdash F \vee G$ implies $\lceil L \rceil \vdash F$ or $\lceil L \rceil \vdash G$; (ii) $\lceil L \rceil \vdash \sim(F \wedge G)$ implies $\lceil L \rceil \vdash \sim F$ or $\lceil L \rceil \vdash \sim G$.*

The property of constructive inference guarantees that, on the basis of definite knowledge, query formulas are decomposable. Obviously, the first condition (of constructible truth) excludes the possibility of certain disjunctive tautologies such as the classical *tertium non datur*, whereas its negative counterpart excludes, for instance, the dual *principle of contradiction*.

The next property (due to Urbas (1990)) excludes the possibility of trivial inferences, i.e. non-tautological inferences which are solely based on the form of a KB and a query and not on their content. For example, $\{s(L), \sim s(L)\} \vdash m(P, S)$ is such a trivial inference which is valid in classical logic, i.e. from contradictory information on Linda being a smoker, we may infer that Peter is married with Susan, and thus we would get (infinitely many) unsensible answers to any query. This is clearly undesirable in a knowledge system.

Definition 13 (Tautology) *$F \in L_{\mathsf{Query}}$ is called a* tautology *in a knowledge system, if $X \vdash F$ for all $X \in L_{\mathsf{KB}}$.*

Definition 14 (Non-Explosive Inference) *An inference relation $\vdash$ is called* non-explosive *if for every non-tautology $F \in L_{\mathsf{Query}}$, and for every knowledge base $X > 0$, there is a variant F' of F (obtained by uniform substitution of propositional constituents) such that $X \not\vdash F'$.*

While most standard logics allow for trivial inferences, their positive fragments and certain paraconsistent logics, such as Belnap's (1977) four-valued, or Nelson's (Almukdad/Nelson 1984) paraconsistent constructive logic, are non-explosive.

The following important property guarantees the freedom of knowledge base evolution.

Definition 15 (Input Completeness) *A KS is called* input complete *if*
$$\forall X_1, X_2 \in L_{\mathrm{KB}}\, \exists A \subseteq L_{\mathrm{Input}} : X_2 = \mathsf{Upd}(X_1, A)$$

Observation 5 *A KS is input complete iff KBs are both input constructible and input destructible, i.e. both of the following conditions hold:*

$$(i)\ \forall X \in L_{\mathrm{KB}}\, \exists A \subseteq L_{\mathrm{Input}} : X = \mathsf{Upd}(0, A)$$
$$(ii)\ \forall X \in L_{\mathrm{KB}}\, \exists A \subseteq L_{\mathrm{Input}} : \mathsf{Upd}(X, A) = 0$$

Further formal properties of knowledge systems are discussed in (Wagner 1994b).

2.2.2 Nonmonotonicity

The following definition captures the idea that a system is considered *monotonic* if all consequences of a KB are preserved after it is updated by some new piece of information.

Definition 16 (Monotonicity) *A KS is called* monotonic *if for all $X \in L_{\mathrm{KB}}$, and all $F \in L_{\mathrm{Input}}$, we have $C(X) \subseteq C(\mathsf{Upd}(X, F))$.*

Though fundamental in the theory of consequence operations due to Tarski, this is too strong a requirement for knowledge systems in general.

There are two 'parameters' on which Monotonicity depends: the update operation may be ampliative or not, and the inference relation may be persistent or not.

Observation 6 *A KS is monotonic if it is ampliative and persistent.*[12]

Proof: For any $X \in L_{\mathrm{KB}}$, and any $F \in L_{\mathrm{Input}}$, we get $X \leq \mathsf{Upd}(X, F)$ by Ampliative Update, and consequently $C(X) \subseteq C(\mathsf{Upd}(X, F))$, by Persistent Inference. □

Practical systems will be nonmonotonic since they will allow for non-persistent queries (by means of negation-as-failure) and for non-ampliative updates (by means of deletion, resp. contraction).

[12] Or, rather exotically, if all inputs are reductive and all queries are 'antipersistent', i.e. preserved under information decrease. It is still an open problem, whether – or under which conditions – the converse holds.

2.3 Vivid Knowledge Systems

The idea of *vividness* as a design principle for knowledge systems was first proposed by Levesque (1986). However, while for Levesque the main issue was to have *complete* information, we have generalized and redefined the notion of vividness based on two fundamental principles: cognitive adequacy and computational feasibility.[13]

In a vivid knowledge system (VKS) we have two kinds of negation: besides a principal negation (called *strong*) expressing explicit falsity, there should be a second negation (called *weak*) being able to express default-implicit negative information in the style of negation-as-failure.

It is desirable for a KS to be robust towards any possible update. This includes inputs which are inconsistent with the current knowledge base. Such inputs may be erroneous, but it might be as well the case that the new input is correct, and some old piece of information in the KB is erroneous or outdated. In any case, it seems important that the main functions of a KS are not corrupted by inconsistent inputs. Thus, a sophisticated inconsistency handling mechanism might prevent that both a query formula and its negation can ever be inferred from a KB even if the KB contains contradictory information.

Definition 17 (Inherently Consistent Inference) *An inference relation $\vdash$ is called* inherently consistent *with respect to a negation operator $\neg$, if for every $X \in L_{\mathrm{KB}}$ and every $F \in L_{\mathrm{Query}}$, it is never the case that $X \vdash \neg F$ and $X \vdash F$.*

Notice that this holds trivially if we make the restriction that L_{KB} admits only of consistent KBs. But such a restriction is not realistic. We shall assume, therefore, that L_{KB} also contains inconsistent KBs. In this case, inference in classical logic is not inherently consistent. In fact, Inherently Consistent Inference is violated by any negation satisfying the classical explosion principle *ex contradictione sequitur quodlibet*, $\{F, \neg F\} \vdash G$. But it is also violated by those paraconsistent logics where contradictions are derivable, $\{F, \neg F\} \vdash F \wedge \neg F$, such as in Belnap's 4-valued logic.

Observation 7 *Inference in $\boldsymbol{F}$ is inherently consistent wrt weak negation:*

$$\forall X \in 2^{\mathrm{Lit}} \forall F \in L(-, \sim, \wedge, \vee) : X \not\vdash F \wedge -F$$

Although we do not make it a strict requirement, it seems desirable for a KS that the following *coherence*[14] property, relating weak with strong negation, holds.

[13] This is further discussed in (Wagner 1994a).
[14] The name is adopted from (Pereira/Alferes 1992).

Definition 18 (Negation Coherence) *A knowledge system with weak and strong negation satisfies* Negation Coherence, *if for any consistent* $X \in L_{\mathrm{KB}}$, *and any* $F \in L_{\mathrm{Query}}$, $X \vdash -F$ *whenever* $X \vdash \sim F$.

Definition 19 (Minimal Change) [15] *Let* $u \in L_{\mathrm{Unit}}$ *be any information unit. We say that* Upd *satisfies the principle of* Minimal Change *if it satisfies both Minimal Change for* Unit Expansion *and for* Unit Contraction: *for any* $X, X' \in L_{\mathrm{KB}}$, *and any* $u \in L_{\mathrm{Unit}}$,

Minimal Change for Unit Expansion $X + u := \mathsf{Upd}(X, u)$ *is the least extension of* X *such that* u *can be inferred, expressed by the conditions* (+1) *and* (+2):

$$(+1)\ X + u \geq X$$
$$(+2)\ X' \geq X\ \&\ X' \vdash u\ \Rightarrow\ X + u \leq X'$$

Minimal Change for Unit Contraction $X - u := \mathsf{Upd}(X, -u)$ *is the greatest reduction of* X *such that* $-u$ *can be inferred, expressed by* (−1) *and* (−2):

$$(-1)\ X - u \leq X$$
$$(-2)\ X' \leq X\ \&\ X' \vdash -u\ \Rightarrow\ X - u \geq X'$$

Definition 20 (VKS) *A knowledge system with weak and strong negation is called a* vivid knowledge system (VKS), *if it satisfies Constructive and Non-Explosive Inference, Inherently Consistent Inference wrt weak negation, and the principle of Minimal Change.*

It is easy to check that $\boldsymbol{F}$, the system of fact bases, is a VKS.

3 Belnap's Epistemic States

A KS, according to Belnap, answers queries by invoking some inference mechanism, and it accepts input from a variety of sources by using an appropriate update mechanism. In such circumstances inconsistency threatens. Mr. X tells the KS that A while Mrs. Y tells it that $\sim A$. Or, in a different environment, an automatic measurement yields that $m > 0.3$ while the subsequent confirmation attempt yields that $m < 0.3$.

What is the KS to do?

Possibility 1: Refuse to accept inconsistent information. However: this is unfair either to Mr. X or to Mrs. Y. Also, contradictions may not lie on the surface.

[15]This principle was already proposed in (Belnap 1977) under the name of *minimal mutilation*. It is also one of the fundamental principles of AGM-style *belief revision*, see e.g. (Gärdenfors 1988).

Possibility 2: Revise current beliefs in the presence of contradictions. However: it seems to be difficult to determine the proper revision policy doing justice to both Mr. X and Mrs. Y, and it seems to be even more difficult to mechanize it in a satisfactory way.

Possibility 3: Just accept contradictions and report them exactly as they were told, so the user can make up her mind.

Belnap advocates possibility 3, but emphasizes that even if the ultimate goal is possibility 2, i.e. revision, possibility 3 is a good first step towards that goal.

In order to be able to process arbitrary input formulas $F \in L(\sim, \wedge, \vee)$, an epistemic state has to consist of a set of possible situation descriptions (corresponding to partial interpretations, called 'set-ups' by Belnap). Thus, an epistemic state Y is a subset of 2^{Lit}, and the elements of Y represent different epistemic alternatives. For instance, if we know that Susan is a nonsmoker and that Peter is married either with Linda or with Susan, we get the following epistemic state:

$$Y_1 = \{\{\sim s(S),\, m(P,L)\},\, \{\sim s(S), m(P,S)\}\}$$

Following Belnap, yet formulated in a different way, inputs to an epistemic state are processed as follows:

$$\begin{aligned} \mathsf{Upd}_B(Y,l) &= \{X \cup \{l\} : X \in Y\} \\ \mathsf{Upd}_B(Y, F \wedge G) &= \mathsf{Upd}_B(\mathsf{Upd}_B(Y,F), G) \\ \mathsf{Upd}_B(Y, F \vee G) &= \mathsf{Upd}_B(Y,F) \cup \mathsf{Upd}_B(Y,G) \end{aligned}$$

All other cases of compound formulas are treated by DeMorgan-style rewriting (see the rewrite rules above). Notice that Upd_B may create inconsistent situation descriptions even if there are consistent alternatives. It seems, however, more appropriate to discard inconsistent situation descriptions if they are not *minimally inconsistent*,[16] i.e. if there are alternatives with 'less inconsistency'. For this purpose we define

$$\begin{aligned} \mathrm{Cons}(Y) &:= \{X \in Y \mid X \cap \widetilde{X} = \emptyset\} \\ \mathrm{MinInc}(Y) &:= \{X \in Y \mid \neg\exists Z \in Y : (Z \cap \tilde{Z}) \subset (X \cap \widetilde{X})\} \end{aligned}$$

[16]The principle of *minimal inconsistency* was proposed in (Priest 1989).

The first operator, Cons, accepts only consistent situation descriptions, while the second one, MinInc, accepts all situation descriptions which are minimally inconsistent. If an epistemic state Y is consistent, i.e. if it contains at least one consistent epistemic alternative, then $\text{Cons}(Y) = \text{MinInc}(Y)$. We can now define

$$\begin{aligned}\mathsf{Upd}_{ex}(Y,F) &:= \text{Cons}(\mathsf{Upd}_B(Y,F))\\ \mathsf{Upd}_{mi}(Y,F) &:= \text{MinInc}(\mathsf{Upd}_B(Y,F))\end{aligned}$$

Upd_{ex} does not accept inconsistent inputs at all. It implements the *ex contradictione sequitur quodlibet (ECSQ)* principle of classical logic by discarding all inconsistent epistemic alternatives. A good compromise between the hypersensitive inconsistency handling mechanism of Upd_{ex} and the too liberal Upd_B is the principle of *minimal inconsistency* proposed in (Priest 1989), and implemented by Upd_{mi}.

A query formula $F \in L(\sim, \wedge, \vee)$ can be inferred from an epstemic state Y if it can be inferred from every possible situation description $X \in Y$:

$$Y \vdash F \quad :\Longleftrightarrow \quad \text{for all } X \in Y : X \vdash F,$$

where $X \vdash F$ is the inference relation of $\boldsymbol{F}^+$.

The process of information growth can be captured by the following notion of informational extension. An epistemic state Y' is called an *(informational) extension* of Y, symbolically $Y \leq Y'$, if every epistemic alternative in Y' extends one in Y:

Definition 21 (Information Ordering) $Y \leq Y' :\Longleftrightarrow \forall X' \in Y' \, \exists X \in Y : X \subseteq X'$

This ordering, used in (Belnap 1977), was also suggested in domain theory for the semantics of parallel processes (see (Plotkin 1976, Smyth 1978)), and is sometimes called 'Smyth ordering'.

Observation 8 $\langle 2^{2^{\text{Lit}}}, \leq \rangle$ *is a preorder with least element* $0 := \{\emptyset\}$.

Notice that while $\{\text{Lit}\}$ is informationally larger than any non-empty epistemic state $Y \subseteq 2^{\text{Lit}}$, the empty epistemic state is still larger:

$$Y \leq \{\text{Lit}\} < \{\}$$

The elementary pieces of information in epistemic states are disjunctions of literals $l_1 \vee \ldots \vee l_m$. The set of all such disjunctions is denoted by $\text{Lit}^{\vee}$. The inconsistency operation Inc_B now collects all definite and indefinite contradictions:

$$\text{Inc}_B(Y) := \{\bigvee L : L \in \text{Min}(\{K \subseteq \text{Lit} \mid \forall X \in Y \exists l \in K : l \in X \cap \tilde{X}\})\}$$

Definition 22 (Belnap's KS)

$$\boldsymbol{B} := \langle \{\emptyset\}, \leq, 2^{2^{\mathrm{Lit}}}, \vdash, L(\sim, \wedge, \vee), \mathsf{Upd}_B, L(\sim, \wedge, \vee), \mathrm{Inc}_B, \mathrm{Lit}^{\vee} \rangle$$

is called Belnap's KS. *Besides* $\boldsymbol{B}$, *we also define*

$\boldsymbol{B}_{mi}$ *with* $L_{\mathrm{KB}} = \{Y \in 2^{2^{\mathrm{Lit}}} : Y = \mathrm{MinInc}(Y)\}$, *and* Upd_{mi}, *and*

$\boldsymbol{B}_{ex}$ *with* $L_{\mathrm{KB}} = \{Y \in 2^{2^{\mathrm{Lit}}} : Y = \mathrm{Cons}(Y)\}$ *and* Upd_{ex}.

We have to show that KS1–KS6 are satisfied.

KS1 and KS2 are obvious.

KS3: Reflexivity is proved by induction on input formulas. Clearly, $\mathsf{Upd}(Y, l) \vdash l$, since l is added to every element of Y. For conjunctive inputs we obtain $\mathsf{Upd}(Y, F \wedge G) = (Y + F) + G \vdash G$, by the induction hypothesis, and since $\boldsymbol{B}$ is persistent and ampliative (see the resp. observations below), it follows from $Y + F \vdash F$ that $(Y + F) + G \vdash F$, and hence $(Y + F) + G \vdash F \wedge G$. For disjunctive inputs, $\mathsf{Upd}(Y, F \vee G) = (Y + F) \cup (Y + G) \vdash F \vee G$, since for every $X \in (Y + F)$, it holds by the induction hypothesis that $X \vdash F$, hence $X \vdash F \vee G$, and similarly for every $X \in (Y + G)$.

KS4: easy to check.

KS5: It suffices to show that $Y_1 \leq Y_2$ iff $C(Y_1) \subseteq C(Y_2)$. The ($\Rightarrow$)-part, i.e. Persistent Inference, is proved below. We show the contraposition of the ($\Leftarrow$)-part. Assume $Y_1 \not\leq Y_2$, i.e. $\exists X_2 \in Y_2 \forall X_1 \in Y_1 \exists l \in X_1 - X_2$. Let X_2' be such an element of Y_2. Then we can construct a formula $F = \bigvee\{l \in \mathrm{Lit} \mid \exists X_1 \in Y_1 : l \in X_1 - X_2'\}$, for which $Y_1 \vdash F$, but $Y_2 \not\vdash F$.

KS6: Follows immediately from the fact that $\boldsymbol{B}$ is persistent and ampliative (see the resp. observations below). □

Observation 9 *Only the minimal elements of an epistemic state count:* $C(Y) = C(\mathrm{Min}(Y))$.

Proof: In $\boldsymbol{F}^+$, for $X, X' \subseteq \mathrm{Lit}$ it holds that $C(X) \subseteq C(X')$ whenever $X \subseteq X'$. Consequently,

$$C(Y) = \bigcap\{C(X) : X \in Y\} = \bigcap\{C(X) : X \in \mathrm{Min}(Y)\} \qquad \square$$

Observation 10 (Persistent Inference) *In* $\boldsymbol{B}$, $\vdash$ *is persistent, i.e. if* $Y_1 \leq Y_2$ *then* $Y_2 \vdash F$ *whenever* $Y_1 \vdash F$ *for all* $F \in L(\sim, \wedge, \vee)$.

Proof: Let $Y_1 \leq Y_2$. Then,

$$\begin{aligned} C(Y_1) &= \bigcap\{C(X_1) \mid X_1 \in Y_1\} \\ &\subseteq \bigcap\{C(X_1) \mid X_1 \in Y_1 \ \& \ \exists X_2 \in Y_2 : X_1 \subseteq X_2\} \\ &\subseteq \bigcap\{C(X_2) \mid X_2 \in Y_2 \ \& \ \exists X_1 \in Y_1 : X_1 \subseteq X_2\} \\ &= \bigcap\{C(X_2) \mid X_2 \in Y_2\} \\ &= C(Y_2) \quad \square \end{aligned}$$

Observation 11 (Ampliative Update) $Y \leq \mathsf{Upd}_B(Y,F) \leq \mathsf{Upd}_{mi}(Y,F) \leq \mathsf{Upd}_{ex}(Y,F)$

Proof: It suffices to show that $Y \leq \mathsf{Upd}_B(Y,F)$ by straightforward induction on the complexity of F. □

Claim 1 *$\boldsymbol{B}$, $\boldsymbol{B}_{mi}$ and $\boldsymbol{B}_{ex}$ satisfy Monotonicity, i.e.* $C(Y) \subseteq C(\mathsf{Upd}(Y,F))$.

Proof: The assertion follows as a corollary from the two previous observations and observation 6. □

Recall that DNS(F) denotes the *disjunctive normal set* corresponding to the disjunctive normal form of F.

Observation 12

1. $\mathsf{Upd}_B(Y,F) = \{X \cup K : X \in Y, K \in \mathrm{DNS}(F)\}$.
2. $\mathsf{Upd}_B(0,F) = \mathrm{DNS}(F)$.

Proof: See (Wagner 1994a). This observation implies that Contraction and Permutation hold in $\boldsymbol{B}$.

Claim 2 (Four-Valued Logic) *Let* $X \in \mathrm{Fin}(2^{L(\sim,\wedge,\vee)})$, *and* $F \in L(\sim,\wedge,\vee)$, *then*

$$X \models_4 F \quad \textit{iff} \quad \mathsf{Upd}_B(0,X) \vdash F$$

Proof: From the previous observation it follows that

$$\mathrm{MinMod}_4(X) \subseteq \mathsf{Upd}_B(0,X) \subseteq \mathrm{Mod}_4(X).$$

This, together with the fact that $\models_4$ is determined by minimal models, i.e.

$$\mathrm{MinMod}_4(F) \subseteq \mathrm{Mod}_4(G) \Rightarrow \mathrm{Mod}_4(F) \subseteq \mathrm{Mod}_4(G)$$

yields the assertion. □

Claim 3 (Three-Valued Logic) *Recall that standard (= Kleene's strong) 3-valued propositional logic corresponds to* $\models_c$.

$$X \models_c F \quad \textit{iff} \quad \mathsf{Upd}_{ex}(0,X) \vdash F$$

and moreover, if X is consistent, then

$$X \models_c F \quad \textit{iff} \quad \mathsf{Upd}_{mi}(0,X) \vdash F$$

Proof: If X is not consistent, i.e. it does not have a coherent model, then $\mathsf{Upd}_{ex}(0,X) = \{\}$, and the assertion holds trivially. Otherwise every minimal coherent model of X corresponds to a consistent element of $\mathsf{Upd}_B(0,X)$, and

$$\mathrm{MinMod}_c(X) \subseteq \mathsf{Upd}_{ex}(0,X) = \mathsf{Upd}_{mi}(0,X) \subseteq \mathrm{Mod}_c(X)$$

This, together with the fact that $\models_c$ restricted to $L(\sim, \wedge, \vee)$ is determined by minimal models, i.e.

$$\mathrm{MinMod}_c(X) \subseteq \mathrm{Mod}_c(F) \Rightarrow \mathrm{Mod}_c(X) \subseteq \mathrm{Mod}_c(F)$$

yields the assertion. □

Claim 4 (Classical Logic) *Let X, F be as before, let $\mathrm{At}(X, F)$ denote the set of all atoms occuring in X and F, and $\models_2$ denote classical propositional logic. Then,*

$$X \models_2 F \quad \textit{iff} \quad \mathsf{Upd}_{ex}(\mathsf{Upd}_{ex}(0, X), \{a \vee \sim a : a \in \mathrm{At}(X, F)\}) \vdash F$$

3.1 Problems with Belnap's KS

We briefly discuss three problems with $\boldsymbol{B}$.

3.1.1 No Unique Representation

In $\boldsymbol{B}$, epistemic states are not unique representations: $C(Y) = C(Y')$ does not imply that $Y = Y'$. For instance,

$$C(\{\{p\}\}) = C(\{\{p\}, \{p, q\}\})$$

A possible remedy consists of admitting only minimal situation descriptions as elements of an epistemic state. An epistemic state Y is called *canonical* if $Y = \mathrm{Min}(Y)$. We denote the set of all canonical elements of a set of epistemic states $\mathcal{Y} \subseteq 2^{2^{\mathrm{Lit}}}$ by $\mathrm{Can}(\mathcal{Y})$.

We obtain the following system:

$$\boldsymbol{B}^m := \langle \{\emptyset\}, \leq, \mathrm{Can}(2^{2^{\mathrm{Lit}}}), \vdash, L(\sim, \wedge, \vee), \mathsf{Upd}^m_{\boldsymbol{B}}, \mathrm{Lit}, L(\sim, \wedge, \vee) \rangle$$

where $\mathsf{Upd}^m_{\boldsymbol{B}}(Y, F) := \mathrm{Min}(\mathsf{Upd}_{\boldsymbol{B}}(Y, F))$. The collection of all canonical epistemic states forms a lattice, as was shown in (Kaluzkuy/Muravetsky 1993).

Observation 13 *$\langle \mathrm{Can}(2^{2^{\mathrm{Lit}}}), \leq \rangle$ is a lattice order. Meet and join can be defined by $Y_1 \sqcap Y_2 = \mathrm{Min}(Y_1 \cup Y_2)$, and $Y_1 \sqcup Y_2 = \mathrm{Min}(\{X_1 \cup X_2 : X_1 \in Y_1, X_2 \in Y_2\})$.*

3.1.2 No Disjunctive Syllogism

For instance,

$$\mathsf{Upd}_{\boldsymbol{B}}(0, (p \vee q) \wedge \sim p) = \{\{p, \sim p\}, \{q, \sim p\}\} \not\vdash q$$

Notice that the resulting epistemic state contains an inconsistent situation description although there is a consistent epistemic alternative.

The Disjunctive Syllogism holds in $\boldsymbol{B}_{ex}$, where all inconsistent epistemic alternatives are discarded. In this system, however, if $Y \vdash l$ then $\mathsf{Upd}(Y, \tilde{l}) = \{\}$ yields an 'exploded' KB in the sense that everything follows from $\{\}$, i.e.

$\boldsymbol{B}_{ex}$ is explosive. A good compromise seems to be $\boldsymbol{B}_{mi}$ which is both non-explosive and satisfies the Disjunctive Syllogism.

Example 4 *If we have the above Y_1, and we then learn that Susan or Linda smokes, the following update is performed in $\boldsymbol{B}_{mi}$:*

$$\begin{aligned} Y_2 &:= \mathsf{Upd}_{mi}(Y_1, s(S) \vee s(L)) \\ &= \text{MinInc}(\{\{s(S), \sim s(S), m(P,L)\}, \{s(S), \sim s(S), m(P,S)\}, \\ &\qquad \{s(L), \sim s(S), m(P,L)\}, \{s(L), \sim s(S), m(P,S)\}\}) \\ &= \{\,\{s(L), m(P,L), \sim s(S)\}, \{s(L), m(P,S), \sim s(S)\}\,\} \\ &\vdash s(L) \end{aligned}$$

3.1.3 No Input Completeness

Finite epistemic states are input constructible in $\boldsymbol{B}$, i.e.

$$\forall Y \in \text{Fin}(2^{\text{Fin}(2^{\text{Lit}})}) \exists F \in L_{\text{Input}} : Y = \mathsf{Upd}_B(0, F)$$

but since they are not input destructible, $\boldsymbol{B}$ does not satisfy Input Completeness.

4 Disjunctive Fact Bases

In a similar way as we have extended the monotonic system $\boldsymbol{F}^+$ to obtain Belnap's monotonic disjunctive KS in section 3, we now want to extend the nonmonotonic system $\boldsymbol{F}$ to obtain a nonmonotonic disjunctive KS. However, such an extension is not straightforward for at least two reasons. First, if we would simply extend the query language of $\boldsymbol{B}$ by adding weak negation, lemmas would be no longer compatible. This is easy to see. Consider the following example:

$$\{\{p\}\} \vdash p \vee q$$

but $\mathsf{Upd}_B(\{\{p\}\}, p \vee q) = \{\{p\}, \{p, q\}\} \not\vdash -q$

while $\{\{p\}\} \vdash -q$ □

Consequently, we should rather choose $\boldsymbol{B}^m$ as the basis of our extension, since

$$\mathsf{Upd}_B^m(\{\{p\}\}, p \vee q) = \{\{p\}\}$$

But then we get another problem: disjunctive information would be always exclusive in the sense that $\mathsf{Upd}_B^m(0, p \vee q) \vdash -p \vee -q$. And this is clearly undesirable. To remedy the problem, we have to use another notion of minimality called *paraminimality* in (Herre/Jaspars/Wagner 1996). We first define $\text{Min}_X(Y) = \{X' \in \text{Min}(Y) : X' \leq X\}$, and by means of it an operator collecting all *paraminimal* elements:

$$\text{PMin(Y)} = \{\text{X} \in \text{Y} \mid \neg\exists \text{X}' \in \text{Y} : \text{X}' < \text{X} \;\;\&\;\; \text{Min}_{\text{X}'}(\text{Y}) = \text{Min}_{\text{X}}(\text{Y})\}$$

An ordered set Y is called *paracanonical* if $Y = \mathrm{PMin}(Y)$. The set of all paracanonical elements of a set of ordered sets $\mathcal{Y}$ is denoted by $\mathrm{PCan}(\mathcal{Y})$.

Disjunctive Fact Bases are paracanonical epistemic states. For every $Y \in \mathrm{PCan}(2^{2^{\mathrm{Lit}}})$, and $F \in L(-, \sim, \wedge, \vee, |, \exists)$, inference is defined elementwise:

$$Y \vdash F :\Longleftrightarrow \text{for all } X \in Y : X \vdash F$$

where $X \vdash F$ is inference in $\boldsymbol{F}$. For example,

$$Y_1 \vdash {\sim} s(S) \wedge [m(P, L) \,|\, m(P, S)]$$

The information ordering between disjunctive fact bases $Y_1, Y_2 \subseteq 2^{\mathrm{Lit}}$ is the same ordering as in $\boldsymbol{B}$. Likewise, information units and the inconsistency operation are the same. Inputs are processed as follows:

$$\begin{array}{lrcl} (Ul) & \mathsf{Upd}_B(Y, l) & = & \{X \cup \{l\} : X \in Y\} \\ (U\wedge) & \mathsf{Upd}_B(Y, F \wedge G) & = & \mathsf{Upd}_B(\mathsf{Upd}_B(Y, F), G) \\ (U\vee) & \mathsf{Upd}_B(Y, F \vee G) & = & \mathrm{PMin}(\mathsf{Upd}_B(Y, F) \cup \\ & & & \mathsf{Upd}_B(Y, G) \cup \mathsf{Upd}_B(Y, F \wedge G)) \end{array}$$

We define the same additional update operations Upd_{mi}, and Upd_{ex}, as in $\boldsymbol{B}$. The only difference with respect to update in $\boldsymbol{B}$ is the definition $(U\vee)$ for disjunctive inputs which renders disjunctive information now explicitly *inclusive*.

Definition 23 (Disjunctive Fact Bases) *The system of disjunctive fact bases is defined as*

$$V_B\boldsymbol{F} := \langle \{\emptyset\}, \leq, \mathrm{PCan}(2^{2^{\mathrm{Lit}}}), \vdash, L(-, \sim, \wedge, \vee, |), \mathsf{Upd}_B, L(\sim, \wedge, \vee), \mathrm{Inc}_B, \mathrm{Lit}^{\vee} \rangle$$

$V_{mi}\boldsymbol{F}$ and $V_{ex}\boldsymbol{F}$ are formed from $V_B\boldsymbol{F}$ in the same way as $\boldsymbol{B}_{mi}$ and $\boldsymbol{B}_{ex}$ are formed from $\boldsymbol{B}$.

We have to show that $V_B\boldsymbol{F}$, $V_{ex}\boldsymbol{F}$ and $V_{mi}\boldsymbol{F}$ satisfy the KS postulates. KS1, KS2 and KS4 are obvious. KS3 (Reflexivity), KS5 (Knowledge Ordering Adequacy), and KS6 (Ampliative=Persistent) are proved in the same way as for $\boldsymbol{B}$, now using

$$L_{\mathrm{PersQ}} = L_{\mathrm{AmplI}} = L_{\mathrm{Input}} = L(\sim, \wedge, \vee) \quad \square$$

Notice that by the addition of weak negation and exclusive disjunction, inference in Disjunctive Fact Bases is no longer persistent.

Observation 14 *$V_B\boldsymbol{F}$ and $V_{mi}\boldsymbol{F}$ are vivid knowledge systems. $V_{ex}\boldsymbol{F}$ is not vivid since it is explosive.*

Conjecture 1 *Disjunctive Fact Bases capture paraminimal reasoning in partial logic. Let X be a finite subset of $L(\sim, \wedge, \vee)$, and $F \in L(-, \sim, \wedge, \vee, |)$. Then,*

$$X \models_4^{pm} F \quad \text{iff} \quad \mathsf{Upd}_B(0, X) \vdash F$$

where $X \models_4^{pm} F$ is defined as $\mathrm{PMin}(\mathrm{Mod}_4(X)) \subseteq \mathrm{Mod}_4(F)$.

Observation 15 *$V_B\boldsymbol{F}$ violates the principle of Negation Coherence: if, e.g., $Y = \{\{\sim p, q\}, \{p, \sim p\}\}$, then $Y \vdash \sim p$, but $Y \not\vdash -p$. On the other hand, Coherence holds in $V_{mi}\boldsymbol{F}$ and in $V_{ex}\boldsymbol{F}$.*

Observation 16 *$V_{mi}\boldsymbol{F}$ is non-explosive and negation coherent. It has (a restricted form of) the Disjunctive Syllogism. E.g.,*

$$\mathsf{Upd}_{mi}(0, (p \vee q) \wedge \sim p) = \{\{q, \sim p\}\} \vdash q, \quad \textit{while}$$
$$\mathsf{Upd}_{mi}(\{\{\sim q\}\}, (p \vee q) \wedge \sim p) = \{\{p, \sim p, \sim q\}, \{\sim p, q, \sim q\}\} \not\vdash q$$

This example also shows that neither Permutation nor Update Monotonicity hold in $V_{mi}\boldsymbol{F}$.

Claim 5 (Collapse of weak and strong negation)
Let $Y \in \mathrm{Cons}(\mathrm{PCan}(2^{2^{\mathrm{Lit}}}))$, and $a \in \mathrm{At}$. Then, whenever $Y \vdash a \vee \sim a$, weak and strong negation coincide for a in $V_{mi}\boldsymbol{F}$ and $V_{ex}\boldsymbol{F}$:

$$Y \vdash \sim a \quad \textit{iff} \quad Y \vdash -a$$

Proof: Since Y is consistent, all $X \in Y$ are consistent, and consequently, $a \vee \sim a$ is derivable from Y iff either $a \in X$ or $\sim a \in X$ for all $X \in Y$, subsuming three cases:

1. If all $X \in Y$ contain a, then neither $Y \vdash \sim a$ nor $Y \vdash -a$.
2. If all $X \in Y$ contain $\sim a$, then both $Y \vdash \sim a$ and $Y \vdash -a$.
3. Otherwise some $X \in Y$ contain a, and all others contain $\sim a$, hence $Y \not\vdash \sim a$, and $Y \not\vdash -a$. □

Notice that this does not hold in $V_B\boldsymbol{F}$.

Claim 6 *A disjunctive fact base is a unique representation, i.e. in $V_*\boldsymbol{F}$ (for $* = B, mi, ex$), it holds that $C(Y) = C(Y')$ implies $Y = Y'$.*

Proof: In (Wagner 1994), the Unique Representation property was shown for arbitrary $Y \subseteq 2^{\mathrm{Lit}}$ (the reason for it is the availability of weak negation in the query language). □

5 Exact Predicates and the Closed-World Assumption

In knowledge systems, three kinds of predicates can be distinguished. The first distinction, proposed by Körner (1966), reflects the fact that many predicates (especially in empirical domains) have truth value gaps: neither $p(c)$ nor $\sim p(c)$ has to be the case for specific instances of such *inexact* predicates, like, e.g., color attributes which can in some cases not be determined because of vagueness.

	F^+	F	B	B_{ex}	B_{mi}	$V_B F$	$V_{ex} F$	$V_{mi} F$
Contraction	√	√	√	√	√	√	√	√
Permutation	√	√	√	√		√	√	
Update Monotonicity	√	√	√	√		√	√	
Cumulativity	√	√	√	√	√	?	?	?
Monotonicity	√		√	√	√			
Unique Representation	√	√				√	√	√
non-explosive	√	√	√		√	√		√
Negation Coherence	–	√	–	–	–		√	√
Disjunctive Syllogism	–	–		√	√		√	√

Table 1: Formal properties of some basic knowledge systems (– denotes *not applicable*, ? denotes *open problem*).

Other predicates, e.g. from legal or theoretical domains, are *exact*, and we then have, for instance, $m(S) \vee \sim m(S)$ and $prime(2^{77}-1) \vee \sim prime(2^{77}-1)$, stating that Sophia is either married or unmarried and that $2^{77}-1$ is either a prime or a non-prime number. Only exact predicates can be totally represented in a knowledge base. Therefore, only exact predicates can be subject to the Closed-World Asssumption.

As in relational database theory, we distinguish between the *schema* Σ and the *instance*, resp. *state*, $X \in L_{\mathrm{KB}}$ of a vivid knowledge base. The schema determines the language of a KB, i.e. the available predicates (relation symbols) together with their domains and their epistemic category (exact or inexact). Since, in general, not all predicates are subject to the Closed-World Assumption (CWA), the schema also stipulates those relation symbols for which the CWA will be assumed. Finally, the schema contains a set of *integrity constraints*, i.e. closed query formulas which have to be satisfied by any evolving state of an associated knowledge base.[17]

Definition 24 (VKB Schema) *A vivid knowledge base schema is a quadruple*

$$\langle Rel, ExRel, CWRel, IC \rangle$$

consisting of a set Rel *of relation schemas, a set* ExRel $\subseteq$ Rel *of exact relation symbols, a set* CWRel $\subseteq$ ExRel *of CWA relation symbols, and a set* $IC \subseteq L_{\mathsf{Query}}$ *of integrity constraints.*

Definition 25 *For a schema* $\Sigma = \langle Rel, ExRel, CWRel, IC \rangle$, *we say that* Y *is a knowledge base over* Σ, *denoted* $Y : \Sigma$, *if*

[17]We shall not discuss integrity constraints in this paper.

1. *Y contains only predicates from* Rel, *such that the resp. domain constraints are satisfied.*

2. *For any exact predicate $p \in ExRel$, and any constant c from its domain, the resp. instance of the* tertium non datur (TND) *holds:* $Y \vdash p(c) \vee \sim p(c)$.

3. *If $q \in CWRel$, then Y does not contain any indefinite information on q, i.e. $Y \vdash q(c)$, or $Y \vdash \sim q(c)$.*

4. *For any $F \in IC$, $Y \vdash F$.*

Definition 26 (Closed-World Assumption) *For vivid knowledge bases $Y : \Sigma$, we have the following additional inference rule for inferring negative conclusions,*[18]

$$Y \vdash_\Sigma \sim p(c) \quad \textit{if} \quad p \in CWRel \;\&\; Y \vdash_\Sigma -p(c)$$

together with the corresponding extension of update,

$$\mathsf{Upd}_\Sigma(Y, \sim p(c)) = \mathsf{Upd}_\Sigma(Y, -p(c)) \quad \textit{if} \quad p \in CWRel$$

Notice that this CWA inference rule refers to both the state and the schema, i.e. it leads to an extension of the basic inference relation $\vdash$ defined by the underlying KS. The connection between them can be expressed by means of the *CWA closure:*
$CWA(Y) := \mathsf{Upd}(Y, \{\sim p(c) : p \in CWRel \;\&\; c \in Const \;\&\; \mathrm{KB} \vdash_\Sigma -p(c)\})$
Obviously, we have $Y \vdash_\Sigma F$ iff $CWA(Y) \vdash F$.

Example 5 (The CWA in Fact Bases) *While we cannot assume the CWA for empirical predicates like* smoker, *we should also not assume it for the relation* married *in some local KB since people may get married all over the world, and thus* married *will not be totaly represented in a local KB. We may assume the CWA, however, for a predicate like* resident, *simply because all residents of a city are registered in the local KB of that city. Thus, from the fact base*

$$X_3 = \{\sim s(S), m(P, L), r(P), r(S)\}$$

over a schema with CWRel $= \{r\}$, *we may infer that Peter is married with a non-resident,*

$$\mathsf{Ans}(CWA(X_3), m(x, y) \wedge \sim r(y)) = \{\langle P, L \rangle\}$$

[18]The Closed-World Assumption, in a less general form, was originally proposed in (Reiter 1978). Notice that our form of the CWA relates explicit with default-implicit falsity, i.e. strong with weak negation: an atomic sentence formed with a totaly represented predicate is (explicitly) false if it is false by default, i.e. its strong negation holds if its weak negation does.

but neither that he is married with a nonsmoker,

$$\mathsf{Ans}(CWA(X_3), m(x,y) \wedge \sim s(y)) = \emptyset$$

nor that he is not married with Susan, $CWA(X_3) \not\vdash \sim m(P,S)$.

A query F is inferable from a disjunctive fact base $Y : \Sigma$ if it can be derived from the closure of Y with respect to *ExRel* and *CWRel*. Formally,

$$Y \vdash_\Sigma F :\Longleftrightarrow \mathrm{Exact}(Y) \vdash F$$

where

$$\mathrm{Exact}(Y) = \mathsf{Upd}(\mathrm{CWA}(Y), \{p(c) \vee \sim p(c) : p \in \mathit{ExRel} - \mathit{CWRel} \ \& \ c \in \mathit{Const}\})$$

Notice that in KBs of definite knowledge systems, like fact bases, it is not possible to declare exact predicates not subject to the CWA. Therefore, in definite knowledge systems, *ExRel* = *CWRel*.

Inputs leading to the violation of integrity constraints have to be rejected. The update operation for disjunctive fact bases has to be modified accordingly:

$$\mathsf{Upd}_*^\Sigma(Y,F) = \{X \in \mathsf{Upd}_*(Y,F) : X \vdash IC\}$$

where $* = B, mi, ex$.

6 Reasoning with Three Kinds of Predicates

Only certain exact predicates can be totally represented in a KB. Totally represented exact predicates are subject to the CWA. For example, the local KB of some city knows all residents of the city, i.e. the CWA holds for *resident*, but it does not have complete information of every resident whether (s)he is married or not because (s)he might have married in another city and this information is not present. Consequently, the CWA does not apply to *married* in this KB.

The CWA helps to reduce disjunctive complexity which is exponential in the number of exact non-CWA predicates: if n is the number of unknown ground atoms which can be formed by means of predicates declared as exact but not subject to the CWA by a VKB, then the VKB contains 2^n possible situation descriptions.

We illustrate these distinctions with an example. Let m, r, s, l denote the predicates *married*, *resident*, *smoker* and *is_looking_at*, and let M, P, S stand for the individuals *Mary*, *Peter* and *Susan*. Let

$$Y = \{\{m(M),\ r(M),\ s(M),\ \sim m(S),\ \sim s(S),\ l(M,P),\ l(P,S)\}\}$$

be a disjunctive fact base over the schema $\Sigma = \langle \{m,r,s,l\},\ \{m,r\},\ \{r\},\ \emptyset \rangle$. The interesting queries we can ask Y and the resp. answers are:

1. Does a married person look at an unmarried one? Yes, but Y does not know who, either Mary at Peter, or Peter at Susan:

$$Y \vdash_\Sigma \exists x, y : l(x,y) \wedge m(x) \wedge \sim m(y)$$
$$\mathsf{Ans}(Y,\ l(x,y) \wedge m(x) \wedge \sim m(y)) = \{\{\langle M, P\rangle,\ \langle P, S\rangle\}\}$$

2. Does a resident look at a non-resident? Yes, Mary at Peter.

$$\mathsf{Ans}(Y,\ l(x,y) \wedge r(x) \wedge \sim r(y)) = \{\{\langle M, P\rangle\}\}$$

since $Y \vdash_\Sigma \sim r(P)$ if $Y \vdash -r(P)$.

3. Does a smoker look at a nonsmoker? No. Y is completely ignorant about Peter being a smoker or not: neither is he a smoker, nor is he a nonsmoker, nor is he a smoker or nonsmoker (he might be neither):

$$\mathsf{Ans}(Y,\ l(x,y) \wedge s(x) \wedge \sim s(y)) = \emptyset$$

Notice that the explicit TND completion of Y wrt *ExRel* would yield $2^3 = 8$ possible situation descriptions, which are reduced to $2^1 = 2$ by the CWA declaration of *resident.*[19]

7 Conclusion

We have presented natural extensions of Belnap's system of epistemic states. First, the addition of weak negation to the query language leads to unique representations and renders it possible to express nonmonotonically exclusive disjunctive information. Second, the adoption of the principle of minimal inconsistency leads to Negation Coherence and the Disjunctive Syllogism. Third, the addition of exactness declarations for specific predicates leads to a kind of local reasoning by cases. In a future paper we shall show that disjunctive fact bases with exactness declarations, in combination with deduction rules, allow for restricted forms of Contraposition, and thus establish a bridge for practical reasoning between partial, resp. constructive, and classical logic.

[19]Without the CWA there are 3 tertium non datur disjunctions formed with $m(P)$, $r(S)$, and $r(P)$, while with the CWA there is only one such disjunction: $m(P) \vee \sim m(P)$.

References

Almukdad, A., & Nelson, D.: Constructible Falsity and Inexact Predicates. *Journal of Symbolic Logic* 49 (1984), 231–233.

Belnap, N.D.: A Useful Four-valued Logic. In: *Modern Uses of Many-valued Logic*, ed. G. Epstein & J.M. Dunn, Dordrecht 1977, 8–37.

Gärdenfors, P.: *Knowledge in Flux.* Cambridge/Ma. 1988.

van Gelder A., & Topor, R.W.: Safety and Translation of Relational Calculus Queries. *ACM Transactions on Database Systems* 16 (1991), 235–278.

Gelfond M., & Lifschitz, V.: Logic Programs with Classical Negation. *Proceedings ICLP 1990*, Cambridge/Ma. 1990.

Gelfond M., & Lifschitz, V.: Classical Negation in Logic Programs and Disjunctive Databases. *Journal of New Generation Computing* 9 (1991), 365–385.

Herre, H., Jaspars, J., & Wagner, G.: Partial Logics with Two Kinds of Negation as a Foundation for Knowledge-Based Reasoning. Forthcoming in: What is Negation?, ed. by D. Gabbay & H. Wansing, Oxford, 1996.

Kaluzkuy Y., & Muravetsky, A.Y.: A Knowledge Representation Based on the Belnap's Four-Valued Logic. *Journal of Applied Non-Classical Logics* 3 (1993), 189–203.

Körner, S.: *Experience and Theory.* London 1966.

Levesque, H.J.: Foundations of a Functional Approach to Knowledge Representation. *Artificial IntelligenceI* 23 (1984), 155–212.

Levesque, H.J.: Making Believers out of Computers. *Artificial Intelligence* 30 (1986), 81–107.

Lobo, J., Minker, J., & Rajasekar, A.: *Foundations of Disjunctive Logic Programming.* Cambridge/Ma. 1992.

Minker, J.: On Indefinite Databases and the Closed-World Assumption. *Proceedings of CADE-82*, Berlin 1982, 292–308.

Pearce, D., & Wagner, G.: Reasoning with Negative Information I – Strong Negation in Logic Programs. LWI Technical Report, Freie Universität Berlin, 1989. Also in: *Language, Knowledge, and Intentionality: Perspectives on the Philosophy of Jaakko Hintikka.*, ed. by L. Haaparanta, M. Kusch & I. Niiniluoto, (Acta Philosophica Fennica 49), Helsinki 1990, 430–453.

Pereira, L.M., & Alferes, J.J.: Wellfounded Semantics for Logic Programs with Explicit Negation. *Proceedings ECAI'92*, New York 1992.

Plotkin, G.D.: A Powerdomain Construction. *SIAM Journal of Computing* 5 (1976), 452–487.

Priest, G.: Reasoning about Truth. *Artificial Intelligence* 39 (1989), 231–244.

Reiter, R.: On Closed-World Databases. In: *Logic and Databases*, ed./ by J. Minker & H. Gallaire, New York 1978.

Ross, K.A., & Topor, R.W.: Inferring Negative Information from Disjunctive Databases. *Journal of Automated Reasoning* 4 (1988), 397–424.

Sakama, C.: Possible Model Semantics for Disjunctive Databases. *Proceedings 1st International Conference on Deductive and Object-Oriented Databases (DOOD'89)*, Amsterdam 1989.

Smyth, M.: Powerdomains. *Journal of Computers and System Sciences* 16 (1978), 23–36.

Urbas, I.: Paraconsistency. *Studies in Soviet Thought* 39 (1990), 343–354.

Wagner, G.: Logic Programming with Strong Negation and Inexact Predicates. *Journal of Logic and Computation* 1 (1991), 835–859.

Wagner, G.: *Vivid Logic – Knowledged-Based Reasoning with Two Kinds of Negation*, LNAI 764. Berlin 1994a.

Wagner, G.: *Disjunctive, Deductive and Active Knowledge Bases.* LWI Report 22/1994b, Freie Universität Berlin; available as **ddakbs.dvi.gz**, respectively **ddakbs.ps.gz**, via ftp from **ftp.inf.fu-berlin.de** in the directory **/pub/reports/wagner** (notice that the file has to be 'gunzip'ed).

Name Index

Subject Index

Structuralist Theory of Science

Focal Issues, New Results

Edited by Wolfgang Balzer and C. Ulises Moulines

1996. Large-octavo. XI, 295 pages. Cloth. ISBN 3-11-014075-6

(Perspektiven der Analytischen Philosophie / Perspectives in Analytical Philosophy, Vol. 6)

A collection of papers on the formal analysis of science and its methodology.

From the contents: Moulines: General introduction to the approach – Diederich, Bartelborth, Sintonen, Kuipers, Lauth: Discussion of general topics relating to the theory of science and epistemology, in which structuralism is compared to other approaches – Balzer, Gähde, Mühlhölzer, Moulines/Polanski: Special questions arising from the structuralist approach and new research findings – Hinst and Mormann: The application of logico-mathematical methods to the foundation of the programme.

Walter de Gruyter · Berlin · New York

Bei Fragen zur Produktsicherheit wenden Sie sich bitte an:
If you have any questions regarding product safety,
please contact:

Walter de Gruyter GmbH
Genthiner Straße 13
10785 Berlin
productsafety@degruyterbrill.com